METHODS IN MOLECULAR BIOLOGY

Series Editor
John M. Walker
School of Life and Medical Sciences
University of Hertfordshire
Hatfield, Hertfordshire, UK

For further volumes:
http://www.springer.com/series/7651

For over 35 years, biological scientists have come to rely on the research protocols and methodologies in the critically acclaimed *Methods in Molecular Biology* series. The series was the first to introduce the step-by-step protocols approach that has become the standard in all biomedical protocol publishing. Each protocol is provided in readily-reproducible step-by-step fashion, opening with an introductory overview, a list of the materials and reagents needed to complete the experiment, and followed by a detailed procedure that is supported with a helpful notes section offering tips and tricks of the trade as well as troubleshooting advice. These hallmark features were introduced by series editor Dr. John Walker and constitute the key ingredient in each and every volume of the *Methods in Molecular Biology* series. Tested and trusted, comprehensive and reliable, all protocols from the series are indexed in PubMed.

Chlamydia trachomatis

Methods and Protocols

Edited by

Amanda Claire Brown

Department of Microbiology and Immunology, Vet School, Cornell University, Ithaca, NY, USA

Humana Press

Editor
Amanda Claire Brown
Department of Microbiology
and Immunology, Vet School
Cornell University
Ithaca, NY, USA

ISSN 1064-3745 ISSN 1940-6029 (electronic)
Methods in Molecular Biology
ISBN 978-1-4939-9693-3 ISBN 978-1-4939-9694-0 (eBook)
https://doi.org/10.1007/978-1-4939-9694-0

Cover caption: A B&W electron micrograph of the different forms of Chlamydia inside the inclusion. Image provided by Isabelle Derré, PhD.

This Humana imprint is published by the registered company Springer Science+Business Media, LLC, part of Springer Nature.
The registered company address is: 233 Spring Street, New York, NY 10013, U.S.A.

Preface

Hello and welcome to *Chlamydia trachomatis: Methods and Protocols. Chlamydia trachomatis* represents the most commonly diagnosed sexually transmitted infection in the world and can both lead to serious health implications and place a huge burden on public health. However, our increasing dependence on antibiotics has led to a nonchalance regarding chlamydia, especially in the West, where it is often viewed as "Oh it's only chlamydia; nothing to worry about; luckily, it's nothing serious." Currently, drug resistance in *Chlamydia trachomatis* has been slow to emerge; however, that may not always be the case. With the development of high-level drug resistance, as seen in other bacterial pathogens, treatment for chlamydia may no longer be as simple as it currently is, and, given its prevalence, this represents a very large concern. Plus, infections often go untreated in asymptomatic patients, which can cause serious effects on fertility and to newborn babies, as well as can have devastating consequences in countries where access to effective treatments is often limited. To combat this, we need to understand the biology of the bacterium as well as to develop new and improved methods to diagnose and treat the infection.

Chlamydia trachomatis is a notoriously difficult pathogen to work with, due to the added complexity of being an obligate intracellular organism, and lags behind other pathogens, such as *E. coli*, and even more technically difficult pathogens, such as *Mycobacterium tuberculosis*, in the repertoire of methods which have been developed to date. Here are described, for the first time in a single volume, comprehensive, step-by-step protocols designed to allow researchers to work at the very forefront of the current available technology. These include methods to identify *Chlamydia trachomatis* in patient samples, ranging from simple point-of-care tests to whole-genome sequencing, methods for propagation of strains in both cell culture and animal models, techniques to manipulate *Chlamydia trachomatis* in molecular genetic methodologies, high-throughput screening methods for testing new potential drugs against intracellular bacteria, screens for antibiotic resistance, methods for labeling and enumeration, and descriptions of genotyping technologies, as well as dual RNA-seq transcriptional profiling: all of which will provide ultimately a better understanding of the pathogen, allowing for the development of improved treatment regimens and the discovery of new drugs.

This volume has been made possible by the contribution and hard work of the authors from all over the world, who are the very best researchers on this pathogen, who readily shared their methodologies, and to whom I am extremely grateful. I hope this book helps to promote and stimulate further research into this highly relevant and important, albeit often over-looked and really quite difficult, form of bacteria. I believe this book will prove to be a most useful resource to the chlamydial research community.

Ithaca, NY, USA *Amanda Claire Brown*

Contents

Contributors

PATIYAN ANDERSSON • *Menzies School of Health Research, Charles Darwin University, Darwin, NT, Australia; Microbiological Diagnostic Unit Public Health Laboratory, Department of Microbiology and Immunology, The Peter Doherty Institute for Infection and Immunology, The University of Melbourne, Melbourne, VIC, Australia*

CHARLES W. ARMITAGE • *Institute of Health and Biomedical Innovation, School of Biomedical Sciences, Queensland University of Technology, Brisbane, QLD, Australia*

ROBERT J. BASTIDAS • *Department of Molecular Genetics and Microbiology, Duke University, Durham, NC, USA*

KENNETH W. BEAGLEY • *Institute of Health and Biomedical Innovation, School of Biomedical Sciences, Queensland University of Technology, Brisbane, QLD, Australia*

AMANDA CLAIRE BROWN • *Oxford Gene Technology, Oxford, UK; Department of Microbiology and Immunology, Vet School, Cornell University, Ithaca, NY, USA*

CAMERON BUCKLEY • *Faculty of Medicine, Centre for Clinical Research, The University of Queensland, Herston, QLD, Australia*

KATALIN BURIÁN • *Institute of Medical Microbiology and Immunobiology, University of Szeged, Szeged, Hungary*

ALISON J. CAREY • *Institute of Health and Biomedical Innovation, School of Biomedical Sciences, Queensland University of Technology, Brisbane, QLD, Australia*

ILADA CHOOPARA • *Program in Biotechnology, Faculty of Science, Chulalongkorn University, Bangkok, Thailand*

METTE T. CHRISTIANSEN • *UCL, London, UK*

ISABELLE DERRÉ • *Department of Microbiology, Immunology, and Cancer Biology, University of Virginia, Charlottesville, VA, USA*

KEVIN E. EBOIGBODIN • *Research and Development, Orion Diagnostica Oy, Espoo, Finland*

RACHEL ENDE • *Department of Microbiology, Immunology, and Cancer Biology, University of Virginia, Charlottesville, VA, USA*

ROBERT FARIS • *Department of Microbiology and Immunology, University of Iowa Carver College of Medicine, Iowa City, IA, USA*

KENNETH A. FIELDS • *Department of Microbiology, Immunology and Molecular Genetics, University of Kentucky College of Medicine, Lexington, KY, USA*

PHILIP M. GIFFARD • *Menzies School of Health Research, Charles Darwin University, Darwin, NT, Australia; College of Health and Human Sciences, Charles Darwin University, Darwin, NT, Australia*

REGAN J. HAYWARD • *The iThree Institute, University of Technology Sydney, Ultimo, NSW, Australia*

DEBORAH C. HOLT • *Menzies School of Health Research, Charles Darwin University, Darwin, NT, Australia; College of Health and Human Sciences, Charles Darwin University, Darwin, NT, Australia*

MARK HOSER • *Molecular Biology, GeneForm Technologies, Broadstairs, UK*

MICHAEL S. HUMPHRYS • *Institute for Genome Sciences, University of Maryland School of Medicine, Baltimore, MA, USA*

MARTINA JELOCNIK • *Genecology Research Centre, University of the Sunshine Coast, Sippy Downs, QLD, Australia*

LISA M. JORGENSON • *Department of Pathology and Microbiology, University of Nebraska Medical Center, Omaha, NE, USA*
MATEUSZ KĘDZIOR • *Department of Molecular Genetics and Microbiology, Duke University, Durham, NC, USA*
NICOLE L. KUSHNER • *Department of Microbiology and Immunology, Vet School, Cornell University, Ithaca, NY, USA*
VASSILI N. LAZAREV • *Federal Research and Clinical Center of Physical-Chemical Medicine of Federal Medical Biological Agency, Moscow, Russia; Moscow Institute of Physics and Technology, Dolgoprudny, Moscow Region, Russia*
STEVEN LIANG • *Department of Pathology and Molecular Medicine, McMaster University, Hamilton, ON, Canada*
SUNČANICA LJUBIN-STERNAK • *Medical Microbiology Department, School of Medicine, University of Zagreb, Zagreb, Croatia; Clinical Microbiology Department, Teaching Institute of Public Health "Dr. Andrija Štampar", Zagreb, Croatia*
JAMES B. MAHONY • *Department of Pathology and Molecular Medicine, McMaster University, Hamilton, ON, Canada*
ANUP MAHURKAR • *Institute for Genome Sciences, University of Maryland School of Medicine, Baltimore, MA, USA*
JAMES W. MARSH • *The iThree Institute, University of Technology Sydney, Ultimo, NSW, Australia; Department of Microbiome Science, Max Planck Institute for Developmental Biology, Tübingen, Germany*
TOMISLAV MEŠTROVIĆ • *University North, University Centre Varaždin, Varaždin, Croatia; Clinical Microbiology and Parasitology Unit, Polyclinic "Dr. Zora Profozić", Zagreb, Croatia*
GARRY S. A. MYERS • *The iThree Institute, University of Technology Sydney, Ultimo, NSW, Australia*
MACY G. OLSON • *Department of Pathology and Microbiology, University of Nebraska Medical Center, Omaha, NE, USA*
YVONNE PANNEKOEK • *Department of Medical Microbiology, Amsterdam UMC, University of Amsterdam, Amsterdam, The Netherlands*
OLEG V. PODGORNY • *Federal Research and Clinical Center of Physical-Chemical Medicine of Federal Medical Biological Agency, Moscow, Russia; Koltzov Institute of Developmental Biology, Russian Academy of Sciences, Moscow, Russia*
NADEZHDA F. POLINA • *Federal Research and Clinical Center of Physical-Chemical Medicine of Federal Medical Biological Agency, Moscow, Russia*
ADAM POLKINGHORNE • *Department of Microbiology and Infectious Diseases, Nepean Hospital, NSW Health Pathology, Penrith, NSW, Australia*
TÍMEA RAFFAI • *Institute of Medical Microbiology and Immunobiology, University of Szeged, Szeged, Hungary*
MOSTAFA RAHNAMA • *Department of Microbiology, Immunology and Molecular Genetics, University of Kentucky College of Medicine, Lexington, KY, USA*
ELIZABETH A. RUCKS • *Department of Pathology and Microbiology, University of Nebraska Medical Center, Omaha, NE, USA*
AMOL SHETTY • *Institute for Genome Sciences, University of Maryland School of Medicine, Baltimore, MA, USA*
NARAPORN SOMBOONNA • *Department of Microbiology, Faculty of Science, Chulalongkorn University, Bangkok, Thailand*

DEZSŐ P. VIROK • *Institute of Medical Microbiology and Immunobiology, University of Szeged, Szeged, Hungary*
JASMINA VRANEŠ • *Medical Microbiology Department, School of Medicine, University of Zagreb, Zagreb, Croatia; Clinical Microbiology Department, Teaching Institute of Public Health "Dr. Andrija Štampar", Zagreb, Croatia*
MARY M. WEBER • *Department of Microbiology and Immunology, University of Iowa Carver College of Medicine, Iowa City, IA, USA*
DAVID M. WHILEY • *Faculty of Medicine, Centre for Clinical Research, The University of Queensland, Herston, QLD, Australia; UQ Child Health Research Centre, The University of Queensland, Herston, QLD, Australia; Pathology Queensland Central Laboratory, Brisbane, QLD, Australia*
RAY E. WIDNER • *Department of Pathology and Microbiology, University of Nebraska Medical Center, Omaha, NE, USA*
KATERINA WOLF • *Department of Microbiology, Immunology and Molecular Genetics, University of Kentucky College of Medicine, Lexington, KY, USA*

Chapter 1

Detection of *Chlamydia trachomatis* and *Neisseria gonorrhoeae* Using Multiplex Strand Invasion Based Amplification (mSIBA)

Kevin E. Eboigbodin and Mark Hoser

Abstract

Chlamydia trachomatis (CT) and *Neisseria gonorrhoeae* (NG) are among the most prevalent causes of sexually transmitted infections (STIs) worldwide. Timely and accurate diagnosis plays an important role in deciding appropriate treatment and preventing the spread of the infection. Strand invasion based amplification (SIBA), is an established isothermal nucleic acid amplification method for the rapid and accurate detection of infectious diseases. SIBA was applied for the simultaneous detection of CT and NG in less than 1 h. The multiplex SIBA (mSIBA) method displayed high analytical sensitivity and specificity for the detection of CT and NG. Since the method is performed at low and constant temperature, it can therefore be run on portable instruments. SIBA enables rapid screening for CT and NG within point-of-care or central laboratory settings.

Key words Detection, Diagnosis, Strand invasion based amplification, Multiplex, Polymerase chain, *Chlamydia trachomatis*, *Neisseria gonorrhoeae*

1 Introduction

Chlamydia trachomatis (CT) and *Neisseria gonorrhoeae* (NG) are among the most common pathogens responsible for sexually transmitted infections (STIs) worldwide [1, 2]. Globally, these pathogens are responsible for over 200 million of new STIs cases annually [3]. Timely and accurate diagnosis plays an important role in deciding appropriate treatment and preventing the spread of the infection. Nucleic acid amplification tests (NAATs) such as real time polymerase chain reaction (PCR) have become the most attractive method used for diagnosis of STIs due to their improved sensitivity and specificity over immunoassays and culture-based methods [4, 5]. Moreover, culture method is laborious and time consuming, typically requiring 2–3 days to get the results. Due to the high analytical sensitivity of NAATs, they can be used for the detection

Amanda Claire Brown (ed.), *Chlamydia trachomatis: Methods and Protocols*, Methods in Molecular Biology, vol. 2042, https://doi.org/10.1007/978-1-4939-9694-0_1, © Springer Science+Business Media, LLC, part of Springer Nature 2019

of CT and NG directly from noninvasive specimens such as urine [6].

Isothermal nucleic acid amplification methods offer significant advantages over polymerase chain reaction (PCR) because they are typically faster and do not require sophisticated instruments needed for PCR thermal cycling. Strand invasion based amplification (SIBA), is an established isothermal nucleic acid amplification method for the rapid and accurate detection of infectious diseases [7–10]. SIBA utilizes a recombinase-coated oligonucleotide, invasion oligonucleotide (IO) for the separation of the target duplex. This allows target-specific primers to bind and extend the target via the action of a DNA polymerase. The continuous invasion and extension of the target subsequently leads to the exponential amplification of the target nucleic acid under isothermal condition (Fig. 1).

SIBA was applied for the simultaneous detection of CT and NG within a single reactions tube. The multiplex assay includes primers, invasion oligonucleotide, internal control, and probes required for the simultaneous detection of CT and NG. The internal control, known as synchronized internal control (SIC), uses the same IO and forward primer as the CT assay (Fig. 2). The SIC enables more accurate assessment of reagent functionality as well as sample-derived inhibition. The multiplex SIBA (mSIBA) assay displayed high analytical sensitivity and specificity for the detection of CT and NG. Since the method is performed at low and constant temperature, it can therefore be run on portable instruments. SIBA enables rapid screening for CT and NG within point-of-care or central laboratory settings.

2 Materials

The reactions are performed using previously described components [7]. General laboratory equipment and consumables, such as micropipettes and sterile pipette tips, as well as required PPE is assumed and therefore not listed here.

1. Biosafety cabinet.
2. Vortex mixer.
3. Heating block.
4. Real-time PCR Instruments or any fluorescence reader with temperature control function (*see* **Note 1**).
5. Centrifuge.
6. 0.2 ml reaction tubes or PCR multiwell plates (96-well plate).
7. 1.5 ml reaction tubes.
8. Magnesium acetate.

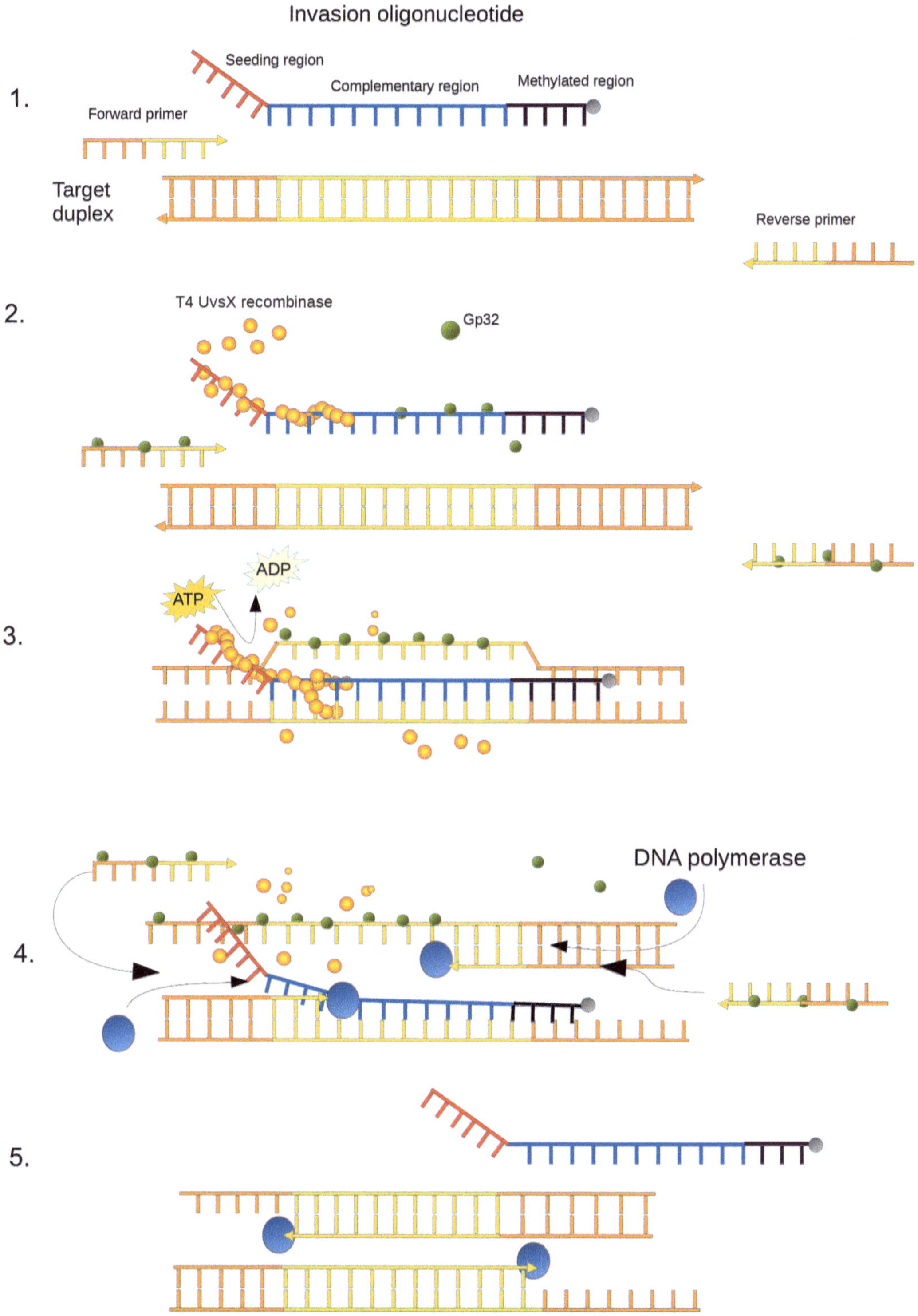

Fig. 1 DNA amplification by strand invasion based amplification method. (1) The SIBA reaction requires two target-specific primers and invasion oligonucleotide (IO). Gp32 binds to all single-stranded DNA. (2) The recombinase protein, UvsX, coats the IO, displacing the bound gp32. The primers are too short to act as

9. DMSO.
10. SYBR Green I (100×) (Thermofisher Scientific, cat. no. S7563).
11. SIBA reagent kit (Orion Diagnostica Oy).
12. T4-gp32 (Orion Diagnostica Oy).
13. T4-UvsX (Orion Diagnostica Oy).
14. Molecular grade water.
15. Oligonucleotides (*see* **Notes 2–5**).

3 Methods

1. Thaw the SIBA reagent kit on ice. The reagent kit contains a substrate buffer (A), enzyme mix (B), T4-gp32, and T4-UvsX (*see* **Note 6**).
2. Prepare a 100× working concentration of SYBR Green 1 in DMSO.
3. Prepare 10× oligo mix by diluting the oligonucleotides in water to give 10× the final concentration used in the reaction. For all reactions, forward and reverse primers and the IO should be used at 200 nM final concentrations, except for the SIC assay, in which the reverse primer should be used at 100 nM final concentration. Probes for CT, NG, and SIC assays should be used at 150, 50 and 100 nM, respectively (*see* **Notes 2–5**, 7, and **8**).
4. Preparation of template in magnesium acetate.

   Prepare 10× working concentration of the magnesium acetate needed per reaction (i.e., for 10 mM magnesium acetate per reaction, prepared a 100 mM of working stock. Dilute the template using the 10× working magnesium acetate stock (for example 2 μl template added to 982 μl of 10× working magnesium acetate stock). Add 2 μl of the diluted template to the SIBA reaction.
5. Preparation of SIBA reaction (20 μl per reaction). Prepare mastermix by sequentially adding the following reagents: 0.98 μl·H_2O, 7 μl Substrate mix A, 7 μl enzyme mix B, 0.5 μl T4-UvsX, 0.5 μl T4-gp32, and 0.02 μl SYBR Green I;

Fig. 1 (continued) substrates for UvsX. (3) The recombinase-coated IO invades the complementary region of the target duplex. The invasion process facilitates the separation of the target duplex, enabling target-specific primers to bind the target. (4) The strand displacement polymerase extends the dissociated target duplex from the primers. (5) This event leads to the production of two copies of the target duplex. Recombinase-mediated target duplex separation and polymerase-mediated extension are the basis for exponential amplification. Image and description were modified from Eboigbodin et al. [11]

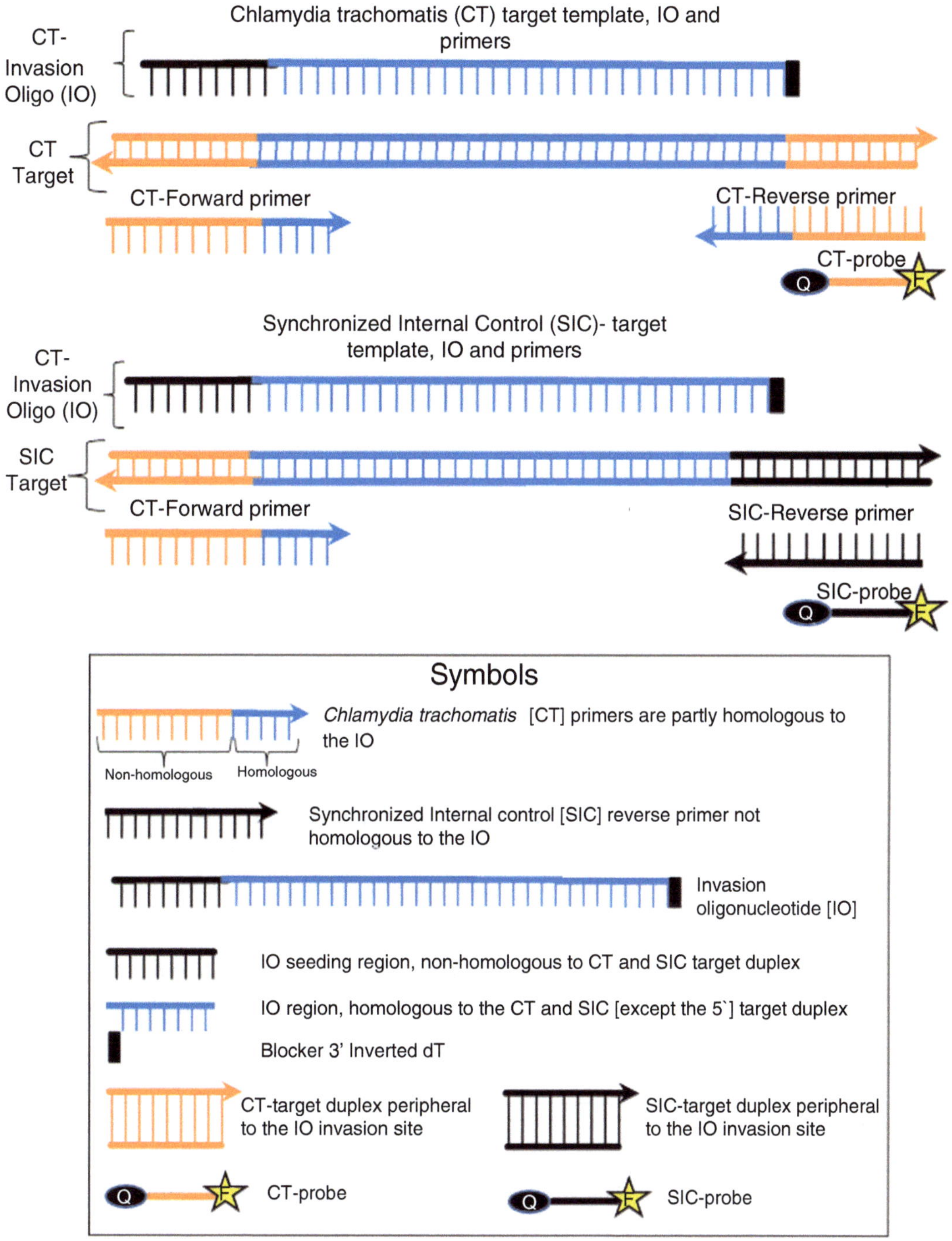

Fig. 2 Configuration of oligonucleotides and probes used in the synchronized internal control (SIC) and (CT) assays. Image and description were modified from Eboigbodin and Hoser [8]

Table 1
Components of SIBA mastermix

Reagent	Unit	Stock concentration	Final concentration	Volume for 1 reaction (μl)	Volume for 50 reactions (μl)
H_2O				0.98	49
Substrate mix A		2.86	1	7	350
Enzyme mix B		2.86	1	7	350
T4-gp32	mg/ml	10	0.25	0.5	25.0
T4-UvsX	mg/ml	10	0.25	0.5	25.0
SYBR Green I (1:100)	×	100	0.1	0.02	1.0
			Total	16.0	800.0

total volume = 16 μl (*see* **Note 7**). The layout for preparation of 50 reactions is shown in Table 1.

6. Distribute 16 μl of mastermix into 96-well plate or 0.2 ml reaction tubes. Add 2 μl of 10× oligo mix into each well/tube. Finally, add 2 μl of magnesium acetate (no template control, NTC) or 2 μl of template prepared in magnesium acetate. Stock concentration of magnesium acetate used is 140 mM (14 mM final concentration). Thus each well should receive the following: 16 μl Master mix, 2 μl 10× oligo mix, 2 μl magnesium acetate (NTC) or template prepared in magnesium acetate.
7. Seal the plate and run on a real-time PCR Instruments or any fluorescence reader with temperature control function. Set instrument to 40 °C for 60 min, and fluorescence readings taken at 1-min intervals (*see* **Notes 9** and **10**). Read FAM, ROX and Cy5 channel. If required, conduct melt curve analysis after the 60 min reaction time.

4 Notes

1. Portable fluorescence reader such as Genie (Optigene, UK) or the ESEQuant Tube Scanner (Qiagen, Germany) are also suitable.
2. The CT and NG SIBA assays was designed to specifically amplify and detect conserved sequence region from the CT cryptic plasmid and *porA* gene, respectively. Forward and reverse primers, probes and the IO were designed according to previously published protocol [8]. The length of the primers should be less than 23 nucleotides (preferably between 16 and

21). Primers of less than 23 nucleotides in length have less affinity for recombinase and are therefore less prone to primer dimers. An internal control assay namely, Synchronized Internal Control (SIC) was also designed to amplify an artificial DNA sequence that is partially identical to same target sequence as the CT assay. The IC used the same forward primer and IO as the CT assay, but the reverse primer and probe differed (Fig. 2).

3. The list of oligonucleotide used for CT, NG and SIC are shown in Table 2. The IO should be PAGE purified. All other oligonucleotides should be HPLC purified.
4. Several primer pairs should be designed and tested in the presence or absence of the target template. The fastest and the most specific primers should be selected and used for subsequent studies.
5. The SIC can also be designed to use the same forward primer and IO as the NG assay. Therefore, two or multiple SIC assays can be used in the same reaction tube (i.e., one per target assay).
6. Store reagents in the −20 °C freezer.

Table 2
Oligonucleotides used for the CT, NG, and SIC multiplex assay

Name	Sequence 5′ → 3′
CT-F primer	TTATCGACTGGGTGATTA
CT-R primer	CTTTCTGGCCAAGAATTAT
CT-IO	**TCCTCCTCTTCCTTG**ATTACAGCAGCTGCGAAAAAGAGACGAAAAmUmUmAmAmCmUmAmAmGmGmAmUmAmAmUmU(InvdT)
CT-probe	(ROX)CC+T+TT+C+TGG+C+C+AA+G(IABkFQ)
NG-F primer	GCAAGTCCGCCTATACGCCTG
NG-R primer	CCGACAACAGCCGGAACT
NG-IO	**TTCTCCTTACACT**CGCCTGCTACTTTCACGCTGGAAAGTAATCAGmAmUmGmAmAmAmCmCmAmGmUmUmCmCmGmA(InvdT)
NG-probe	(CY5)+C+C+G+A+C+A+AC+AG+C+C(IAbRQSp)
SIC-R-primer	TCTCGGTCAATATATTTA
SIC-probe	(FAM)+T+C+G+G+T+C+A+A+T+A+T(DABCYL)
SIC-template	ATTTATCGACTGGGTGATTACAGCAGCTGCGAAAAAGAGACGAAAATTAACTAATAAATATATTGACCGAGA

For invasion oligonucleotide (IO), bold sequences denote nonhomologous seeding regions. mA, mC, mG, and mU denote 2′-*O*-methyl RNA nucleotides. *F* forward, *R* reverse, *IABkFQ* Iowa black FQ quencher, *IAbRQSp* Iowa black RQ quencher, *InvdT* inverted dT, + lock nucleic acid bases, *CT Chlamydia trachomatis, NG Neisseria gonorrhoeae, SIC* Synchronized Internal Control

7. Avoid leaving mastermix at room temperature for a long period. Use cold racks or place on ice.
8. The 5′-end of the IO contains sequence that are noncomplementary to the target region. This noncomplementary region facilitates optimal coating of the IO by the recombinase [7, 11, 12]. In addition, the 3′-end of the IO contains 2′-*O*-methyl RNA nucleotides and a blocker, inverted dT, that prevent the IO from acting as a primer or being extended by the reverse primer. Probes are used to detect and distinguish CT and NG assay in the same reaction. The probes are short single-stranded dual-labeled probes (12–14 nucleotides) containing a mixture of DNA and LNA bases. These probes are designed to be complementary to the downstream region of the target DNA and are homologous to the 5′-end of the reverse primer. The CT probe was labeled with ROX, Iowa black FQ quencher, NG probe was labeled with Cy5 and Iowa black RQ quencher, and SIC probe was labeled with FAM and Dabcyl respectively. The sequences of the oligonucleotides used for the CT, NG, and SIC assay are listed in Table 2.
9. The CT, NG, SIC assays can either be performed as a multiplex CT/NG/SIC or as individual singleplex assay reactions. All reactions are performed with 14 mM magnesium acetate, except for the NG singleplex reactions, in which 10 mM magnesium acetate was used. SIBA reactions were incubated at 40 °C for 60 min, and fluorescence readings were taken at 60-s intervals. PCR device or other fluorescence readers with temperature control capability can be used to run SIBA reactions.
10. Optionally two units of restriction endonuclease (MlyI) can be added to the NG singleplex or the CT/NG/SIC multiplex assay in order to reduce the detection time of *Neisseria gonorrhoeae*. For multiplex assay, an additional 75 mM sucrose and 0.25 U sucrose phosphorylase should be included into the SIBA mastermix.

References

1. Centers for Disease Control and Prevention (2014) Sexually transmitted disease surveillance 2013. U.S. Department of Health and Human Services, Atlanta, p 2014
2. Baugh LR, Hill AA, Brown EL, Hunter CP (2001) Quantitative analysis of mRNA amplification by in vitro transcription. Nucleic Acids Res 29(5):e29. https://doi.org/10.1093/nar/29.5.e29
3. Newman L, Rowley J, Vander Hoorn S, Wijesooriya NS, Unemo M, Low N, Stevens G, Gottlieb S, Kiarie J, Temmerman M (2015) Global estimates of the prevalence and incidence of four curable sexually transmitted infections in 2012 based on systematic review and global reporting. PLoS One 10(12): e0143304. https://doi.org/10.1371/journal.pone.0143304
4. Papp JR, Schachter J, Gaydos CA, Van Der Pol B (2014) Recommendations for the laboratory-based detection of *Chlamydia trachomatis* and *Neisseria gonorrhoeae* — 2014. MMWR Recomm Rep 63(RR-02):1–19

5. Van Dyck E, Ieven M, Pattyn S, Van Damme L, Laga M (2001) Detection of *Chlamydia trachomatis* and *Neisseria gonorrhoeae* by enzyme immunoassay, culture, and three nucleic acid amplification tests. J Clin Microbiol 39 (5):1751–1756. https://doi.org/10.1128/jcm.39.5.1751-1756.2001
6. Gaydos CA, Quinn TC, Willis D, Weissfeld A, Hook EW, Martin DH, Ferrero DV, Schachter J (2003) Performance of the APTIMA combo 2 assay for detection of *Chlamydia trachomatis* and *Neisseria gonorrhoeae* in female urine and endocervical swab specimens. J Clin Microbiol 41(1):304–309. https://doi.org/10.1128/jcm.41.1.304-309.2003
7. Hoser MJ, Mansukoski HK, Morrical SW, Eboigbodin KE (2014) Strand invasion based amplification (SIBA®): a novel isothermal DNA amplification technology demonstrating high specificity and sensitivity for a single molecule of target analyte. PLoS One 9(11): e112656. https://doi.org/10.1371/journal.pone.0112656
8. Eboigbodin KE, Hoser MJ (2016) Multiplex strand invasion based amplification (mSIBA) assay for detection of *Chlamydia trachomatis* and *Neisseria gonorrhoeae*. Sci Rep 6:20487. https://doi.org/10.1038/srep20487
9. Eboigbodin KE, Moilanen K, Elf S, Hoser M (2017) Rapid and sensitive real-time assay for the detection of respiratory syncytial virus using RT-SIBA® BMC Infectious Diseases 17 (1):134. https://doi.org/10.1186/s12879-017-2227-x
10. Eboigbodin KE, Brummer M, Ojalehto T, Hoser M (2016) Rapid molecular diagnostic test for Zika virus with low demands on sample preparation and instrumentation. Diagn Microbiol Infect Dis 86(4):369–371. https://doi.org/10.1016/j.diagmicrobio.2016.08.027
11. Eboigbodin K, Filén S, Ojalehto T, Brummer M, Elf S, Pousi K, Hoser M (2016) Reverse transcription strand invasion based amplification (RT-SIBA): a method for rapid detection of influenza A and B. Appl Microbiol Biotechnol:1–9. https://doi.org/10.1007/s00253-016-7491-y
12. Formosa T, Alberts BM (1986) Purification and characterization of the T4 bacteriophage uvsX protein. J Biol Chem 261 (13):6107–6118

Chapter 2

Point-of-Care *Chlamydia trachomatis* Detection Using Loop-Mediated Isothermal Amplification and Hydroxynaphthol Blue

Naraporn Somboonna and Ilada Choopara

Abstract

Loop-mediated isothermal amplification (LAMP) is a new nucleic acid amplification technique that provides high sensitivity and specificity compatible to conventional polymerase chain reaction (PCR) and simplicity, without the requirement of an expensive thermal cycler. Additionally, LAMP coupled hydroxynaphthol blue (HNB) allowed for visual color detection by the naked eye. Here we describe the methodology of *Chlamydia trachomatis* detection of vaginal swab from crude DNA lysis to LAMP-HNB reaction. The result can be read by the naked eye through color change from violet (negative) to sky blue (positive). The result can be confirmed by UV spectra and agarose gel electrophoresis. This assay uses all nonhazardous chemical reagents and is hence safe to the users, and requires little specialist training or knowledge.

Key words Loop-mediated isothermal amplification (LAMP), Hydroxynaphthol blue, Color detection, Local detection

1 Introduction

Loop-mediated isothermal amplification (LAMP) is a recent technique for targeted nucleotide amplification at very high sensitivity and specificity. LAMP is superior to PCR in terms of a very short reaction time (less than 1 h) and only one incubation temperature is required [1, 2]. LAMP uses *Bst* DNA polymerase and a set of 4–6 target-specific primers for an autocycling and strand displacement DNA synthesis at a constant temperature of 60–65 °C in 40–60 min. Thus the only necessary equipment for LAMP is a heating block or hot water bath. LAMP offers high sensitivity (as low as 1–10 genome copy), high specificity, rapidity, simplicity and inexpensiveness, and thus is appropriate for local disease detection [3–5]. Several point-of-care diagnostic kits based on LAMP have thereby been developed for bacteria, viruses, fungi and parasites, and in forensic sciences [5–9].

Amanda Claire Brown (ed.), *Chlamydia trachomatis: Methods and Protocols*, Methods in Molecular Biology, vol. 2042,
https://doi.org/10.1007/978-1-4939-9694-0_2,

Originally, the LAMP product was analyzed by a precipitate of magnesium pyrophosphate appearing as a white turbidity, that can be observed by the naked eye [10]. While the DNA product is amplified, a pyrophosphate is release and coupled with a magnesium ion in a buffer to form an insoluble magnesium pyrophosphate. However, this level of turbidity depends on the number of amplified products, so small genome copy number could potentially yield a false negative or questionable result. To improve detection, the decreased concentration of magnesium ion in the buffer during successful LAMP amplification can be titrated by a metal ion indicator, such as hydroxynaphthol blue (HNB). LAMP-HNB replaces the requirement of agarose gel electrophoresis and is more specific for detection of the LAMP product than this initial identification of the white precipitation. Further, HNB can be added together with LAMP reagents in one step, limiting a possibility of contamination and increasing to the ease of use/low technical requirements of the method [11, 12]. This chapter describes the methods required for a of crude DNA lysis and LAMP-HNB, using primers that are specific to *C. trachomatis* outer membrane A gene (*ompA*) [4].

2 Materials

Prepare all solutions using molecular grade reagents and with aseptic technique. Store LAMP reagents at ≤−20 °C (unless indicated otherwise). Diligently follow all waste disposal regulations when disposing waste materials.

2.1 Crude DNA Extraction

1. TE (Tris–EDTA) buffer: 0.01 M Tris–HCl and 0.001 M EDTA, sterilize by autoclaving. Store at room temperature.
2. 0.2 mL thin-wall PCR tubes.
3. NanoDrop spectrophotometer, or Qubit®3.0 fluorometer (*see* **Note 1**).

2.2 LAMP Reaction

1. Primers as given in Table 1 [4].
2. 0.2 mL thin-wall PCR tubes.
3. dNTP mix.
4. Betaine.
5. $MgSO_4$.
6. *Bst* DNA polymerase (large fragment).
7. 10× ThermoPol Reaction Buffer (200 mM Tris–HCl, 100 mM $(NH_4)_2SO_4$, 100 mM·KCl, 20 mM·$MgSO_4$, 1% Triton X-100).

Table 1
Names and sequences of primers and probes for LAMP reaction

Primer name	Sequence 5′→3′
F3	GAACAGA(A/T)GC(T/A)GCGACAG
B3	(C/T)GGGTTTAGAGTAGT(G/A)A(C/T)ATC
FIP	TTAACTCCAATGTA(A/G)GGAGTGAACA-ATGCCTCTATTGA(C/T)TACCAT
BIP	GGTCT(A/C)GAG(T/C)AAGTTTTGATGCCG–CAAGAT(T/A)GCTT(C/T) AGCCAATT
LF	CATATT(T/C)A(AAT(C/T)CGTATAGCTCAGCC
LB	AT(C/T)CGTATAGCTCAGCC

8. Hydroxynaphthol blue (HNB): Store at room temperature and protect from light.
9. DNA template (sample).
10. Positive control (*C. trachomatis* DNA).

2.3 Agarose Gel Electrophoresis (10× TBE Buffer)

1. 0.5 M EDTA: Weight 93.05 g EDTA and dissolve in 200 mL purified water with magnetic stirrer. Then, add NaOH until the solution becomes clear, and adjust the volume to 500 mL with purified deionized water. Store at room temperature.
2. 10× TBE Buffer: 108 g Tris, 55 g boric acid, 40 mL 0.5 M Na_2-EDTA (pH 8.0), and adjust volume to 1 L with purified water. Store at room temperature. (10× TBE Buffer may take some time to dissolve, even with stirring.)
3. Sample loading buffer.
4. DNA ladder/marker.
5. Agarose.

3 Methods

Carry out all procedures on ice unless otherwise specified.

3.1 Crude DNA Extraction

1. Mix 40 μL of clinical sample (e.g., vaginal swab in collection medium, such as M4RT [4]) with 20 μL of TE Buffer in 0.2 mL thin-wall PCR tube (*see* **Note 2**).
2. Heat at 95 °C for 5 min, then let it cool on ice. If debris precipitates at the bottom of the tube, use the supernatant.
3. Measure DNA concentration (*see* **Note 1**).
4. Store crude DNA lysate at −20 °C.

3.2 LAMP Reaction

1. Mix 0.2 μM each of primers named F3 and B3, 1.6 μM each of FIP and BIP primers, 1.4 μM each of LF and LB primers in 0.2 mL thin-wall PCR tube (*see* **Note 2**).
2. Add 1.4 mM of dNTP mix, 0.3 M betaine, 6 mM $MgSO_4$, 1× ThermoPol Reaction Buffer, 100 ng of DNA sample, 8 U of *Bst* DNA polymerase large fragment, and mix very well. Then add 133 μM HNB, in a final volume of 25 μL (*see* **Note 3**). Include a positive (confirmed *C. trachomatis* DNA) and negative control (sterile double distilled water in place of DNA sample) with each reaction set.
3. Mix all reaction tubes very well, and incubate at 61 °C for 40–60 min (*see* **Note 4**).
4. Move reaction tubes to 80 °C for 4 min to terminate the reaction and let it cool suddenly.
5. Analyze LAMP products by the naked eye, observing a violet (negative) to sky blue (positive) color change (Fig. 1); compare color visually with the positive and negative control tubes (*see* **Note 5**).
6. Color result can also be determined by UV-Vis spectra reading at approximately 580 and 650 nm wavelength. A positive result often shows a ratio of A650/A580 more than 1.0 (*see* **Note 6**).

3.3 (Optional) Verification of LAMP Product by Agarose Gel Electrophoresis

1. Weigh 2 g agarose powder into 100 mL 1× TBE Buffer.
2. Boil until the agarose is completely dissolved. Then, let agarose solution cool down to about 50 °C.
3. Add ethidium bromide (EtBr) to a final concentration of approximately 0.2–0.5 μg/mL (*see* **Note** 7).
4. Pour the agarose onto a gel tray with attached combs in place. Leave at room temperature for the agarose to solidify (*see* **Note 8**).
5. Fill an electrophoretic chamber with 1× TBE Buffer as to prepare the agarose gel.
6. Mix an appropriate volume of sample, for the gel comb used, with loading buffer, and load.
7. Run the gel at 100 V for 45–60 min (until the dye in a sample reaches approximately 80% down the gel).
8. Visualize the gel in a UV transilluminator: LAMP product will show several ladder bands (Fig. 1a).

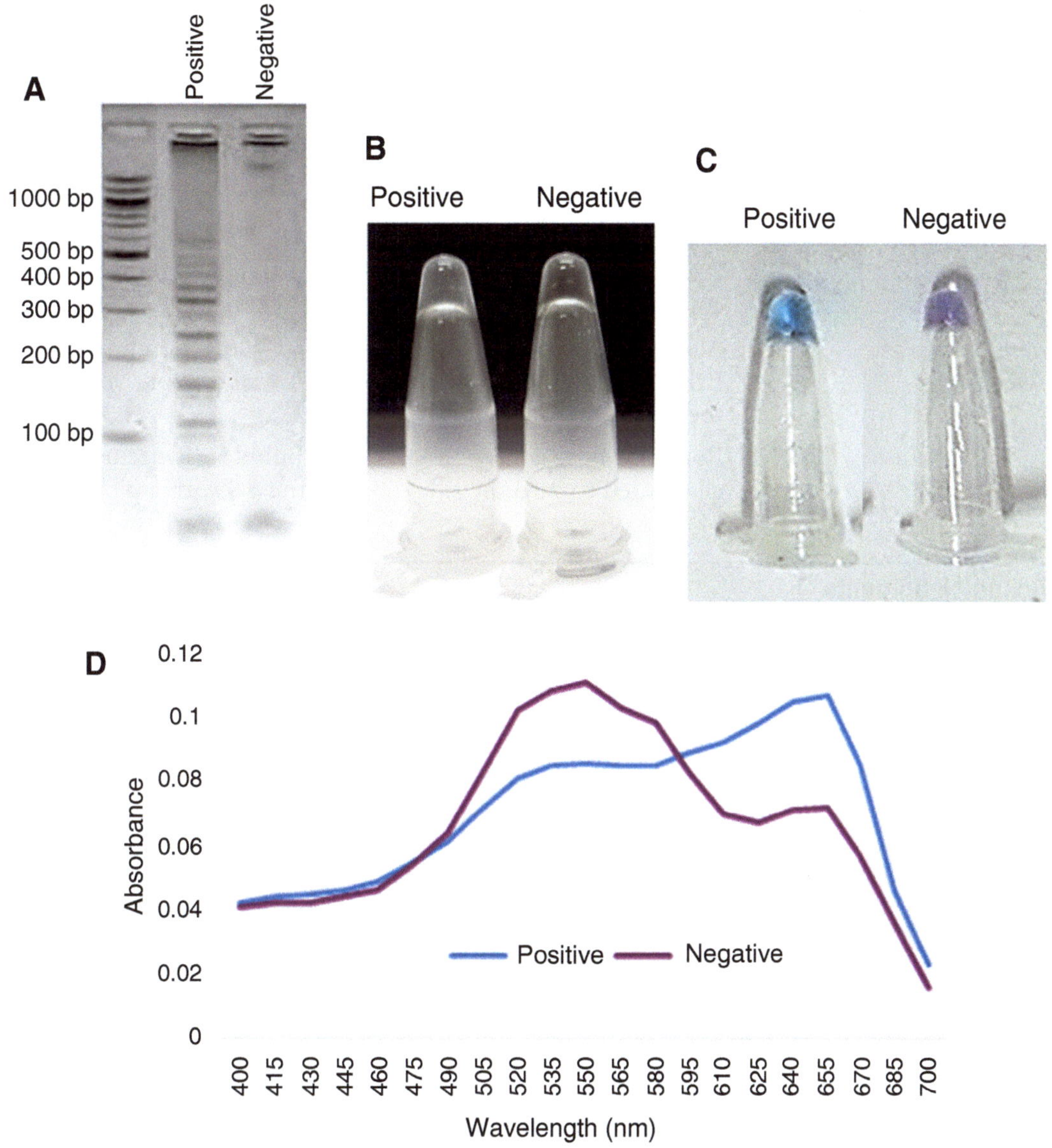

Fig. 1 Analysis of LAMP product using (**a**) agarose gel electrophoresis, (**b**) turbidity, (**c**) HNB, and (**d**) UV spectra. Positive reaction contained *C. trachomatis* genome. Negative reaction contained DNA-free water in place of *C. trachomatis* genome

4 Notes

1. For double-stranded DNA analysis, use Qubit®3.0 Fluorometer to measure DNA concentration at 260 nm wavelength. A ratio of A260/A280 represents a quality of DNA, which should range between 1.80 and 2.0.
2. Use 0.2 mL thin-wall PCR tubes that are of DNase, RNase, and nucleic acid free. A thin-wall tube allows an efficient transfer of temperature between outside and inside of the tube.

3. Always add HNB at the final component. For example, adding HNB prior to $MgSO_4$ may cause the solution to remain as a blue color because of no magnesium ions (thus false positive). To minimize false reading error, the reaction components should be mixed very well before adding HNB.
4. Forty minutes is the minimum time for LAMP incubation.
5. Visualize the color against white background for clarity.
6. If the ratio of A650/A580 is around 1.0, it is recommended to continue to Subheading 3.3.
7. Alternatively, you may stain an electrophoretic gel in EtBr solution, and rinse it in purified water before you image the gel.
8. Pour slowly to avoid bubbles. Any bubbles (if occur) can be eliminated toward the side/edge of the gel with a pipette tip.

Acknowledgments

This work was supported by Research Assistantship Funding from Faculty of Science, Chulalongkorn University (to I. Choopara).

References

1. Mori Y, Kanda H, Notomi T (2013) Loop-mediated isothermal amplification (LAMP): recent progress in research and development. J Infect Chemother 19:404–411
2. Notomi T, Okayama H, Masubuchi H, Yonekawa T, Watanabe K, Amino N et al (2000) Loop-mediated isothermal amplification of DNA. Nucleic Acids Res 28:e63
3. Lin GZ, Zheng FY, Zhou JZ, Gong XW, Wang GH, Cao XA et al (2011) Loop-mediated isothermal amplification assay targeting the *omp25* gene for rapid detection of *Brucella* spp. Mol Cell Probes 25:126–129
4. Choopara I, Arunrut N, Kiatpathomchai W, Dean D, Somboonna N (2017) Rapid and visual *Chlamydia trachomatis* detection using loop-mediated isothermal amplification and hydroxynaphthol blue. Lett Appl Microbiol 64:51–56
5. Somboonna N, Choopara I, Arunrut N, Sukhonpan K, Sayasathid J, Dean D et al (2018) Rapid and sensitive detection of *Chlamydia trachomatis* sexually transmitted infections in resource-constrained settings in Thailand at the point-of-care. PLoS Negl Trop Dis 12:e0006900. https://doi.org/10.1371/journal.pntd.0006900
6. Nimitphak T, Kiatpathomchai W, Flegel TW (2008) Shrimp hepatopancreatic parvovirus detection by combining loop-mediated isothermal amplification with a lateral flow dipstick. J Virol Methods 154:56–60
7. Poon LL, Wong BW, Ma EH, Chan KH, Chow LM, Abeyewickreme W et al (2006) Sensitive and inexpensive molecular test for falciparum malaria: detecting *Plasmodium falciparum* DNA directly from heat-treated blood by loop-mediated isothermal amplification. Clin Chem 52:303–306
8. Sun J, Najafzadeh MJ, Vicente V, Xi L, de Hoog GS (2010) Rapid detection of pathogenic fungi using loop-mediated isothermal amplification, exemplified by *Fonsecaea* agents of chromoblastomycosis. J Microbiol Methods 80:19–24
9. Watthanapanpituck K, Kiatpathomchai W, Chu E, Panvisavas N (2014) Identification of human DNA in forensic evidence by loop-mediated isothermal amplification combined with a colorimetric gold nanoparticle hybridization probe. Int J Legal Med 128:923–931
10. Mori Y, Notomi T (2009) Loop-mediated isothermal amplification (LAMP): a rapid, accurate, and cost-effective diagnostic method for

infectious diseases. J Infect Chemother 15:62–69

11. Goto M, Honda E, Ogura A, Nomoto A, Hanaki K (2009) Colorimetric detection of loop-mediated isothermal amplification reaction by using hydroxynaphthol blue. BioTechniques 46:167–172
12. Goto M, Shimada K, Sato A, Takahashi E, Fukasawa T, Takahashi T et al (2010) Rapid detection of *Pseudomonas aeruginosa* in mouse feces by colorimetric loop-mediated isothermal amplification. J Microbiol Meth 81:247–252

Chapter 3

Application of Loop-Mediated Isothermal Amplification Assay for the Detection of *Chlamydia trachomatis* and *Neisseria gonorrhoeae*

Kevin E. Eboigbodin

Abstract

The loop-mediated isothermal amplification (LAMP) is one of the most widely used isothermal nucleic acid amplification techniques due to it its simplicity and adaptability within limited resource or point-of-care settings. Here, LAMP was utilized for the rapid amplification and detection of *Chlamydia trachomatis* and *Neisseria gonorrhoeae*.

Key words Amplification, Detection, Diagnosis, Loop-mediated isothermal amplification (LAMP), *Chlamydia trachomatis*, *Neisseria gonorrhoeae*

1 Introduction

Isothermal nucleic acid amplification techniques (iNAATs) are an alternative to real-time polymerase chain reaction (qPCR) methods. iNAAT techniques are performed at low and constant temperature and therefore do not require expensive and sophisticated instruments, such as are needed for thermal cycling of qPCR protocols. Furthermore, iNAAT are often more tolerant to specimens derived inhibitors when compared to qPCR. Consequently, iNAATs are able to amplify nucleic acids from a crude sample matrix without the need for nucleic acid purification. Several iNAAT techniques have been reported in literature, including strand invasion based amplification (SIBA) [1, 2, 3], recombinase polymerase amplification (RPA) [4, 5], rolling circle replication (RCA) [6], helicase-dependent amplification (HDA) [7], strand displacement amplification (SDA) [8], nicking enzyme amplification reaction (NEAR) [9], and loop-mediated isothermal amplification (LAMP) [10].

The loop-mediated isothermal amplification (LAMP) is one of the most widely used iNAAT techniques due to it its simplicity and

Amanda Claire Brown (ed.), *Chlamydia trachomatis: Methods and Protocols*, Methods in Molecular Biology, vol. 2042,
https://doi.org/10.1007/978-1-4939-9694-0_3,

adaptability within limited resource or point-of-care settings. LAMP was first published by Notomi et al. in 2000 for the detection of hepatitis virus B DNA [10]. To date LAMP has now been applied for the detection of DNA/RNA from a wide variety of applications, such as infectious diseases, cancer biomarkers, and food and environmental contaminants [11]. LAMP uses a strand displacing polymerase and 4–6 primers for the rapid amplification and detection of a target nucleic acid sequence (Fig. 1). LAMP requires a minimum of four primer sets, known as FIP (forward inner primer), BIP (backward inner primer), F3 (forward primer) and B3 (backward primer). These primers are designed to recognize six distinct regions of the target sequences. In addition, two optional primers, LF (loop forward) and LB (loop backward can be included in order to enhance LAMP assay reaction times. LAMP reactions can be monitored in real time and/or by visual inspection using intercalating dyes, turbidometric, colorimetric, and lateral flow methods [12, 13]. Lamp reactions can also be detected using lateral flow biosensor or gel electrophoresis [14].

To date, *Chlamydia trachomatis* (CT) and *Neisseria gonorrhoeae* (NG) still remain the leading pathogens responsible for sexually transmitted infections (STIs) worldwide [15, 16]. Here, LAMP assay reactions were set up for the rapid detection of CT and NG. LAMP primers sets for NG were designed to amplify and detect the porA gene sequence using PrimerExplorer Version 4 (http://primerexplorer.jp/e/). For Lamp CT assay, primers that amplify and detect sequence from CT cryptic plasmid were used [17].

Typical general laboratory equipment is not listed below.

2 Materials

1. LAMP reagent kit, OptiGene Isothermal Mastermix ISO-100 (OptiGene, UK) (*see* **Note 1**).
2. Biosafety cabinet.
3. Vortex mixer.
4. Heating block.
5. Real-time PCR Instruments or any fluorescence reader with temperature control function (*see* **Note 2**).
6. Centrifuge.
7. 0.2 ml reaction tubes or PCR multiwell plates (96-well plate).
8. 1.5 ml reaction tubes.
9. Molecular grade water.
10. Primers (*see* **Note 3**).
11. Positive template (5 μl).

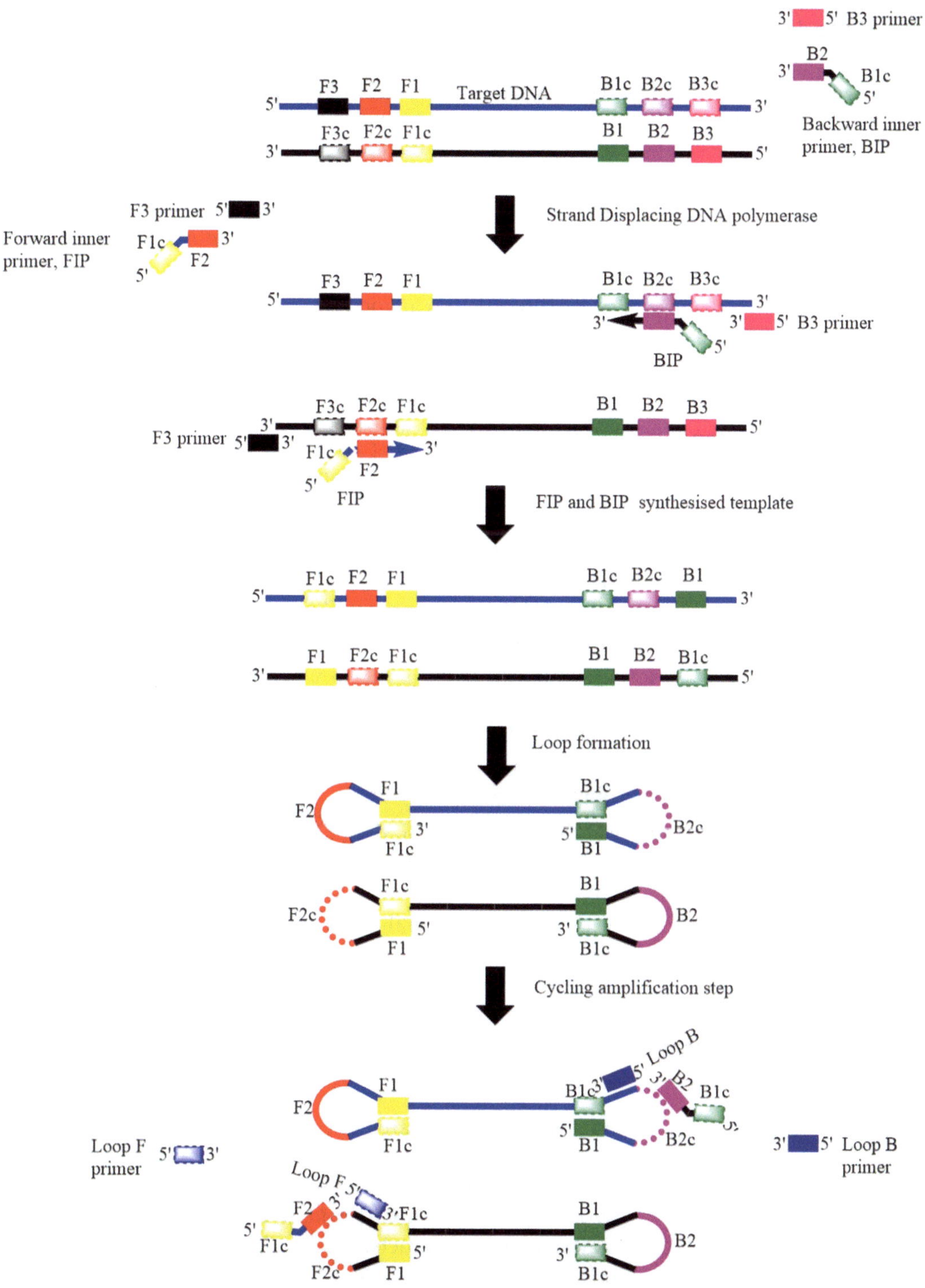

Fig. 1 Schematic description of loop-mediated isothermal amplification (LAMP). The location of four primers that recognize six distinct regions of the target sequence. Exponential amplification is mediated via a strand displacing polymerase

3 Methods

1. Preparation of template. Add an appropriate amount (0.5 ml) of urine to 0.5 ml of water, and heat at 90 °C for 10 min. Dilute heated lysate by tenfold in water (*see* **Note 4**). 5 μl of the diluted lysate can be added directly to LAMP reactions; prepare template in the same manner from *Chlamydia trachomatis* elementary bodies (EB) and/or *Neisseria gonorrhoeae* cells for use as the positive control template.
2. For each required reaction (# samples, plus positive and negative controls), prepare 10× CT or NG primer mix by diluting the oligonucleotides in water to give 10× the final concentration used in the reaction. For CT LAMP 10× primer mix, prepare 4 μM of LF and LB, 2 μM of F3 and B3, and 8 μM of BIP and FIP. For NG LAMP 10× primer mix, prepare 2 μM of F3 and B3, and 8 μM of BIP and FIP. The sequences of the LAMP primers are shown in Table 1.
3. Prepare LAMP reagent mastermix by sequentially adding the following reagents: water, LAMP reaction mix and primer mix to give a reaction volume of 20 μl per reaction. The layout for the preparation of a 10 reaction mastermix is shown in Table 2.
4. Distribute 20 μl of mastermix per well of a 96-well plate or 0.2 ml reaction tubes. Finally, add 5 μl of water (no template control, NTC) or 5 μl of template (DNA extracted from sample) or 5 μl positive control template to give a final reaction volume of 25 μl per reaction.
5. Seal the plate (or close the 0.2 ml reaction tube) and run on a Real-Time PCR Instruments or any fluorescence reader with temperature control capabilities. Set instrument to 64 °C for 60 min, and fluorescence readings taken at one-minute intervals (*see* **Note 5**). Measure fluorescence using the FAM channel.

4 Notes

1. The LAMP reactions are performed using a commercial LAMP reagent kit. The LAMP reagent kit already contains optimized DNA polymerase, intercalating dye (FAM), buffer inorganic pyrophosphatase, $MgCl_2$, and dNTPs for the detection of LAMP reaction products.
2. Portable fluorescence readers such as Genie (Optigene, UK) or the ESEQuant Tube Scanner (Qiagen, Germany) are also suitable.
3. All primers should be HPLC purified.

Table 1
Oligonucleotides used for the CT and NG LAMP assays

Target	Primer/probes/IO	Sequence	Reference
CT-plasmid	LoopB	5′-CGAGCAGCAAGCTATATT-3′	[17]
	LoopF	5′-AAACTCTTGCAGATTCATA-3′	
	LampB	5′GACCGAAGGTACTAAACAAGTTTTTTTGTTTAGGAATCTT-3′	
	LampF	5′-GTTAAGGGCGCATCTAGGATTAGATTAGATTTTATTGGTCTATTGTCCTTGG-3′	
	B3	5′-TATTCCTTGAGTCATCC-3′	
	F3	5′-GATCATATCGAGGATCTT-3′	
NG-*porA*	LampB	5′-TCCCCCGGATTTTCCGGTTTCAGTAGCAGGCGTATAGGCG-3′	[18]
	LampF	5′-GTAACGCACGGAAACCGGCAGTGGCTTCGCAATTGGGTA-3′	
	B3	5′-CCGGAACTGGTTTCATCTGA-3′	
	F3	5′-TTGATCCTTGGGACAGCAAT-3′	

Table 2
Components of LAMP mastermix

Reagent	Stock concentration	Final concentration	Volume for 1 reaction (μl)	Volume for 10 reactions (μl)
Water			2.5	25
LAMP reaction mix	1.67×	1	15	150
10× primer mix	10×	1	2.5	25
Total			20.0	200.0

4. Lower fold dilution (e.g., two- or fourfold dilution) could be used in the absence of sample derived inhibition.
5. LAMP reaction can further be optimized by varying the concentration of primers and magnesium, and the reaction temperature (e.g., 55–67 °C).

References

1. Hoser MJ, Mansukoski HK, Morrical SW, Eboigbodin KE (2014) Strand invasion based amplification (SIBA®): a novel isothermal DNA amplification technology demonstrating high specificity and sensitivity for a single molecule of target analyte. PLoS One 9. https://doi.org/10.1371/journal.pone.0112656
2. Eboigbodin K, Filén S, Ojalehto T, Brummer M, Elf S, Pousi K, Hoser M (2016) Reverse transcription strand invasion based amplification (RT-SIBA): a method for rapid detection of influenza A and B. Appl Microbiol Biotechnol 100(12):5559–5567. https://doi.org/10.1007/s00253-016-7491-y
3. Eboigbodin KE, Moilanen K, Elf S, Hoser M (2017) Rapid and sensitive real-time assay for the detection of respiratory syncytial virus using RT-SIBA®. BMC Infectious Diseases 17(1):134. https://doi.org/10.1186/s12879-017-2227-x
4. Ali MM, Li F, Zhang Z, Zhang K, Kang D, Ankrum JA, Le XC, Zhao W (2014) Rolling circle amplification: a versatile tool for chemical biology, materials science and medicine. Chemical Society Reviews 43(10):3324
5. Piepenburg O, Williams CH, Stemple DL, Armes NA (2006) DNA detection using recombination proteins. PLoS Biol 4(7):e204. https://doi.org/10.1371/journal.pbio.0040204
6. Demidov VV (2002) Rolling-circle amplification in DNA diagnostics: the power of simplicity. Expert Rev Mol Diagn 2(6):542–548. https://doi.org/10.1586/14737159.2.6.542
7. Vincent M, Xu Y, Kong H (2004) Helicase-dependent isothermal DNA amplification. EMBO Rep 5(8):795–800. http://www.nature.com/embor/journal/v5/n8/suppinfo/7400200_S1.html
8. Walker GT, Fraiser MS, Schram JL, Little MC, Nadeau JG, Malinowski DP (1992) Strand displacement amplification—an isothermal, in vitro DNA amplification technique. Nucleic Acids Res 20(7):1691–1696. https://doi.org/10.1093/nar/20.7.1691
9. Van Ness J, Van Ness LK, Galas DJ (2003) Isothermal reactions for the amplification of oligonucleotides. Proc Natl Acad Sci 100(8):4504–4509. https://doi.org/10.1073/pnas.0730811100
10. Notomi T, Okayama H, Masubuchi H, Yonekawa T, Watanabe K, Amino N, Hase T (2000) Loop-mediated isothermal amplification of DNA. Nucleic Acids Res 28(12):e63. https://doi.org/10.1093/nar/28.12.e63
11. Kundapur RR, Nema V (2016) Loop-mediated isothermal amplification: beyond microbial identification. Cogent Biology 2(1):1137110. https://doi.org/10.1080/23312025.2015.1137110
12. Notomi T, Mori Y, Tomita N, Kanda H (2015) Loop-mediated isothermal amplification (LAMP): principle, features, and future

prospects. J Microbiol 53(1):1–5. https://doi.org/10.1007/s12275-015-4656-9
13. Goto M, Honda E, Ogura A, Nomoto A, Hanaki K-I (2009) Colorimetric detection of loop-mediated isothermal amplification reaction by using hydroxy naphthol blue. BioTechniques 46(3):167–172. https://doi.org/10.2144/000113072
14. Wang Y, Li H, Wang Y, Zhang L, Xu J, Ye C (2017) Loop-mediated isothermal amplification label-based gold nanoparticles lateral flow biosensor for detection of *Enterococcus faecalis* and *Staphylococcus aureus*. Front Microbiol 8:192–192. https://doi.org/10.3389/fmicb.2017.00192
15. Prevention CfDCa (2014) Sexually transmitted disease surveillance 2013. U.S. Department of Health and Human Services; 2014. Centers for Disease Control and Prevention, Atlanta
16. Baugh LR, Hill AA, Brown EL, Hunter CP (2001) Quantitative analysis of mRNA amplification by in vitro transcription. Nucleic Acids Res 29(5):e29. https://doi.org/10.1093/nar/29.5.e29
17. Gandelman OA, Church VL, Moore CA, Kiddle G, Carne CA, Parmar S, Jalal H, Tisi LC, Murray JAH (2010) Novel bioluminescent quantitative detection of nucleic acid amplification in real-time. PLoS One 5(11):e14155. https://doi.org/10.1371/journal.pone.0014155
18. Eboigbodin KE, Hoser MJ (2016) Multiplex strand invasion based amplification (mSIBA) assay for detection of *Chlamydia trachomatis* and *Neisseria gonorrhoeae*. Sci Rep 6:20487. https://doi.org/10.1038/srep20487. https://www.nature.com/articles/srep20487#supplementary-information

Chapter 4

Simultaneous Detection of *Chlamydia trachomatis* and *Neisseria gonorrhoeae* Using Real-Time Multiplex qPCR Assay

Kevin E. Eboigbodin

Abstract

Real-time polymerase chain reaction (qPCR) has become a prominent technique in life science research particularly for the detection and monitoring of biomarkers, pathogens, and environmental contaminants. *Chlamydia trachomatis* (CT) and *Neisseria gonorrhoeae* (NG) are among the major pathogens responsible for sexually transmitted diseases (STDs). Here, multiplex qPCR was utilized for the amplification and detection *Chlamydia trachomatis* and *Neisseria gonorrhoeae* within the same reaction tube.

Key words Amplification, Detection, Diagnosis, Multiplex, qPCR, *Chlamydia trachomatis*, *Neisseria gonorrhoeae*

1 Introduction

Sexually transmitted infections (STIs) are among the leading cause of infection worldwide and carry an enormous economic and healthcare burden [1]. *Chlamydia trachomatis* (CT) and *Neisseria gonorrhoeae* (NG) are among the major pathogens responsible for STIs [2, 3]. Infections caused by *Chlamydia trachomatis* (*CT*) and *Neisseria gonorrhoeae* (*NG*) are often treatable particularly when detected early. Consequently, prompt and accurate diagnosis is important for deciding appropriate treatment and preventing the spread of the infection. Real-time polymerase chain reaction (qPCR) is the most widely used nucleic acid amplification technique (NAAT) for the diagnosis of STIs. qPCR offer superior sensitivity and specificity over immunoassays and traditional culture based methods [4, 5].

Despite the emergence of non-PCR based nucleic acid amplification techniques, PCR still remains the gold standard NAAT [6–11]. Due to the high analytical sensitivity displayed by qPCR, CT and NG can be directly detected from noninvasive specimens such

Amanda Claire Brown (ed.), *Chlamydia trachomatis: Methods and Protocols*, Methods in Molecular Biology, vol. 2042,
https://doi.org/10.1007/978-1-4939-9694-0_4,

as urine and vaginal swabs [9]. qPCR uses a Taq polymerase and two flanking primers for the amplification of a target nucleic acid sequence. qPCR reactions are performed under thermal cycling conditions allowing for repeated cycles of dissociation and extension of a target duplex. The orchestrated cycles of dissociation and extension lead to an exponential amplification of the target duplex. The amplification reaction can be detected using either intercalating dyes or a target specific probe.

Coinfection of CT and NG is often common (and can be as high as 50%), therefore it is often more efficient and cost effective to simultaneously process and detect CT and NG from suspected patient specimens [12, 13]. Multiplex qPCR allows CT and NG to be simultaneously detected and distinguish within a single reaction tube. In multiplex qPCR, two sets of primers and probes are designed to detect specific sequence of CT and NG respectively. The probe for CT and NG are labeled with distinct fluorophores that enable each pathogen to be detected and distinguished from each other. Several probe chemistries have been developed for performing multiplex qPCR reactions [14]. These include TaqMan, Scorpions, Lux, and molecular beacon probes, allowing for more than one target to be detected in a single qPCR reaction tube [15, 16].

Here, a TaqMan probed based multiplex qPCR assay is described for the detection of CT and NG in the same reaction tube [6, 17]. The multiplex qPCR assay also includes an internal control (IC) for assessing potential sample-related inhibition. The qPCR primers and probe sets for CT are used to amplify and detect the conserved region of *porA*. A set of qPCR primers and probe were also used to amplify and detect an artificial DNA sequence which served as the internal control. CT, NG, and IC probes were labeled with FAM, ROX, and Cy5 reported dyes. This allowed CT, NG, and IC to be detected and distinguish within the same reaction tube.

Typical general laboratory equipment is not listed below.

2 Materials

1. iTaq Universal Probes Supermix reagent (Bio-Rad, UK) (*see* **Note 1**).
2. DNA extraction kit (*see* **Note 2**).
3. Biosafety cabinet.
4. Vortex mixer.
5. Heating block.
6. Real-time PCR Instrument (*see* **Note 3**).
7. Centrifuge.
8. PCR plates (96-well plate).

9. PCR sealing film.
10. 1.5 ml reaction tubes.
11. Molecular grade water.
12. Primers and probes (*see* **Note 4**).
13. Synthetic internal control template.
14. Positive template (e.g., quantitated CT cryptic plasmid and NG DNA).

3 Methods

1. Extract DNA from patient specimens such as urine and vaginal swabs using commercial nucleic acid extraction kit. The DNA can be extracted from *Chlamydia trachomatis* cells/elementary bodies (EB) and *Neisseria gonorrhoeae* cells which serves as the positive control for the qPCR reactions. Alternatively, commercially quantitated CT cryptic plasmid and NG DNA can be used as the positive control. Use water and/or nucleic acids from unrelated microbes as the negative control.
2. For each required reaction (# samples, plus positive and negative controls), prepare 10× CT primer mix, 10× NG primer mix and 20× IC primer mix by diluting the oligonucleotides in water to give 10× (20× for the IC primer mix) the final concentration used in the reaction. For CT 10× primer mix, prepare 2.5 μM of CT forward primer, 2.5 μM of CT reverse primer and 1 μM of CT probe. For NG 10× primer mix, prepare 2.5 μM of NG forward primer, 2.5 μM of NG reverse primer and 1 μM of NG probe. For IC 20× primer mix, prepare 2 μM of IC forward primer, 2 μM of IC reverse primer, 2 μM of NG probe, and 1200 copies per μl of IC template. The sequences of the multiplex qPCR oligonucleotides are shown in Table 1.
3. Prepare multiplex qPCR mastermix by sequentially adding the following reagents into a 1.5 ml reaction tube: water, itAQ Universal probe Supermix, 10× CT primer mix, 10× NG primer mix, and 20× IC primer mix to give a reaction volume of 20 μl per reaction. Mix gently but thoroughly and briefly centrifuge at 1000 × *g* for 30 s. The layout for the preparation of a 10 reaction mastermix is shown in Table 2.
4. Distribute 20 μl of multiplex qPCR mastermix per well on a qPCR plate. Finally, add 5 μl of positive control template (DNA extracted from cells or EB), 5 μl of water (no template control, NTC) or 5 μl DNA extracted from specimens to give a final reaction volume of 25 μl per reaction.

Table 1
Oligonucleotides used for the multiplex qPCR CT/NG assay

Target	Name	Sequence	Reference
CT-plasmid	F-primer	5′-AACCAAGGTCGATGTGATAG-3′	[17]
	R-primer	5′-TCAGATAATTGGCGATTCTT-3′	
	Probe-FAM	5′-(FAM)CGAACTCATCGGCGATAAGG(BHQ1)-3′	
NG-*porA*	F-primer	5′-CAGCATTCAATTTGTTCCGAGTC-3′	
	R-primer	5′-GAACTGGTTTCATCTGATTACTTTCCA-3′	
	Probe-ROX	5′-(ROX)CGCCTATACGCCTGCTACTTTCACGC(BHQ2)-3′	
Internal control	F-primer	5′-GTGCTCACACCAGTTGCCGC-3′	
	R-primer	5′-GCTTGGCAGCTCGCATCTCG-3′	
	Probe-CY5	5′-(CY5)ATTGTGTGGGTGTGGTGTGGGTGTGTGC(BHQ3)-3′	
	Template	5-GTGCTCACACCAGTTGCCGCGGAAAGTATGTGGAATGTT AACACACCCACACCACACCCACACACGTGTTGGATCAA TTTCGAGATGCGAGCTGCCAAGC-3′	

Table 2
Components of multiplex qPCR mastermix

Reagent	Stock concentration	Final concentration	Volume for 1 reaction (μl)	Volume for 10 reactions (μl)
Water			1.25	12.5
itAQ Universal probe supermix	2×	1	12.5	125
10× CT primer mix	10×	1	2.5	25
10× NG primer mix	10×	1	2.5	25
20× IC primer mix	20×	1	1.25	12.5
Total			20.0	180.0

5. Seal the plate using the manufacturer's recommended qPCR plate sealing film and briefly centrifuge at 1000 × *g* for 30 s. Place plate and run on a real-time PCR Instruments. Set instrument to 5 min denaturation step at 95 °C, then 45 cycles of 95 °C for 15 s, 60 °C for 60 s, and 72 °C for 1 s. Run a final incubation step at 40 °C for 30 s.

4 Notes

1. The multiplex qPCR reactions are performed using a Bio-Rad qPCR probe reagent kit. Other commercial qPCR probe reagent kit may also be suitable.
2. Several commercial nucleic acid extraction kits are available for the extraction of DNA from cells and specimens. For example Qiagen extraction kits.
3. The multiplex qPCR reactions are performed using a Bio-Rad CFX95 PCR instrument. However, the qPCR reaction can be performed using other qPCR instruments such as the Roche LightCycler LC480, Applied Biosystems real-time PCR instruments, Qiagen Rotor-Gene, and Agilent Mx real-time PCR.
4. All primers should be HPLC-purified. Dissolve primer stock in Tris–EDTA buffer (10 mM Tris, 1 mM EDTA pH 8.0).

References

1. Newman L, Rowley J, Vander Hoorn S, Wijesooriya NS, Unemo M, Low N, Stevens G, Gottlieb S, Kiarie J, Temmerman M (2015) Global estimates of the prevalence and incidence of four curable sexually transmitted infections in 2012 based on systematic review and global reporting. PLoS One 10(12): e0143304. https://doi.org/10.1371/journal.pone.0143304
2. Prevention CfDCa (2014) Sexually Transmitted Disease Surveillance 2013. U.S. Department of Health and Human Services, Atlanta. Centers for Disease Control and Prevention
3. Baugh LR, Hill AA, Brown EL, Hunter CP (2001) Quantitative analysis of mRNA amplification by in vitro transcription. Nucleic Acids Res 29(5):e29. https://doi.org/10.1093/nar/29.5.e29
4. Papp JR, Schachter J, Gaydos CA, Van Der Pol B (2014) Recommendations for the laboratory-based detection of *Chlamydia trachomatis* and *Neisseria gonorrhoeae* — 2014. MMWR Recomm Rep 63(RR-02):1–19
5. Van Dyck E, Ieven M, Pattyn S, Van Damme L, Laga M (2001) Detection of *Chlamydia trachomatis* and *Neisseria gonorrhoeae* by enzyme immunoassay, culture, and three nucleic acid amplification tests. J Clin Microbiol 39 (5):1751–1756. https://doi.org/10.1128/jcm.39.5.1751-1756.2001
6. Eboigbodin KE, Hoser MJ (2016) Multiplex strand invasion based amplification (mSIBA) assay for detection of *Chlamydia trachomatis* and *Neisseria gonorrhoeae*. Sci Rep 6:20487. https://doi.org/10.1038/srep20487. https://www.nature.com/articles/srep20487#supplementary-information
7. Kundapur RR, Nema V (2016) Loop-mediated isothermal amplification: beyond microbial identification. Cogent Biology 2 (1):1137110. https://doi.org/10.1080/23312025.2015.1137110
8. Eboigbodin K, Filén S, Ojalehto T, Brummer M, Elf S, Pousi K, Hoser M (2016) Reverse transcription strand invasion based amplification (RT-SIBA): a method for rapid detection of influenza A and B. Appl Microbiol Biotechnol 100(12):5559–5567. https://doi.org/10.1007/s00253-016-7491-y
9. Gaydos CA, Quinn TC, Willis D, Weissfeld A, Hook EW, Martin DH, Ferrero DV, Schachter J (2003) Performance of the APTIMA combo 2 assay for detection of *Chlamydia trachomatis* and *Neisseria gonorrhoeae* in female urine and endocervical swab specimens. J Clin Microbiol 41(1):304–309. https://doi.org/10.1128/jcm.41.1.304-309.2003
10. Eboigbodin K, Filén S, Ojalehto T, Brummer M, Elf S, Pousi K, Hoser M (2016) Reverse transcription strand invasion based amplification (RT-SIBA): a method for rapid detection of influenza A and B. Appl Microbiol

Biotechnol 1–9. https://doi.org/10.1007/s00253-016-7491-y

11. Eboigbodin KE, Moilanen K, Elf S, Hoser M (2017) Rapid and sensitive real-time assay for the detection of respiratory syncytial virus using RT-SIBA®. BMC Infectious Diseases 17(1):134. https://doi.org/10.1186/s12879-017-2227-x
12. Forward KR (2010) Risk of coinfection with *Chlamydia trachomatis* and *Neisseria gonorrhoeae* in Nova Scotia. Can J Infect Dis Med Microbiol 21(2):e84–e86
13. Lim RBT, Wong ML, Cook AR, Brun C, Chan RKW, Sen P, Chio M (2015) Determinants of chlamydia, gonorrhea, and coinfection in heterosexual adolescents attending the national public sexually transmitted infection clinic in Singapore. Sex Transm Dis 42(8):450–456. https://doi.org/10.1097/olq.0000000000000316
14. Navarro E, Serrano-Heras G, Castaño MJ, Solera J (2015) Real-time PCR detection chemistry. Clin Chim Acta 439:231–250. https://doi.org/10.1016/j.cca.2014.10.017
15. Besson G, Kazanji M (2009) One-step, multiplex, real-time PCR assay with molecular beacon probes for simultaneous detection, differentiation, and quantification of human T-cell leukemia virus types 1, 2, and 3. J Clin Microbiol 47(4):1129–1135. https://doi.org/10.1128/JCM.02006-08
16. Cura CI, Duffy T, Lucero RH, Bisio M, Péneau J, Jimenez-Coello M, Calabuig E, Gimenez MJ, Valencia Ayala E, Kjos SA, Santalla J, Mahaney SM, Cayo NM, Nagel C, Barcán L, Málaga Machaca ES, Acosta Viana KY, Brutus L, Ocampo SB, Aznar C, Cuba Cuba CA, Gürtler RE, Ramsey JM, Ribeiro I, VandeBerg JL, Yadon ZE, Osuna A, Schijman AG (2015) Multiplex real-time PCR assay using TaqMan probes for the identification of *Trypanosoma cruzi* DTUs in biological and clinical samples. PLoS Negl Trop Dis 9(5):e0003765. https://doi.org/10.1371/journal.pntd.0003765
17. Hopkins MJ, Ashton LJ, Alloba F, Alawattegama A, Hart IJ (2010) Validation of a laboratory-developed real-time PCR protocol for detection of *Chlamydia trachomatis* and *Neisseria gonorrhoeae* in urine. Sex Transm Infect 86(3):207–211. https://doi.org/10.1136/sti.2009.040634

Chapter 5

Antimicrobial Resistance Screening in *Chlamydia trachomatis* by Optimized McCoy Cell Culture System and Direct qPCR-Based Monitoring of Chlamydial Growth

Tomislav Meštrović, Dezső P. Virok, Sunčanica Ljubin-Sternak, Tímea Raffai, Katalin Burián, and Jasmina Vraneš

Abstract

Obligate intracellular localization of *Chlamydia trachomatis* (*C. trachomatis*) complicates antimicrobial sensitivity testing efforts that we are so accustomed to in routine bacteriology. Cell culture systems with immunofluorescence staining, to identify cellular inclusions in the presence of various concentrations of antimicrobial drugs, are still the most pervasive techniques, but more specific and sensitive nucleic acid concentration measuring methods are increasingly being used. Here we describe how to approach antimicrobial susceptibility/resistance screening in *C. trachomatis* by using a McCoy cell culture system, optimized by a research group from Croatia, and direct qPCR-based monitoring of chlamydial growth, optimized by a research group from Hungary.

Key words *Chlamydia trachomatis*, Antimicrobial susceptibility testing, Antimicrobial resistance, Cell culture, qPCR

1 Introduction

Antimicrobial susceptibility testing and screening for resistance in *Chlamydia trachomatis* (*C. trachomatis*) significantly differs from standard procedures in bacteriology, due to the pathogen's obligate intracellular localization in the human host [1–3]. Cell culture systems with immunofluorescence staining, to identify cellular inclusions in the presence of various concentrations of antimicrobial drugs, are still most commonly used [2]. McCoy cells, derived from mouse fibroblasts, offer the most comparable and consistent results, albeit HeLa (human cervical adenocarcinoma), HL (human epithelial cells) and HEp-2 (human epidermoid laryngeal carcinoma) can also be utilized [1]. A cell culture method that uses

Tomislav Meštrović and Dezső P. Virok contributed equally to this work.

Amanda Claire Brown (ed.), *Chlamydia trachomatis: Methods and Protocols*, Methods in Molecular Biology, vol. 2042, https://doi.org/10.1007/978-1-4939-9694-0_5, © Springer Science+Business Media, LLC, part of Springer Nature 2019

McCoy cells was optimized by a Croatian research group, according to the recommendations of Suchland et al. [1, 4–6] and other research groups in this field [7–10].

Still, although the fluorescent inclusion labeling and optic counting is the gold standard of chlamydial growth monitoring, there are several problems with this approach—including the extended time required for labeling and manual reading, subjectivity of inclusion calling, subjectivity of judging atypical inclusions, low dynamic range and eventually low-throughput. There are immunofluorescence based approaches to circumvent the manual reading, including automatic microscopy [11], fluorescent plate reading [12] and DNA-chip based inclusion scanning and counting [13], but these methods are medium-throughput approaches, with typically low dynamic ranges, and, in the case of automatic microscopy, the cost of the hardware is prohibitive. Also, it should be noted, inclusion count does not provide information of the number of bacteria in the inclusion, and a second step is needed to measure the concentration of at least the infectious elementary bodies (EB)—the so-called recoverable inclusion forming unit (IFU).

To circumvent these problems, nucleic acid concentration measuring methods as a correlate to chlamydial growth were developed where, instead of inclusion counting, the accumulation of chlamydial genome or transcript (e.g., rRNA) is quantitated by preferentially quantitative PCR (qPCR) [14]. These methods are objective, and due to the nature of PCR have high dynamic range. The only significant problem is the required nucleic acid purification step, which is a relatively expensive multistep process and therefore significantly reduces the throughput. A research group from Hungary developed a method where, instead of purification, the infected host cells are simply lysed by two freeze–thaw cycles and then a direct EvaGreen based qPCR of the cell lysates can be performed with *Chlamydia*-specific primers [14]. This method has minimal hands-on time, high dynamic range and specificity, and was successfully applied to measure the effect of various antibiotics [15], antiseptics [16] and *Chlamydia* growth promoting compounds [17]. It should be noted that this method measures genomic DNA, therefore the viability of the bacterium cannot be evaluated. If viability is the question in general, or the relative concentrations of EB, reticulate bodies (RB) and aberrant bodies (AB), than reverse transcription-qPCR (RT-qPCR) should be applied with general primers (e.g., rRNA) or EB, RB, and AB specific primers to measure state-specific gene expression [14, 18]. In these cases, due to the instability of the RNA in cell lysates, the RNA purification likely cannot be omitted. Here we describe the methodology for the direct qPCR-based monitoring of Chlamydial growth.

2 Materials

Standard laboratory equipment (e.g., centrifuge, freezer, inverse fluorescent microscope, adequate safety cabinet) is essential for using this kind of methodology. Prepare and keep all reagents at room temperature (unless indicated otherwise). Conscientiously follow all waste disposal regulations when discarding waste materials, and local biosafety rules when working with potential pathogens (*see* **Note 1**).

2.1 McCoy Cell Culture System

1. 10% heat-inactivated fetal bovine serum.
2. 4 Sucrose Phosphate (4SP) buffer for 250 mL: dissolve 34.23 g sucrose in 150 mL of sterile water; separately add 0.567 g Na_2HPO_4 to 50 mL of sterile water, add to sucrose solution, mix, adjust pH to 7.1 and volume to 250 mL. Filter sterilize and store at 4 °C.
3. Antimicrobial powders for MEM supplementation: cycloheximide, gentamicin and vancomycin.
4. Antimicrobial powders for susceptibility testing.
5. Buffered glycerol mounting medium.
6. Dulbecco's Modified Eagle's Medium (DMEM): 4 mM L-glutamine, 4500 mg/L glucose, 1 mM sodium pyruvate, and 1500 mg/L sodium bicarbonate.
7. Eagle's Minimum Essential Medium (MEM) with Earle's balanced salts, L-glutamine, nonessential amino acids, and sodium pyruvate.
8. Fluorescein isothiocyanate-conjugated monoclonal antibodies against *C. trachomatis* species-specific major membrane outer protein (MOMP) or chlamydial genus-specific lipid polysaccharide (LPS).
9. McCoy cells (ATCC CRL-1696).
10. Methanol fixative.
11. Sterile microtiter plates with flat bottom specified for cell culture growth.
12. CO_2 incubator.
13. Phosphate buffer saline (PBS): 0.13 M NaCl, 2.68 μM KCl, 8.74 μM $Na_2PO_4 \times 12H_2O$ and 1.47 μM KH_2PO_4.
14. Pipette tips, sterile.
15. Plastic container.
16. 4-mm-diameter sterile glass beads.
17. Sterile pipettes and multichannel pipettes.

2.2 Direct qPCR-Based Monitoring of Chlamydial Growth

1. HeLa 229 cells (ATCC).
2. Cell culture flasks.
3. Eppendorf tubes.
4. Filter tips, sterile.
5. Minimal Essential Medium (MEM) with Earle salts supplemented with 10% heat-inactivated fetal bovine serum (FBS), 2 mmol/L L-glutamine, 1× MEM vitamins, 1× nonessential amino acids, 0.005% Na pyruvate, 25 μg/mL gentamycin, 1 μg/mL fungisone.
6. 96-well microtiter plates with flat bottom specified for cell culture growth, sterile.
7. Milli-Q water.
8. Motorized pipette filler.
9. Oligonucleotide primers, designed to Chlamydia genes (*see* **Note 2**).
10. Phosphate buffer saline (PBS): 0.13 M NaCl, 2.68 μM KCl, 8.74 μM $Na_2PO_4 \times 12H_2O$ and 1.47 μM KH_2PO_4.
11. Pipette tips, sterile.
12. Plastic container.
13. qPCR Master Mix with SYBR Green or EvaGreen; the master mix is a concentrated solution of DNA Polymerase, dNTPs, and all of the components required for PCR, except DNA template and primers.
14. Strip PCR tubes and caps.
15. Sterile, conical, polystyrene test tubes.
16. Sterile pipette tips, single and multichannel pipettes.
17. Sterile serological pipettes.
18. Sucrose-Phosphate-Glutamate Buffer (SPG) 218 mM Sucrose, 3.8 mM KH_2PO_4, 7.2 mM K_2HPO_4, 4.9 mM L-glutamate, pH 7.2.
19. Trypsin protease.

3 Methods

3.1 McCoy Cell Culture System

1. Seed 100 μL of McCoy cell suspension containing 50,000 cells into wells of a 96-well tissue culture plate 24 h before infecting them with *C. trachomatis* and incubate overnight at 37 °C with 5% CO_2. On the day of infection remove the medium and replace it with 100 μL inoculum containing *C. trachomatis* isolates, to give an MOI of 0.5–1 (*see* **Notes 1**, **3**, and **4**).

2. Centrifuge the plate at 750 × *g* for 30 min, and then replace the inoculum with 100 μL DMEM containing 1 μg/mL of cycloheximide, 20 μg/mL of gentamicin and 10 μg/mL of vancomycin. Incubate the infected cells at 37 °C with 5% CO_2 for 48 h (*see* **Note 5**).
3. Two days after infection, harvest the cells from infected wells using a sterilized pipette tip, and transfer to a sterile screw-capped tube containing 4-mm-diameter sterile glass beads to an equivalent of 3 mL. Using a vortex agitate for 1 min to release EBs from the cells, centrifuge at 110 × *g* for 5 min to remove cell debris, and remove the supernatant and add to an equal volume of 4 Sucrose Phosphate (4SP). Store it at −70 °C.
4. After thawing, amplify chlamydial isolates designated for antimicrobial susceptibility testing by multiple passages through 24 h old McCoy cell monolayers in MEM without cycloheximide and antibiotics, so that they infect approximately 10–50% of the cells to achieve sufficient amount and inoculum size of 5000–10,000 inclusion forming units (IFU) per well of the microtiter plate.
5. Reconstitute antimicrobial agents according to the manufacturer's instructions. Prepare each drug as a stock solution at a concentration of 1024 μg/mL, according to its potency calculated from assay purity by high-performance liquid chromatography (HPLC), water content and active fraction of each substance, as specified by the manufacturer (*see* **Note 6**) [18]. Use the following formula for potency: ***potency = assay purity × active fraction × (1—water content)***.
6. Aspirate and discard the growth medium from the fresh microtiter plate (where McCoy cells were seeded 24 h prior to its use) and inoculate each well with 100 μg of the previously amplified chlamydial test strain.
7. Incubate two batches of plates in parallel, and centrifuge them for 1 h at 1200 × *g* at 37 °C. After centrifugation immediately transfer plates into the incubator at 37 °C with 5% CO_2 for an additional 2 h.
8. During this period serially dilute previously prepared antimicrobial stock solutions from the initial concentration of 1024–0.008 μg/mL in MEM containing 10% heat-inactivated fetal bovine serum and 1 μg/mL of cycloheximide.
9. At the end of a 2 h incubation period, remove the microtiter plates from the incubator and aspirate the supernatants with multichannel pipette. Add 100 μL of the appropriately diluted antimicrobial drug to the appropriate wells for a minimum final concentration range of 0.008–8 μg/mL (and preferably up to 128 μg/mL). Always use antibiotic-free controls.

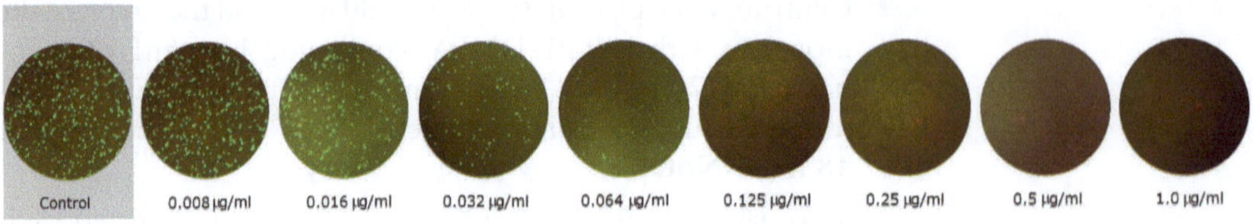

Fig. 1 Photomicrographs of growth pattern of *C. trachomatis* at various concentrations of antibiotic azithromycin showing determination of minimal inhibitory concentration (MIC), defined as the lowest concentration of antibiotic without visible inclusions (0.125 µg/mL)—one dilution higher than MIC_{TP} (0.064 µg/mL) (Reproduced from ref. 4 with permission from Springer)

10. Incubate the plates for 72 h at 37 °C in 5% CO_2, and then fixate the cultures in just one microtiter plate with methanol. Stain the inclusions in that plate with either genus specific or species-specific fluorescein conjugated monoclonal antibodies (*see* **Note 7**). Freeze the second plate at −70 °C in order to make a passage of cultures by employing a freeze–thaw method.
11. Transfer the plate with stained inclusions to a moist chamber and incubate it protected from light for 30 min at room temperature (23 ± 3 °C). Upon washing the wells twice with PBS, fix them with buffered glycerol mounting medium and analyze under the invert fluorescent microscope to ascertain minimal inhibitory concentration (MIC) values.
12. Determine MIC by using the transition point MIC (MIC_{TP}), defined as the concentration of drug where there is a significantly lower number of inclusions and at least 90% of them show altered morphology and size (Fig. 1) [1, 8]. Interpret MIC as the drug concentration that is one twofold dilution higher than the MIC_{TP} (*see* **Note 8**).
13. Thaw the second plate, and repeat the procedure as for the first plate—that is, pass cultures onto new cells, centrifuge, overlay with antibiotic-free medium (MEM with cycloheximide only), incubate for 72 h, fixate with methanol, stain, incubate in moist chamber, fixate with glycerol. Finally, analyze the plate under the invert fluorescent microscope to ascertain minimal chlamydicidal concentration (MCC) values (*see* **Note 9**).

3.2 Direct qPCR-Based Monitoring of Chlamydial Growth

1. Seed HeLa 229 cells (ATCC) in 96-well plates at a density of 4×10^4 cells/well in 100 µL of minimal essential medium (MEM) with Earle's salts supplemented with 10% heat-inactivated fetal bovine serum, 2 mmol/L of L-glutamine, 1× MEM vitamins, 1× nonessential amino acids, 0.005% Na pyruvate, 25 µg/mL of gentamycin, and 1 µg/mL of fungisone.
2. Incubate the transferred cells for 1 h at room temperature to avoid edge formation, and then overnight at 37 °C, 5% CO_2 to obtain a 90% confluent cell layer.

3. The next day, before the infection, wash the cell monolayer twice with 100 μL/well with sterile phosphate buffer saline (PBS (0.13 M NaCl, 2.68 μM KCl, 8.74 μM $Na_2PO_4 \times 12H_2O$, and 1.47 μM KH_2PO_4)) at pH 7.4 and at room temperature.
4. Suspend the *Chlamydia* inoculum in 0.5% (w/v) MEM containing 0.5% glucose and add to the cell layers at the given multiplicity of infection (MOI). qPCR has been shown to accurately detect *C. trachomatis* growth between 0.004 and 64 MOI [15]. Perform the infection by incubation for 60 min at 37 °C, 5% CO_2, or by centrifugation for 60 min at 600 × *g* (*see* **Note 10**).
5. Following the infection, wash the cells twice with 100 μL sterile PBS, and add MEM culture medium containing 1 μg/mL of cycloheximide and the tested antimicrobial compound (e.g., ciprofloxacin by twofold dilutions in the concentration range from 16 to 0.0078 μg/mL). Incubate the cells at 37 °C, 5% CO_2 for 48 h (incubation time is dependent on the *Chlamydia* strain) (*see* **Note 11**).
6. After 48 h of incubation remove the MEM culture medium and wash the HeLa cells with 100 μL PBS twice, and finally add 100 μL PBS (or MQ water) to the cells before lysis.
7. Accomplish the cell lysis by 2× freeze–thaw cycles: 45 min at −80 °C followed by thawing on a plate shaker at room temperature, once thawed repeat. At the end of the cell lysis thoroughly mix the lysed cells (including the edges of the wells) using a multichannel pipette (*see* **Note 12**).
8. Use the mixed lysates as a template for the qPCR analysis without further dilution. Perform the qPCR in a 20 μL final volume, containing 4 μL cell lysate, 2–2 μL primers (10 pmol/μL) and 10 μL 2× qPCR MasterMix (SsoFast EvaGreen Supermix, Bio-Rad) (*see* **Notes 2** and **13–15**).
9. To objectively determine MIC, compare the cycle threshold (Ct) values of the samples replicates ($n \geq 3$) with the highest antibiotic concentration to each of the sample replicates ($n \geq 3$) with the lower antibiotic concentrations. We consider the samples with the highest antibiotic concentration as the inoculum. Figure 2 shows the qPCR analysis of the ciprofloxacin dose-response curve on *C. trachomatis* growth. For visibility purposes, the *y* axis shows the average negative Ct values of the samples ($n = 3$ for each antibiotic concentrations), therefore the *y* axis showing a ~log2 scale of the chlamydial genome concentration (growth). Since the *x* axis shows the 1:2 dilutions of ciprofloxacin, the *x* axis is also on log2 scale.
10. Perform Student's *t*-tests between the Ct levels of the replicates of the inoculum and the replicates with a given antibiotic

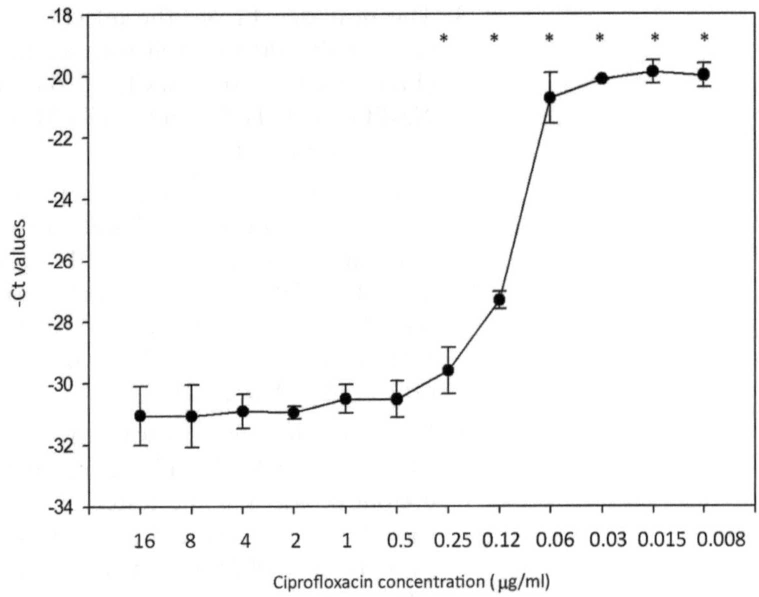

Fig. 2 qPCR based evaluation of ciprofloxacin MIC on *C. trachomatis*. HeLa cells were infected with *C. trachomatis* (MOI 0.2) in the presence of various concentrations of ciprofloxacin. 48 h post infection chlamydial genome concentrations were measured by qPCR, and ciprofloxacin MIC was calculated as described in the text. $^{*}P < 0.05$

concentration. Consider the antibiotic concentration of the last sample replicates where the Ct levels do not differ significantly ($P < 0.05$) when compared to the inoculum as the MIC value. As Fig. 2. shows, the 0.5 μg/mL ciprofloxacin concentration was the last that did not differ significantly from the inoculum (the 16 μg/mL ciprofloxacin concentration sample), therefore the ciprofloxacin MIC was 0.5 μg/mL.

4 Notes

1. As *C. trachomatis* represents a microorganism from the biosafety level 2 (BSL 2) group of agents, be sure to follow local required rules and practices that include personal protective measures and equipment, decontamination and adequate biohazardous material waste.
2. Any primers specific for a chlamydial gene theoretically can be used since any primer pairs can detect genomic DNA. We advise to try a few primer pairs because there can be difference in sensitivity, primer-dimer formation and general specificity between them [15].

3. Prior to inoculating with *C. trachomatis*, check the McCoy cells under the microscope to assess the adequacy of growth and to rule out eventual presence of infection.
4. Eagle's minimal essential medium (MEM) can also be used in the isolation step, with a formulation that usually incorporates either Hanks' or Earle's balanced salts in order to provide adequate buffering of the medium [19]. However, Dulbecco's modified Eagle's medium (DMEM) has approximately 4× as much amino acids and vitamins, as well as 2–4× as much glucose when compared to the original formula, and it also contains iron and phenol red, making it suitable for most cell types.
5. After 48 h chlamydial inclusions should be large enough and observable under the light microscope, so that you can determine the extent of infection visually [20].
6. Keep such stock solutions at −70 °C for a maximum period of 2 weeks, as there is a potential loss of potency of antimicrobial agents after that time period.
7. High specificity of cell culture detection is achieved when a *C. trachomatis* MOMP-specific stain is used. Although monoclonal antibodies directed against the family-specific LPS of *Chlamydiaceae* are less costly, bacteria that share LPS antigens may be stained as well. Thus LPS stains might be suitable for routine use, but a species-specific (MOMP) stain is advised in situations that necessitate increased specificity [21].
8. The reproducibility of *C. trachomatis* antimicrobial sensitivity testing in cell culture systems relies heavily on the endpoint utilized for defining MIC, as differentiating aberrant from normal inclusions can be cumbersome, and subjective interpretation may result in several-fold variation in the obtained MICs. In that regard, MIC_{TP} provides a consistent endpoint at which practically all chlamydial inclusions are inhibited; thus, using a twofold concentration above the MIC_{TP} is considered a reasonable standardized approach to define the MIC (Fig. 1) [1, 5, 8].
9. The MCC is defined as the lowest concentration of drug with no visible inclusions after one passage from the cell culture containing antimicrobial drugs to the cell culture without them. It represents the lowest dilution where the culture has been sterilized completely [1, 5, 8].
10. Due to the high sensitivity of the qPCR lower MOIs can be used without centrifugation. The lack of centrifugation is necessary when the influence of a compound on the attachment/entry of the *Chlamydia* is being studied.

11. Washing to remove the remnants of the inoculum is important because the stock *Chlamydia* contains debris and chlamydial DNA even after purification. This remnant DNA increases the base level of chlamydial DNA, making detection of small changes of chlamydial DNA accumulation difficult to follow.
12. Instead of −80, −20 °C also can be used for cell lysis. The lysed samples in 96-well plates are usually stored in −20 °C freezer up to a few weeks. As was described before [15], DNA purification does not increase the sensitivity of the qPCR compared to the direct qPCR, possibly due to the fact that during the purification fraction of the chlamydial DNA is unavoidably being lost. However, a disadvantage of the direct qPCR is the possible presence of PCR inhibitors in the cell lysates. While we rarely notice this effect, the potential inhibition should be monitored by performing a qPCR with chlamydial DNA diluted, for example, 1:2 by untreated (e.g., antibiotic-free) cell lysate and chlamydial DNA diluted 1:2 with cell lysate containing the highest concentration of the applied compound (e.g., antibiotic). If there is a significant inhibition (or promotion) due to the applied compound, there should be a noticeable difference between the Ct levels of the two sample types.
13. We find that fusion polymerases generally work better with crude cell lysates, resulting in lower Ct levels, but regular Taq polymerases also can be used [15]. Importantly, the shape of the dose-response curve did not change due to lower sensitivity; therefore, for example, the minimal inhibitory concentration (MIC) value is not dependent on the type of the DNA polymerase used.
14. Instead of SYBR Green we prefer EvaGreen because of the higher fluorescent signal intensity, but SYBR Green qPCR mixes also can be used.
15. We performed the qPCR in 10 μL final volume with similar results as with the 20 μL.

Acknowledgments

Tomislav Meštrović and Dezső P. Virok contributed equally to this work.

References

1. Suchland RJ, Geisler WM, Stamm WE (2003) Methodologies and cell lines used for antimicrobial susceptibility testing of *Chlamydia* spp. Antimicrob Agents Chemother 47:636–642
2. Meštrović T, Ljubin-Sternak S (2018) Molecular mechanisms of *Chlamydia trachomatis* resistance to antimicrobial drugs. Front Biosci (Landmark Ed) 23:656–670

3. Ljubin-Sternak S, Meštrović T (2014) *Chlamydia trachomatis* and genital mycoplasmas: pathogens with an impact on human reproductive health. J Pathog 2014:183167
4. Ljubin-Sternak S, Meštrović T, Vilibić-Čavlek T, Mlinarić-Galinović G, Sviben M, Markotić A, Škerk V (2013) *In vitro* susceptibility of urogenital *Chlamydia trachomatis* strains in a country with high azithromycin consumption rate. Folia Microbiol (Praha) 58:361–365
5. Meštrović T, LjubinSternak S, Bedenić B (2015) Technical aspects of *Chlamydia trachomatis* antimicrobial susceptibility testing in cell culture system. Technical Journal 9:136–141
6. Meštrović T, Ljubin-Sternak S, Sviben M, Bedenić B, Vraneš J, Markotić A, Škerk V (2016) Antimicrobial sensitivity profile of *Chlamydia trachomatis* isolates from Croatia in McCoy cell culture system and comparison with the literature. Clin Lab 62:357–364
7. Samra Z, Rosenberg S, Soffer Y, Dan M (2001) In vitro susceptibility of recent clinical isolates of *Chlamydia trachomatis* to macrolides and tetracyclines. Diagn Microbiol Infect Dis 39:177–179
8. Donati M, Di Francesco A, D'Antuono A, Delucca F, Shurdhi A, Moroni A, Baldelli R, Cevenini R (2010) In vitro activities of several antimicrobial agents against recently isolated and genotyped *Chlamydia trachomatis* urogenital serovars D through K. Antimicrob Agents Chemother 54:5379–5380
9. Bhengraj AR, Vardhan H, Srivastava P, Salhan S, Mittal A (2010) Decreased susceptibility to azithromycin and doxycycline in clinical isolates of *Chlamydia trachomatis* obtained from recurrently infected female patients in India. Chemotherapy 56:371–377
10. Kai S, Wada K, Sadahira T, Araki M, Ishii A, Watanabe T, Monden K, Uno S, Araki T, Nasu Y (2017) Antimicrobial susceptibilities of *Chlamydia trachomatis* isolated from the urethra and pharynx of Japanese males. J Infect Chemother 23:512–516
11. Osaka I, Hills JM, Kieweg SL, Shinogle HE, Moore DS, Hefty PS (2012) An automated image-based method for rapid analysis of *Chlamydia* infection as a tool for screening antichlamydial agents. Antimicrob Agents Chemother 56:4184–4188
12. Southern T, Bess L, Harmon J, Taylor L, Caldwell H (2012) Fluorometric high-throughput assay for measuring chlamydial neutralizing antibody. Clin Vaccine Immunol 19:1864–1869
13. Bogdanov A, Endrész V, Urbán S, Lantos I, Deák J, Burián K, Önder K, Ayaydin F, Balázs P, Virok DP (2014) Application of DNA chip scanning technology for automatic detection of *Chlamydia trachomatis* and *Chlamydia pneumoniae* inclusions. Antimicrob Agents Chemother 58:405–413
14. Peuchant O, Duvert JP, Clerc M, Raherison S, Bébéar C, Bébéar CM, de Barbeyrac B (2011) Effects of antibiotics on *Chlamydia trachomatis* viability as determined by real-time quantitative PCR. J Med Microbiol 60:508–514
15. Eszik I, Lantos I, Önder K, Somogyvári F, Burián K, Endrész V, Virok DP (2016) High dynamic range detection of *Chlamydia trachomatis* growth by direct quantitative PCR of the infected cells. J Microbiol Methods 120:15–22
16. Párducz L, Eszik I, Wagner G, Burián K, Endrész V, Virok DP (2016) Impact of antiseptics on *Chlamydia trachomatis* growth. Lett Appl Microbiol 63:260–267
17. Bogdanov A, Janovák L, Lantos I, Endrész V, Sebők D, Szabó T, Dékány I, Deák J, Rázga Z, Burián K, Virok DP (2017) Nonactivated titanium-dioxide nanoparticles promote the growth of *Chlamydia trachomatis* and decrease the antimicrobial activity of silver nanoparticles. J Appl Microbiol 123:1335–1345
18. Belland RJ, Nelson DE, Virok D, Crane DD, Hogan D, Sturdevant D, Beatty WL, Caldwell HD (2003) Transcriptome analysis of chlamydial growth during IFN-gamma-mediated persistence and reactivation. Proc Natl Acad Sci U S A 100:15971–15976
19. Atlas RM (ed) (2010) Handbook of microbiological media, 4th edn. CRC Press, London
20. Labiran C, Clarke IN, Cutcliffe LT, Wang Y, Skilton RJ, Persson K, Bjartling C, Herrmann B, Christerson L, Marsh P (2012) Genotyping markers used for multi locus VNTR analysis with ompA (MLVA-ompA) and multi sequence typing (MST) retain stability in *Chlamydia trachomatis*. Front Cell Infect Microbiol 2:68
21. Barnes RC (1989) Laboratory diagnosis of human chlamydial infections. Clin Microbiol Rev 2:119–136

Chapter 6

Whole-Genome Sequencing of *Chlamydia trachomatis* Directly from Human Samples

Amanda Claire Brown and Mette T. Christiansen

Abstract

Whole-genome sequencing is a powerful, high-resolution tool that can be used to generate accurate data on bacterial population structure, phylogeography and mutations associated with antimicrobial resistance. The ability to sequence pathogen genomes directly from clinical specimens, without the requirement for in vitro culturing, is attractive in terms of time- and labor-saving, especially in the case of slow growing, or obligate intracellular pathogens, such as *Chlamydia trachomatis*. However clinical samples typically contain too low levels of pathogen nucleic acid, plus relatively high levels of human and natural microbiota DNA/RNA, to make this a viable option. Using a combination of whole-genome enrichment and deep sequencing, which has been proven to be a nonmutagenic approach, we can capture all known variation found within *C. trachomatis* genomes. The method is a consistent and sensitive tool that enables rapid whole-genome sequencing of *C. trachomatis* directly from clinical samples and has the potential to be adapted to other pathogens with a similar clonal nature.

Key words Whole-genome enrichment, Whole-genome sequencing, *Chlamydia trachomatis*, Clinical samples, SureSelect^{XT}

1 Introduction

The introduction of relatively affordable, desktop next generation sequencing (NGS) machines, such as Illumina's MiSeq and Next-Seq, and Life Technologies' Ion PGM and Ion Torrent, have revolutionized the ability to perform whole-genome sequencing (WGS) on small-sized genomes, such as those of bacterial and viral pathogens. It is now becoming possible to introduce WGS into diagnostic microbiology laboratories [1–3]. However, clinical samples typically contain too low levels of pathogen nucleic acid, plus relatively high levels of human DNA/RNA as well as DNA/RNA from the natural microbiota, to allow sequencing without an in vitro culture. Adding a culturing step can add significant delay to data acquisition, especially in the case of organisms that are difficult to culture, such as *Chlamydia trachomatis*, and

Amanda Claire Brown (ed.), *Chlamydia trachomatis: Methods and Protocols*, Methods in Molecular Biology, vol. 2042,
https://doi.org/10.1007/978-1-4939-9694-0_6,

organisms that are slow-growing, such as *Mycobacterium tuberculosis*.

C. trachomatis is the most common bacterial agent in sexually transmitted infections (STI), globally accounting for more than 100 million infections per year [4, 5]. *C. trachomatis* causes pelvic inflammatory disease in women, which has severe long-term consequences including ectopic pregnancy, infertility and chronic pelvic pain; in addition serovar A–C causes the blinding disease, trachoma, which affects millions of people worldwide [6, 7]. The loss of productivity and morbidity resulting from *C. trachomatis* infection has been recognized as placing a heavy economic burden on society [8, 9].

C. trachomatis strains are classified into two biovars: the ocular/urogenital biovar and the lymphogranuloma venereum (LGV) biovar [10]. The two biovars can be subdivided into 15–19 different serovars. Further genotypic classification is based on nucleotide sequencing of the *ompA* gene, which encodes the major outer membrane protein and is the target of serotype classification. The ocular/urogenital biovar consists of the ocular serovars A–C and the urogenital serovars D–K, all of which are usually confined to mucosal epithelia whereas the LGV biovar, consisting of serotypes L1–L3, is more invasive and can disseminate to other tissues and the draining lymphatic system. It has previously been demonstrated that genotyping of the *ompA* gene is insufficient for exploring *C. trachomatis* population structure and performing molecular epidemiological studies on transmission, as this region undergoes high levels of recombination [11]. Also variation within the *ompA* gene differs among serovars and sexual networks can be predominated by a single serovar, making strain distribution and evolutionary studies impractical [12, 13]. In this context, whole-genome sequencing (WGS) has been used to generate accurate data on bacterial population structure and phylogeography [14]. In addition whole-genome sequencing can also facilitate the identification of mutations associated with antimicrobial resistance [3, 15].

Clinical samples often contain low numbers of pathogens and to obtain sufficient material for WGS of *C. trachomatis*, in vitro culture is usually required; however, as *C. trachomatis* is an obligate intracellular pathogen it is labor intensive to grow in vitro [16]. For this reason methods that allow sequencing directly from *C. trachomatis* positive samples are particularly attractive. An antibody-based enrichment method, which targets intact *C. trachomatis* cells, followed by whole-genome amplification of the total DNA within the sample was recently described [17]. This method proved useful for sequencing *C. trachomatis* from complex clinical samples but showed only a 15–30% success rate, which underlines the need for a more reliable methodology.

The method described in this chapter utilizes the SureSelect^{XT} Target-Enrichment protocol (Agilent Technologies), which uses custom designed 120-mer RNA oligonucleotides. These RNA "baits" span the entire genome, and can recover (by hybridization) low copy number pathogens from clinical samples with sufficiently high sensitivity and specificity to enable ultra-deep whole-genome sequencing. Using this method we were able to obtain full length *C. trachomatis* genomes (>95–100% coverage of a reference genome) from eight of ten clinical samples tested (seven vaginal swabs and three urine samples). We also tested nine cultured *C. trachomatis* samples, representing different serovars, and obtained full length genomes for all. The proportion of reads mapping to *C. trachomatis* and the mean read depth across each genome were strongly linked to the number of bacterial copies within the original sample. Phylogenetic analysis confirmed the known population structure and the data showed potential for identification of minority variants and mutations associated with antimicrobial resistance. The sensitivity of this method was >tenfold higher than other reported methodologies, and offers the opportunity for gaining a wider understanding of the *C. trachomatis* population structures, transmission patterns and of the evolution of antimicrobial resistance. We have successfully used this approach on not just *C. trachomatis*, but also herpesviruses (including cytomegalovirus (CMV)), and *M. tuberculosis* [18–21], hepatitis B, hepatitis C, human immunodeficiency virus (HIV), influenza A and norovirus (publications currently in preparation).

2 Materials

In addition to the materials listed below sterile, nuclease-free, aerosol barrier pipette tips should be used for all liquid manipulations, and standard molecular biology protocol, such as wearing of gloves and taking care to not cross-contaminate samples, should be performed.

2.1 DNA Extraction and Quantification

1. QIAmp Mini Kit (Cat. No. 51304), Qiagen (*see* **Note 1**).
2. DNA LoBind 1.5 mL tubes (022431021, Eppendorf).
3. PBS (sterile).
4. Vortex mixer.
5. 1.5 mL tube heat block set at 56 °C.
6. Microfuge.
7. 96–100% ethanol, molecular biology grade.
8. Nuclease-free water (P1193, Promega).
9. Qubit dsDNA HS Assay Kit (Q32854), and Qubit 3.0 Flurometer (Q33216), Life Technologies.
10. Qubit Assay Tubes (Q32856), Life Technologies.
11. 50 mL sterile Falcon tubes (352070), Corning.

2.2 DNA Shearing

1. Covaris E210 Focused-ultrasonicator.
2. Covaris microTUBE with AFA fiber and snap cap (520045, Covaris).
3. TE_{LOW} buffer (10 mM Tris–HCl, pH. 8.0, 0.1 mM EDTA, 12090-015, Life Technologies).
4. Human genomic DNA (G3041, Promega) (*see* **Note 2**).
5. DNA LoBind 1.5 mL tubes (022431021, Eppendorf).
6. Microfuge.

2.3 Postshearing Sample Purification

1. Agencourt AMPureXP beads (A63880, Beckman Coulter).
2. DynaMag-2 magnetic stand (12321D, Life Technologies).
3. 70% ethanol (*see* **Note 3**).
4. Nuclease-free water (P1193, Promega).
5. DNA LoBind 1.5 mL tubes (022431021, Eppendorf).
6. Microfuge.
7. Heat block set at 37 °C.
8. 2200 TapeStation Nucleic Acid System (G2965AA, Agilent).
9. D1000 ScreenTape (5067-5582, Agilent) and D1000 reagents (5067-5583, Agilent).
10. Thin-wall 8-strip tubes (3148, Thermo Scientific) and caps (3148C, Thermo Scientific).

2.4 End Repair, A-Tailing, and Adapter Ligation

1. SureSelectXT Reagent Kit, MSQ, 16 (G9612A, Agilent) (*see* **Note 4**).
2. End Repair mix (52 μL per sample): 35.2 μL nuclease-free water, 10 μL 10× End Repair Buffer, 1.6 μL dNTP mix, 1 μL T4 DNA polymerase, 2 μL Klenow DNA polymerase, and 2.2 μL T4 polynucleotide kinase.
3. A-Tailing Mix (20 μL per sample): 11 μL nuclease-free water, 12 μL 10× Klenow DNA polymerase buffer, 1 μL ATP, and 3 μL Exo (−) Klenow DNA polymerase.
4. Adapter Ligation Mix (37 μL per sample): 15.5 μL nuclease-free water, 10 μL 5× T4 DNA ligase buffer, 10 μL Adaptor Oligo mix, and 1.5 μL T4 DNA ligase.
5. Agencourt AMPureXP beads (A63880, Beckman Coulter).
6. DynaMag-96 Side magnet (12331D, Life Technologies).
7. 70% ethanol (*see* **Note 2**).
8. Nuclease-free water (P1193, Promega).
9. Thin-wall 8-strip tubes (3148, Thermo Scientific) and caps (3148C, Thermo Scientific).
10. 2200 TapeStation Nucleic Acid System (G2965AA, Agilent).

11. D1000 ScreenTape (5067-5582, Agilent) and D1000 reagents (5067–5583, Agilent).
12. Thermal cycler (e.g., SureCycler 8800 (G8800A, Agilent)), with 96-well module (G8810A, Agilent).
13. Microfuge.
14. 5–50 μL 8-channel pipette (e.g., P4808-50, Labnet), and 20–200 μL 8-channel pipette (e.g., P4508-200, Labnet).

2.5 Amplification of Adapter-Ligated Library

1. SureSelectXT Reagent Kit, MSQ, 16 (G9612A, Agilent).
2. Herculase II Fusion DNA Polymerase (600,675, Agilent).
3. Pre-Capture PCR Mix (35 μL per sample): 21 μL Nuclease-free water, 1.25 μL SureSelect Primer (forward), 1.25 μL SureSelect Primer (forward), 10 μL Herculase II 5× reaction buffer, 0.5 μL dNTP (100 mM), and 1 μL Herculase II polymerase.
4. Agencourt AMPureXP beads (A63880, Beckman Coulter).
5. DynaMag-96 Side magnet (12331D, Life Technologies).
6. 70% ethanol (*see* **Note 3**).
7. Nuclease-free water (P1193, Promega).
8. Thin-wall 8-strip tubes (3148, Thermo Scientific) and caps (3148C, Thermo Scientific).
9. 2200 TapeStation Nucleic Acid System (G2965AA, Agilent).
10. D1000 ScreenTape (5067-5582, Agilent) and D1000 reagents (5067-5583, Agilent).
11. Thermal cycler (e.g., SureCycler 8800 (G8800A, Agilent)), with 96-well module (G8810A, Agilent).
12. 550 μL 8-channel pipette (e.g., P4808-50, Labnet), and 20–200 μL 8-channel pipette (e.g., P4508-200, Labnet).

2.6 Hybridization of Adapter-Ligated Library to RNA Baits

1. SureSelectXT Reagent Kit, MSQ, 16 (G9612A, Agilent).
2. SureSelect Block Mix (5.6 μL per sample): 2.5 μL SureSelect block #1 (green cap), 2.5 μL SureSelect block #2 (blue cap), and 0.6 μL SureSelect block #3 (brown cap).
3. Hybridization Buffer (20 μL per sample): 12.5 μL SureSelect Hyb #1 (orange cap), 0.5 μL SureSelect Hyb #2 (red cap), 5 μL SureSelect Hyb #3 (yellow cap), and 6.5 μL SureSelect Hyb #4 (black cap).
4. *C. trachomatis* RNA baits (Oxford Gene Technologies) (*see* **Note 5**).
5. Vacuum concentrator (DNA SpeedVac 120, Thermo Scientific).
6. Thin-wall 96-well PCR Plates (3146, Thermo Scientific) and caps (3148C, ThermoScientific).

7. 5–50 μL 8-channel pipette (P4808-50, Labnet).
8. Thermal cycler (e.g., SureCycler 8800 (G8800A, Agilent)), with 96-well module (G8810A, Agilent).
9. 1.5 mL tube heat block set at 65 °C.
10. Dynabeads MyOne Streptavidin T1 (65,601, Life Technologies).
11. DynaMag-2 magnetic stand (12321D, Life Technologies).
12. Nutator (421,105, BD).
13. Nuclease-free water (P1193, Promega).
14. Vortex mixer.
15. Microfuge.
16. DNA LoBind 1.5 mL tubes (022431021, Eppendorf).

2.7 Addition of Index Tags by Posthybridization Amplification

1. SureSelectXT Reagent Kit, MSQ, 16 (G9612A, Agilent).
2. Herculase II Fusion DNA Polymerase (600675, Agilent).
3. Post-Capture PCR mix (34 μL per sample): 22.5 μL nuclease-free water, 1 μL SureSelect Indexing Post-Capture PCR (forward) primer (orange cap), 10 μL Herculase II 5× reaction buffer, 0.5 μL dNTP (100 mM), and 1 μL Herculase II polymerase.
4. Agencourt AMPureXP beads (A63880, Beckman Coulter).
5. DynaMag-96 Side magnet (12331D, Life Technologies).
6. 70% ethanol (*see* **Note 3**).
7. Nuclease-free water (P1193, Promega).
8. Thin-wall 8-strip tubes (3148, ThermoScientific) and caps (3148C, ThermoScientific).
9. 2200 TapeStation Nucleic Acid System (G2965AA, Agilent).
10. D1000 ScreenTape (5067-5582, Agilent) and D1000 reagents (5067-5583, Agilent).
11. Thermal cycler (e.g., SureCycler 8800 (G8800A, Agilent)), with 96-well module (G8810A, Agilent).
12. 5–50 μL 8-channel pipette (e.g., P4808-50, Labnet), and 20–200 μL 8-channel pipette (e.g., P4508-200, Labnet).

2.8 Pooling Samples for Multiplexed Sequencing

1. Nuclease-free water (P1193, Promega).
2. DNA LoBind 1.5 mL tubes (022431021, Eppendorf).
3. Vortex mixer.

2.9 Illumina Sequencing

1. MiSeq desktop sequencer (Illumina).
2. MiSeq Reagent Kit v2 (300 cycle, MS-102-2002, Illumina).
3. 10 M NaOH (72068-100 mL, Sigma) (*see* **Note 6**).

4. PhiX Control (FC-110-3001, Illumina).
5. 10 mM Tris–HCl, pH 8.5 with 0.1% Tween 20.
6. Nuclease-free water (P1193, Promega).
7. Vortex mixer.

2.10 Data Analysis

1. CLC Genomics Workbench (version 6.5.0/6.5.1) including the CLC Microbial Genome Finishing Module (version 1.2.1/1.3.0) (Qiagen).

3 Methods

3.1 DNA Extraction and Quantification

1. Using the QIAamp DNA extraction kit, pipet 20 μL Qiagen Protease into the bottom of a 1.5 mL DNA LoBind tube.
2. For vaginal swabs suspend cells in 200 μL sterile PBS and add to the 1.5 mL tube, for urine add 200 μL; if the sample volume is lower than 200 μL make up to 200 μL with PBS.
3. Add 200 μL Buffer AL and mix by pulse-vortexing for 15 s. Incubate at 56 °C for 10 min. Briefly spin down tube to collect sample.
4. Add 200 μL ethanol (96–100%) to the sample, mix by pulse-vortexing for 15 s, and briefly centrifuge to collect.
5. Insert a QIAamp Mini spin column into a 2 mL collection tube, carefully add the sample to the column without wetting the rim. Close the cap and centrifuge at 6000 × *g* for 1 min. Remove QIAamp column to a fresh 2 mL collection tube and discard the tube containing the flow-through (*see* **Note 7**).
6. Add 500 μL Buffer AW1 to the column, spin at 6000 × *g* for 1 min, remove QIAamp column to a fresh 2 mL collection tube, discard the tube containing the flow-through (*see* **Note 7**).
7. Add 500 μL Buffer AW2 to the column, spin at full speed for 1 min, remove QIAamp column to a fresh 2 mL collection tube, and spin at top speed for 1 min; discard the tube containing the flow-through.
8. Place QIAamp column in a new DNA LoBind tube, add 100 μL Buffer AE, and incubate at room temperature for 5 min, then spin at 6000 × *g* for 1 min.
9. Label lids of Qubit assay tubes, number of sample plus 2 standards. Prepare Qubit working solution by diluting Qubit dsDNA HS reagent 1:200 with Qubit dsDNA HS Buffer in a 50 mL plastic tube, allow for 200 μL/sample or standard.
10. Add 190 μL of Qubit working solution to each standard tube, add 199 μL of Qubit working solution to each sample tube.

11. Add 10 μL of each standard to the appropriate tube, and 1 μL of each sample. Mix by vortexing for 2–3 min, take care not to create bubbles. Briefly spin tubes down in a microfuge. Incubate at room temperature for 2 min.
12. Select "DNA" on the Qubit 3.0 Fluorometer, and then "ds High Sensitivity" as the assay. Press "Read Standards" to proceed; insert tube containing standard #1 and press "Read standard," remove tube when read is complete and insert tube containing standard #2, press "Read standard". When complete press "Run samples," select the sample volume (1 μL) and units (ng/μL), insert a sample tube and press "Read tube," remove tube when reading is complete and insert next sample, until all are read.

3.2 DNA Shearing

1. Dilute samples to 3 μg DNA/130 μL using TE_{LOW} in a DNA LoBind tube, where total DNA concentration is lower than 3 μg use human DNA (Promega) to bulk to a final concentration of 3 μg (*see* **Note 3**).
2. Using a pipette tip introduce 130 μL DNA sample to Covaris microTube directly through the presplit septa, do not introduce bubbles.
3. Secure the microTube in the tube holder and shear for 6 × 60 s (duty cycle 10%, intensity 5, 200 cycles per burst with frequency sweeping) (*see* **Note 8**).
4. Remove microTube from Covaris, spin briefly in microfuge.
5. Remove sample, using a pipette tip, and transfer into a new DNA LoBind tube.

3.3 Postshearing Sample Purification

1. Add 180 μL of homogeneous AMPureXP beads (*see* **Note 9**) to each sheared sample tube (~130 μL), mix by vortexing for 5 s, incubate at room temperature for 5 min.
2. Briefly spin to collect sample to bottom of the tube.
3. Place tube on magnetic stand and allow the solution to clear (3–5 min).
4. Keeping the tube on the magnetic stand, carefully remove the cleared supernatant without disturbing the beads.
5. With the tube still in place on the magnetic stand add 500 μL 70% ethanol, incubate for 1 min or when solution clears. Remove ethanol wash and repeat.
6. After second wash step use a P10 tip to remove any residual ethanol from the bottom of the tube.
7. Dry samples on heat block at 37 °C for no more than 5 min (*see* **Note 10**).

8. Add 52 µL nuclease-free water directly to bead pellet. Mix by vortexing and briefly spin tube to collect contents. Incubate for 2 min at room temperature.
9. Put tube on magnetic rack, leave for 2–3 min to clear and collect the supernatant (~50 µL), transfer to a new DNA LoBind tube.
10. Add 3 µL D1000 Sample Buffer and 1 µL D1000 ladder to tube 1 using an 8-tube strip (*see* **Note 11**).
11. Add 3 µL D1000 Sample Buffer and 1 µL cleaned, sheared sample to tubes 2–16 using 8-tube strips (*see* **Note 12**), add caps and vortex, spin down for 1 min.
12. Load samples into TapeStation, and run D1000 tape. Assess quality and quantity, with a peak height between 150 and 200 nucleotides.
13. Samples can be stored overnight in the fridge at 4 °C, or at −20 °C for longer term.

3.4 End Repair, A-Tailing, and Adapter Ligation

1. Prepare End Repair Mix for each sample, on ice.
2. Pipet 52 µL End Repair Mix into thin-wall 8-strip tubes, add 48 µL of each sheared and cleaned DNA sample, mix by gentle pipetting, and close strip caps.
3. Incubate on a thermocycler at 20 °C for 30 min; do not use a heated lid.
4. Add 180 µL of homogeneous AMPureXP beads (*see* **Note 9**) to each sheared sample tube (~100 µL), mix by pipetting until homogeneous, incubate at room temperature for 5 min.
5. Place strip tubes on DynaMag-96 Side magnet and allow the solution to clear (3–5 min).
6. Keeping the tube on the magnetic stand, carefully remove the cleared supernatant without disturbing the beads.
7. With the tube still in place on the magnetic stand add 200 µL 70% ethanol, incubate for 1 min or when solution clears. Remove ethanol wash and repeat.
8. After second wash step use a P10 tip to remove any residual ethanol from the bottom of the tube.
9. Dry samples on thermocycler at 37 °C for no more than 5 min (*see* **Note 10**).
10. Add 32 µL nuclease-free water directly to bead pellet. Mix by vortexing and briefly spin tube to collect contents. Incubate for 2 min at room temperature.
11. Put tube on magnetic rack, leave for 2–3 min to clear and collect the supernatant (~30 µL), transfer to a new 8-strip tube.

12. Samples can be stored overnight in the fridge at 4 °C, or at −20 °C for longer term.
13. Prepare A-Tailing Mix for each sample, on ice.
14. Add 20 μL A-Tailing Mix to each 30 μL End Repaired Sample in 8-strip tube, mix by gentle pipetting, and close strip caps.
15. Incubate on a thermocycler at 37 °C for 30 min; do not use a heated lid.
16. Add 90 μL of homogeneous AMPureXP beads (*see* **Note 9**) to each sheared sample tube (~50 μL), mix by pipetting until homogeneous, incubate at room temperature for 5 min.
17. Place strip tubes on DynaMag-96 Side magnet and allow the solution to clear (3–5 min).
18. Keeping the tube on the magnetic stand, carefully remove the cleared supernatant without disturbing the beads.
19. With the tube still in place on the magnetic stand add 200 μL 70% ethanol, incubate for 1 min or when solution clears. Remove ethanol wash and repeat.
20. After second wash step use a P10 tip to remove any residual ethanol from the bottom of the tube.
21. Dry samples on thermocycler at 37 °C for no more than 5 min (*see* **Note 10**).
22. Add 15 μL nuclease-free water directly to bead pellet. Mix by vortexing and briefly spin tube to collect contents. Incubate for 2 min at room temperature.
23. Put tube on magnetic rack, leave for 2–3 min to clear and collect the supernatant (~13 μL), transfer to a new 8-strip tube.
24. Proceed immediately to adaptor ligation step.
25. Prepare Adaptor Ligation Mix for each sample, on ice.
26. Add 37 μL Adaptor Ligation Mix to each 13 μL A-Tailed Sample in a 8-tube strip, mix by gentle pipetting, and close strip caps.
27. Incubate on a thermocycler at 20 °C for 15 min, do not use a heated lid.
28. Add 90 μL of homogeneous AMPureXP beads (*see* **Note 9**) to each sheared sample tube (~50 μL), mix by pipetting until homogeneous, incubate at room temperature for 5 min.
29. Place strip tubes on DynaMag-96 Side magnet and allow the solution to clear (3–5 min).
30. Keeping the tube on the magnetic stand, carefully remove the cleared supernatant without disturbing the beads.

31. With the tube still in place on the magnetic stand add 200 μL 70% ethanol, incubate for 1 min or when solution clears. Remove ethanol wash and repeat.
32. After second wash step use a P10 tip to remove any residual ethanol from the bottom of the tube.
33. Dry samples on thermocycler at 37 °C for no more than 5 min (*see* **Note 10**).
34. Add 47 μL nuclease-free water directly to bead pellet. Mix by vortexing and briefly spin tube to collect contents. Incubate for 2 min at room temperature.
35. Put tubes on magnetic rack, leave for 2–3 min to clear and collect the supernatant (~45 μL), transfer to a new 8-tube strip.
36. Add 3 μL D1000 Sample Buffer and 1 μL D1000 ladder to tube 1 using an 8-tube strip (*see* **Note 11**).
37. Add 3 μL D1000 Sample Buffer and 1 μL sample to tubes 2–16 using 8-tube strips (*see* **Note 12**), add caps and vortex, spin down for 1 min, carefully remove caps.
38. Load samples into TapeStation, and run D1000 tape. Assess quality and quantity; the peak height should have increased to ~250–275 bp.
39. Samples can be stored overnight in the fridge at 4 °C, or at −20 °C for longer term.

3.5 Amplification of Adapter-Ligated Library

1. Adjust, if required, adaptor ligated samples to 20 ng/μL using nuclease-free water.
2. Prepare Pre-Capture PCR Mix.
3. Add 35 μL Pre-Capture PCR Mix to each 15 μL Adaptor Ligated Sample in 8-strip tube, mix by gentle pipetting, and close strip caps.
4. Cycle on a thermocycler using the parameters in Table 1, repeating **steps 2–4** for 6–12 cycles (*see* **Note 13**).
5. Add 90 μL of homogeneous AMPureXP beads (*see* **Note 9**) to each sheared sample tube (~50 μL), mix by pipetting until homogeneous, incubate at room temperature for 5 min.
6. Place strip tubes on DynaMag-96 Side magnet and allow the solution to clear (3–5 min).
7. Keeping the tubes on the magnetic stand, carefully remove the cleared supernatant without disturbing the beads.
8. With the tube still in place on the magnetic stand add 200 μL 70% ethanol, incubate for 1 min or when solution clears. Remove ethanol wash and repeat.
9. After second wash step use a P10 tip to remove any residual ethanol from the bottom of the tube.

Table 1
Thermocycler parameters for amplification of adapter-ligated library (From Sect. 3.5)

Step	Temperature (°C)	Time
1. Initial denaturing	98	2 min
2. Denaturing	98	30 s
3. Annealing	65	30 s
4. Extension	72	1 min
5. Final extension	72	10 min
6. Cool down	4	Hold

10. Dry samples on thermocycler at 37 °C for no more than 5 min (*see* **Note 10**).
11. Add 32 μL nuclease-free water directly to bead pellet. Mix by vortexing and briefly spin tube to collect contents. Incubate for 2 min at room temperature.
12. Put tubes on magnetic rack, leave for 2–3 min to clear and collect the supernatant (~30 μL), transfer to new 8-tube strip.
13. Add 3 μL D1000 Sample Buffer and 1 μL D1000 ladder to tube 1 using an 8-tube strip (*see* **Note 11**).
14. Add 3 μL D1000 Sample Buffer and 1 μL sample to tubes 2–16 using 8-tube strips (*see* **Note 12**), add caps and vortex, spin down for 1 min, remove caps gently.
15. Load samples into TapeStation, and run D1000 tape. Assess quality and quantity; the peak height should be ~250–275 bp.
16. Samples can be stored overnight in the fridge at 4 °C, or at −20 °C for longer term.

3.6 Hybridization of Adapter-Ligated Library to RNA Baits

1. Concentrate Pre-Capture PCR samples using vacuum concentrator as required so they reach 147–220 ng/μL in nuclease-free water (*see* **Note 14**).
2. Using a 96-well plate (PLATE A) add 500–750 ng in 3.4 μL of each sample into separate wells of column 2, (*see* **Note 15**).
3. Prepare sufficient SureSelect Block Mix for all samples being hybridized.
4. Add 5.6 μL SureSelect Block Mix to each sample well in column 2 of the 96-well plate.
5. Add strip caps to column 2 and seal well, carefully check each to ensure a complete seal.

6. Incubate plate(s) with samples in column 2 for 95 °C for 5 min, and then hold at 65 °C, using a thermocycler with a heated lid (105 °C).
7. Prepare sufficient Hybridization Buffer for the samples being hybridized.
8. Whilst maintaining PLATE A at 65 °C load 20 μL Hybridization Buffer into the required wells in column 1.
9. Add strip caps to column 1 and seal well, carefully check each to ensure a complete seal.
10. Close thermocycler lid and incubate at 65 °C for a least 5 min before proceeding to next step.
11. Prepare SureSelect RNA Capture Baits by diluting with 10% RNase Block, for each sample: 2 μL RNA baits (<3.0 MB), 0.5 μL RNase Block, 4.5 μL nuclease-free water; keep on ice.
12. In a new 96-well plate (PLATE B), on ice, add SureSelect RNA Capture Baits (7 μL) to column 3.
13. Using a multichannel pipette transfer 7 μL SureSelect RNA Capture Baits from column 3 of PLATE B, to column 3 of PLATE A, still in place on the thermocycler at 65 °C.
14. Add strip caps to column 3 of PLATE A and seal well, carefully check each to ensure a complete seal.
15. Close thermocycler lid and incubate at 65 °C for 2 min.
16. Whilst maintaining PLATE A at 65 °C, transfer 13 μL Hybridization Buffer from column 1 to column 3, using a multichannel pipette (*see* **Note 16**).
17. Whilst continuing to maintain PLATE A at 65 °C, transfer the entire contents of column 2 to column 3, using a multichannel pipette, still set at 13 μL (*see* **Note 16**).
18. Mix column 3 well by careful pipetting 8–10 times.
19. Add new strip caps to column 3 of PLATE A and seal well, carefully check each lip to ensure a complete seal (*see* **Note 17**).
20. Add strip caps to columns 1, 2, and 12 of PLATE A, to ensure the plate is balanced under the heated lid.
21. Continue to incubate the hybridization mixture, on the thermocycler for 16–24 h at 65 °C.
22. Prepare 50 μL Dynabeads MyOne Streptavidin T1 beads for each sample hybridized, add to a DNA LoBind tube and place on magnetic rack for 1–2 min.
23. Remove the cleared supernatant.
24. Add 200 μL SureSelect Binding Buffer.
25. Vortex for 5 s, spin down to collect tube contents and return to magnetic rack for 1–2 min.

26. Repeat **steps 23–25** for a total of three washes.
27. Resuspend beads in 200 μL SureSelect Binding Buffer and label each LoBind tube with sample identifier.
28. Open the lid to the thermocycler containing PLATE A, keeping the plate at 65 °C, open the first cap for column 3 (A3), remove the contents (typically 27–29 μL, *see* **Note 18**) and place in corresponding LoBind tube containing resuspended MyOne Streptavidin T1 beads. Invert 3–5 times to mix. Repeat for other samples (B3, C3...).
29. Mix all tubes containing hybridized samples and MyOne Streptavidin T1 beads well on a vortex for 5 s, then transfer to a Nutator. Incubate on the Nutator for 30 min at room temperature (*see* **Note 19**).
30. Briefly spin the tubes in a microfuge to collect the contents, and then add to a magnetic rack, and allow to clear for 1–3 min.
31. Remove the cleared supernatant and discard.
32. Add 500 μL SureSelect Wash Buffer #1 to each tube, close caps and vortex for 5 s.
33. Incubate at room temperature for 15 min, mix every 5 min using a vortex.
34. Briefly spin the tubes in a microfuge to collect the contents, and then add to a magnetic rack, and allow to clear for 1–3 min.
35. Remove the cleared supernatant and discard.
36. Add 500 μL prewarmed 65 °C SureSelect Wash Buffer #2 (*see* **Note 20**) to each tube, close caps and vortex for 5 s.
37. Incubate on a heat block at 65 °C for 10 min, mix every 3 min using a vortex.
38. Repeat **steps 34–37** for a total of three SureSelect Wash Buffer #2 washes.
39. After the final wash briefly spin tube before returning to the magnetic rack and use a P10 tip to ensure all residual wash buffer is removed.
40. Resuspend bead pellet in 45 μL nuclease-free water, leave beads remaining in tube.
41. Samples should be stored at 4 °C.

3.7 Addition of Index Tags by Post-Hybridization Amplification

1. Prepare Post-Capture PCR Mix.
2. Add 34 μL Post-Capture PCR Mix to required number of tubes in 8-strip tube.
3. Resuspend beads in hybridized samples, pipette 15 μL of bead mix into tube containing Post-Capture PCR Mix.

Table 2
Thermocycler parameters for addition of index tags by posthybridization amplification (From Sect. 3.7)

Step	Temperature (°C)	Time
1. Initial denaturing	98	2 min
2. Denaturing	98	30 s
3. Annealing	57	30 s
4. Extension	72	1 min
5. Final extension	72	10 min
6. Cool down	4	Hold

4. Add 1 μL of the appropriate PCR Index Primer to each reaction tube (*see* **Note 21**), mix by gentle pipetting, and close strip caps.
5. Cycle on a thermocycler using the parameters in Table 2, repeating **steps 2–4** for 12–18 cycles (*see* **Note 22**).
6. Add 90 μL of homogeneous AMPureXP beads (*see* **Note 9**) to each sheared sample tube (~50 μL), mix by pipetting until homogeneous, incubate at room temperature for 5 min.
7. Place strip tubes on DynaMag-96 Side magnet and allow the solution to clear (3–5 min).
8. Keeping the tubes on the magnetic stand, carefully remove the cleared supernatant without disturbing the beads.
9. With the tube still in place on the magnetic stand add 200 μL 70% ethanol, incubate for 1 min or when solution clears. Remove ethanol wash and repeat.
10. After second wash step use a P10 tip to remove any residual ethanol from the bottom of the tube.
11. Dry samples on thermocycler at 37 °C for no more than 5 min (*see* **Note 10**).
12. Add 32 μL nuclease-free water directly to bead pellet. Mix by vortexing and briefly spin tube to collect contents. Incubate for 2 min at room temperature.
13. Put tubes on magnetic rack, leave for 2–3 min to clear and collect the supernatant (~30 μL), transfer to new 8-tube strip.
14. Add 3 μL D1000 Sample Buffer and 1 μL D1000 ladder to tube 1 using an 8-tube strip (*see* **Note 11**).
15. Add 3 μL D1000 Sample Buffer and 1 μL sample to tubes 2–16 using 8-tube strips (*see* **Note 12**), add caps and vortex, spin down for 1 min, carefully remove caps.

16. Load samples into TapeStation, run D1000 tape. Assess quality and quantity, recording the peak height (*see* **Note 23**).
17. Label lids of Qubit assay tubes, number of sample plus 2 standards. Prepare Qubit working solution by diluting Qubit dsDNA HS reagent 1:200 with Qubit dsDNA HS Buffer in a 50 mL plastic tube, allow for 200 μL/sample or standard.
18. Add 190 μL of Qubit working solution to each standard tube, add 199 μL of Qubit working solution to each sample tube.
19. Add 10 μL of each standard to the appropriate tube, and 1 μL of each sample. Mix by vortexing for 2–3 min, take care not to create bubbles. Briefly spin tubes down in a microfuge. Incubate at room temperature for 2 min.
20. Select "DNA" on the Qubit 3.0 Fluorometer, and then "ds High Sensitivity" as the assay. Press "Read Standards" to proceed; insert tube containing standard #1 and press "Read standard," remove tube when read is complete and insert tube containing standard #2, press "Read standard." When complete press "Run samples," select the sample volume (1 μL) and units (ng/μL), insert a sample tube and press "Read tube," remove tube when reading is complete and insert next sample, until all are read.
21. Samples can be stored overnight in the fridge at 4 °C, or at −20 °C for longer term.

3.8 Pooling Samples for Multiplexed Sequencing

1. Use the concentration as determined by Qubit, and the size, as determined using TapeStation to calculate the molarity of each Post-Index Amplified sample (*see* **Notes 23** and **24**).
2. Combine the Post-Index Amplified samples so that each is representative in equimolar amounts in the library pool, using the formula below to determine the volume of each sample to use:

$$\text{Volume of post}-\text{index amplified sample} = \frac{V(f)\ X\ C(f)}{\#X\ C(i)}$$

where $V(f)$ is the final desired volume of the pool, $C(f)$ is the desired concentration of the pool (e.g., 2 nM for MiSeq v2), # is the total number of indexes in the pool, and $C(i)$ is the initial concentration of each Post-Index Amplified sample.

3. Mix pool well by vortexing and store at 4 °C.
4. Create a MiSeq sample sheet, using either Illumina Experiment Manager or Excel (*see* **Note 25**).

3.9 Illumina Sequencing

1. Thaw MiSeq Reagent Cartridge and HT1 buffer, overnight at 4 °C or at least 1 h at room temperature in a dish of cold water.

2. Once thawed keep Reagent Cartridge and HT1 buffer at 4 °C until needed.
3. Prepare a fresh dilution of 0.2 N NaOH (combine 980 μL of nuclease-free water and 20 μL 10 M NaOH in a LoBind tube).
4. Prepare PhiX control (*see* **Note 26**) by adding 2 μL 10 nM PhiX library with 3 μL 10 mM Tris–HCl, pH 8.5 with 0.1% Tween 20 in a LoBind tube. Mix well. Then add 5 μL 0.2 N NaOH, briefly vortex and then centrifuge at 280 × *g* for 1 min. Incubate for 5 min at room temperature, then add 990 μL prechilled HT1 buffer. This is 20 pM denatured PhiX, for v2 reagents 12.5 pM is required: combine 375 μL 20 pM denatured PhiX with 225 μL prechilled HT1 (*see* **Notes 27** and **28**).
5. Denature the 5 μL of the 2 nM library pool by combining with 5 μL 0.2 N NaOH, briefly vortex and then centrifuge at 280 × *g* for 1 min. Incubate for 5 min at room temperature, then add 990 μL prechilled HT1 buffer. This is a 10 pM denatured library pool; dilute to 8 pM by combining 800 μL 10 pM denatured library pool with 190 μL prechilled HT1 buffer, and 10 μL 12.5pM denatured PhiX. Mix well.
6. Add 600–1000 μL of the denatured and diluted library pool + PhiX mix to the designated reservoir (labeled as LOAD SAMPLES) on the thawed reagent cartridge.
7. From the MiSeq Welcome Screen select "Sequence" and then follow the run set up steps.
8. Remove flow cell from storage buffer using plastic forceps. Rinse the flow cell under running distilled water to remove excess salts, dry thoroughly using a lint free tissue, clean the flow cell glass with an alcohol wipe to ensure the glass is free for streaks, fingerprints, or fibers. Avoid touching the flow cell port gasket (black rubber area at the base of the flow cell). Allow to dry and load onto MiSeq following the screen prompts.
9. Load fresh PR2 bottle, and ensure the waste bottle is empty, following the screen prompts.
10. Load reagent cartridge into the chiller, following the screen prompts.
11. Change the sample sheet name to match bar code on reagent cartridge, review the run parameters and prerun check results.
12. Select Start Run (*see* **Note 29**).
13. When run is complete perform postrun washes, as per standard MiSeq maintenance.

3.10 Data Analysis

Genome mapping using CLC Genomics Workbench from Qiagen (*see* **Note 30**).

1. Import data. For each sample, transfer the read_1 and read_2 FASTQ files from the MiSeq to your local computer (keep the file names). Import the files in CLC Genomics Workbench (keep the fastq format) - File > Import > Illumina.

   Select the following settings:

   - Select files (.fastq).
   - General options.
     - Paired reads.
   - Discard read names.
   - Illumina options.
     - Remove failed reads.
   - Paired reads information.
     - Paired-end (forward-reverse) minimum distance 1 and maximum distance 800.
   - Quality scores.
     - Select: NCBI/Sanger or Illumina pipeline 1.8 and later.

   When imported the read_1 and read_2 files for each sample will be merged into one single file with (paired) added to the file name.

2. Quality Control (QC). Trim all read pairs for the presence of ambiguous nucleotides and base call quality - Toolbox > NGS core tools > Trim Sequences.

   Select the following settings:

   - Select file (paired).
   - Trim using quality scores.
     - Limit 0.001 (= Phred score of 30).
   - Trim ambiguous nucleotides.
     - Maximum numbers of ambiguities 2.
   - (Adapter trimming was performed in the MiSeq so no further adapter trimming is necessary).
   - Filter on length.
     - Maximum number of nucleotides in reads 1000.
     - Minimum number of nucleotides in reads 15.
   - Create report for inspection.

3. Reference mapping. Map all trimmed reads to a *C. trachomatis* reference genome:

   - Identify the best matching (identical biovar/genotype) complete reference genome from GenBank NCBI nucleotide database (*see* **Note 31**). Select the FASTA format and download the full GenBank file to your local computer.

Import the reference genome as a GenBank file (.gb) - File > Standard Import > Automatic import.

- Map each sample file containing the trimmed sequence reads to the selected reference(s) - Toolbox > NGS core tools > Map Reads to Reference.

  Select the following settings:

- Select imported reference.
- No masking.
- Read alignment.
  - Mismatch cost 2.
  - Linear gap cost.

    Insertion cost 3.

    Deletion cost 3.
  - Length fraction 0.5.
  - Similarity fraction 0.8.
  - Auto-detect paired distances.
- Nonspecific map handling.
  - Map randomly.
- Create read track and report.

  The mean read depth and fraction of the reference genome covered can be identified from the mapping report (*see* **Note 32**).

4. Consensus sequence. From each sample extract the consensus (majority) sequence based on the mapping file containing the trimmed sequence reads mapped to a reference.

   Select the following setting:

- Select the mapped read track.
- Threshold 5.
- Insert "N" ambiguity symbols.
- Conflict resolution "Vote".

For each sample the end result will be a complete (or close to complete) *C. trachomatis* genome generated based on reference mapping and majority base calling. Examples of how such genomes can be used for studies on population structure, in silico genotyping, identification of resistance mutation and variant detection can be found in Christiansen et al. [18].

4 Notes

1. After extraction *C. trachomatis* DNA can be quantified by qPCR, targeting the *C. trachomatis* plasmid and the genomic *omcB* gene, and human RNase-P as an endogenous control.
2. This protocol describes the use of the Agilent SureSelect^{XT} 3 μg input methodology, in the majority of cases less than 3 μg of *C. trachomatis* DNA is obtained following extraction from clinical samples, we therefore use human DNA as a carrier to bulk the sample to 3 μg; however users may prefer to use the 200 ng method, which uses a smaller shearing volume (50 μL), and dilution of adapters 1:10, *see* http://www.chem.agilent.com/library/usermanuals/Public/G7530-90000.pdf for full details.
3. 70% ethanol should be made fresh daily using 100% molecular grade ethanol and ultrapure, molecular biology grade water.
4. This protocol describes the use of the Agilent SureSelect^{XT} 3 μg methodology; however, since we first established this protocol other kits have come to market, including the SureSeq NGS Library Preparation Kit (500,070, Oxford Gene Technology) which requires less bead cleanup steps.
5. 120-mer RNA baits spanning the length of the positive strand of 74 GenBank *C. trachomatis* reference genomes were designed using an in-house PERL script developed by the PATHSEEK consortium. The specificity of the baits was verified by BLASTn searches against the Human Genomic Transcript database. The custom designed *C. trachomatis* bait library was uploaded to SureDesign and synthesized by Agilent Technologies.
6. 10 M NaOH should be diluted to 0.2 N fresh, just before use.
7. Flow through contains Buffer AL or Buffer AW1 and is not compatible with bleach.
8. Add fresh deionized water to the Covaris tank and turn on chiller, allow to reach 5 °C before using.
9. Remove AMPureXP beads from fridge at least 30 min before use. Just before use shake beads vigorously to ensure homogeneous and consistent color.
10. Bead pellet is completely dry when fine cracks can be observed. Do not over dry as this will decrease yield.
11. Remove TapeStation reagents from fridge 30 min prior to use.
12. If sample number is greater than 15 a 96-well plate can be used and multiple tapes.
13. Using a lower number of cycles will reduce the amount of duplication in the final sequencing library. The number of

cycles required to achieve enough product for hybridization should be determined empirically.

14. 500–750 ng DNA from the Pre-Cap PCR is required for each hybridization, if there is not sufficient product from one PCR, repeat with another 15 μL aliquot of the Adapter Ligated sample, and after cleaning pool products.
15. If more than eight samples are being hybridized use multiple plates and thermocycler blocks.
16. This step should be performed as quickly as possible to prevent excessive evaporation.
17. Use new strip caps to seal column 3; the structural integrity of the caps can be compromised during the previous incubation steps.
18. Excessive evaporation, such as the sample volume less than 20 μL following hybridization, can indicate suboptimal capture performance.
19. Visually ensure that sample is mixing on the Nutator platform.
20. Prewarm the required volume of SureSelect Wash Buffer #2 at 65 °C for at least 1 h before use.
21. Choose index combinations as per Illumina guidelines to ensure that indexes are balanced.
22. Using a lower number of cycles will reduce the amount of duplication in the final sequencing library. The number of cycles required to achieve enough product for sequencing should be determined empirically.
23. We have found it to be more accurate to use the TapeStation data to establish the size of the DNA fragments, and the Qubit to measure the DNA concentration in order to calculate the molarity.
24. Calculate molarity of each Post-Index Amplified sample as:

$$\mu g \ \text{DNA} \times \frac{\text{pmol}}{660 \ \text{pg}} \times \frac{10^6 \ \text{pg}}{1 \ \mu g} \times \frac{1}{N} = \text{pM DNA}$$

where N is the average fragment size and $\frac{\text{pmol}}{660 \ \text{pg}}$ is the average molecular weight of a nucleotide pair.

25. If creating a sample sheet in Excel *see* https://support.illumina.com/content/dam/illumina-support/documents/documentation/system_documentation/miseq/miseq-sample-sheet-quick-ref-guide-15028392-j.pdf for the parameters required.
26. PhiX control is helpful when sequencing low diversity samples, however the choice to add it as a spike in is down to the individual user. We found that it improved our cluster formation.

27. If you are using v3 chemistry then further dilution to 12.5 pM is not required.
28. Denatured PhiX libraries can be stored for up to 3 weeks at −20 °C.
29. A 300 cycle run takes ~24 h to complete, monitor run for cluster number and performance.
30. http://www.clcbio.com/products/clc-genomics-workbench-direct-download/.
31. http://www.ncbi.nlm.nih.gov/nucleotide/.
32. When defining the *C. trachomatis* mean read depth obtained from vaginal swab samples, mask the duplicated rRNA regions as these regions have been found to have a significantly higher read depth compared to the rest of the *C. trachomatis* genome and will skew the results.

Acknowledgments

The PATHSEEK consortium received funding from the European Union's Seventh Programme for research, technological development and demonstration under grant agreement No 304875. We acknowledge all the help from the other members of the consortium involved with the establishment of this methodology; particular thanks are due to Helena Tutill, UCL.

References

1. Köser CU, Ellington MJ, Cartwright EJP et al (2012) Routine use of microbial whole genome sequencing in diagnostic and public health microbiology. PLoS Pathog 8: e1002824
2. Köser CU, Bryant JM, Becq J et al (2013) Whole-genome sequencing for rapid susceptibility testing of *M. tuberculosis*. N Engl J Med 369:290–292
3. Olsen RJ, Long SW, Musser JM (2012) Bacterial genomics in infectious disease and the clinical pathology laboratory. Arch Pathol Lab Med 136:1414–1422
4. WHO. Global incidence and prevalence of selected curable sexually transmitted infections - 2008,
5. WHO. Prevalence and incidence of selected sexually transmitted infections,
6. Mylonas I (2012) Female genital *Chlamydia trachomatis* infection: where are we heading? Arch Gynecol Obstet 285:1271–1285
7. Mariotti SP, Pascolini D, Rose-Nussbaumer J (2009) Trachoma: global magnitude of a preventable cause of blindness. Br J Ophthalmol 93:563–568
8. Blandford JM, Gift TL (2006) Productivity losses attributable to untreated chlamydial infection and associated pelvic inflammatory disease in reproductive-aged women. Sex Transm Dis 33:S117–S121
9. Burton MJ, Mabey DCW (2009) The global burden of trachoma: a review. PLoS Negl Trop Dis 3:e460
10. Pedersen LN, Herrmann B, Møller JK (2009) Typing *Chlamydia trachomatis*: from egg yolk to nanotechnology. FEMS Immunol Med Microbiol 55:120–130
11. Millman KL, Tavaré S, Dean D (2001) Recombination in the *ompA* gene but not the *omcB* gene of *Chlamydia* contributes to serovar-specific differences in tissue tropism, immune surveillance, and persistence of the organism. J Bacteriol 183:5997–6008
12. Psarrakos P, Papadogeorgakis E, Sachse K et al (2011) *Chlamydia trachomatis ompA* genotypes in male patients with urethritis in Greece:

conservation of the serovar distribution and evidence for mixed infections with *Chlamydophila abortus*. Mol Cell Probes 25:168–173

13. Stothard DR, Boguslawski G, Jones RB (1998) Phylogenetic analysis of the *Chlamydia trachomatis* major outer membrane protein and examination of potential pathogenic determinants. Infect Immun 66:3618–3625
14. Harris SR, Clarke IN, Seth-Smith HMB et al (2012) Whole-genome analysis of diverse *Chlamydia trachomatis* strains identifies phylogenetic relationships masked by current clinical typing., nature genetics. Nat Genet 44:413–9S1
15. O'Neill CE, Seth-Smith HMB, Van Der Pol B et al (2013) *Chlamydia trachomatis* clinical isolates identified as tetracycline resistant do not exhibit resistance in vitro: whole-genome sequencing reveals a mutation in *porB* but no evidence for tetracycline resistance genes. Microbiology (Reading England) 159:748–756
16. Seth-Smith HMB, Harris SR, Scott P et al (2013) Generating whole bacterial genome sequences of low-abundance species from complex samples with IMS-MDA. Nat Protoc 8:2404–2412
17. Seth-Smith HMB, Harris SR, Skilton RJ et al (2013) Whole-genome sequences of *Chlamydia trachomatis* directly from clinical samples without culture. Genome Res 23:855–866
18. Christiansen MT, Brown AC, Kundu S et al (2014) Whole-genome enrichment and sequencing of *Chlamydia trachomatis* directly from clinical samples. BMC Infect Dis 14:591
19. Depledge DP, Palser AL, Watson SJ et al (2011) Specific capture and whole-genome sequencing of viruses from clinical samples. PLoS One 6:e27805
20. Depledge DP, Kundu S, Jensen NJ et al (2014) Deep sequencing of viral genomes provides insight into the evolution and pathogenesis of varicella zoster virus and its vaccine in humans. Mol Biol Evol 31:397–409
21. Brown AC, Bryant JM, Einer-Jensen K et al (2015) Rapid whole genome sequencing of *M. tuberculosis* directly from clinical samples. J Clin Microbiol 53(7):2230–2237

Chapter 7

Multilocus Sequence Typing (MLST) of *Chlamydiales*

Martina Jelocnik, Adam Polkinghorne, and Yvonne Pannekoek

Abstract

Developed two decades ago as a molecular method to provide definite characterization of a bacterial isolate, Multilocus Sequence Typing (MLST) is today globally adopted as a universal fine-detailed molecular typing tool and has been applied to numerous pathogenic and nonpathogenic bacterial as well eukaryotic organisms. MLST utilizes DNA sequence of several conserved housekeeping (HK) genes which are assigned an allelic number, which then collectively constitute an allelic profile or sequence type (ST), a "molecular barcode" of the interrogated bacterial strain or a eukaryotic organism. Here, we describe the principles and molecular approaches for generating MLST data for an analysis of a bacteria in the order *Chlamydiales*, using a *Chlamydia pecorum*-specific MLST scheme as an example.

Key words Multi locus sequence typing, *Chlamydia*, Genetic diversity, Population structure, Sequence types, PubMLST

1 Introduction

Bacteria of the order *Chlamydiales* are characterized by a unique biphasic obligate intracellular developmental cycle. Despite this seeming limitation, chlamydiae can infect an astonishing diversity of hosts, ranging from humans to terrestrial and marine animals, insects and protozoa [1]. The family *Chlamydiaceae* is perhaps the best described of the 13 described or proposed families within this order [2]. At present, this family is comprised of 16-member species, including 13 well described species and three *Candidatus* species, all belonging to a single genus, *Chlamydia*. The recent description of the latter bacteria suggests that further diversity within this genus is yet to be revealed [2, 3].

Members of the genus *Chlamydia* are commonly recognized as pathogens, responsible for a variety of diseases in humans and animals [1–3]. Several of these are veterinary pathogens that also have a well described zoonotic potential [4]. *Chlamydia trachomatis* is the most common globally widespread sexually transmitted bacteria [5] and major causative agent of preventable ocular

Amanda Claire Brown (ed.), *Chlamydia trachomatis: Methods and Protocols*, Methods in Molecular Biology, vol. 2042,
https://doi.org/10.1007/978-1-4939-9694-0_7,

diseases in humans [6]. *Chlamydia pneumoniae* is the second commonly recognized human pathogen, causing community-acquired pneumonias as well as a range of other clinical manifestations. *C. pneumoniae* has also been associated with infection in several animal hosts, including marsupials and reptiles [7]. In terms of the remaining *Chlamydia* species, at least five species infect a broad range of animal hosts, while the rest have only been documented in a niche host. *Chlamydia abortus* is a significant pathogen of sheep and cattle, known for its role in ovine enzootic abortions [8]. *Chlamydia psittaci*, primarily an avian pathogen, also infects livestock and other animal hosts [9]. Both pathogens can cause serious zoonotic disease [8, 10]. *Chlamydia pecorum* is a globally distributed pathogen of domesticated livestock and wild ruminants and, notably, a serious cause of debilitating disease in the Australian native marsupial, the koala [11, 12]. *Chlamydia avium* and *Chlamydia gallinacea* [13] have been detected in a variety of domesticated bird species as well as wild birds. While documented in other hosts, the bacteria *Chlamydia suis*, *Chlamydia felis*, *Chlamydia muridarum*, and *Chlamydia caviae* are generally considered to have pigs, cats, mice, and guinea pigs as preferred hosts, respectively [4, 14]. A source of untapped diversity in the genus *Chlamydia* appears to be in reptile hosts with several of the most recently described (*Chlamydia poikilothermis* and *Chlamydia serpentis*) [3] and proposed species (Ca. *Chlamydia sanzinia* and Ca. *Chlamydia corallus*) first described in a range of captive reptiles [15].

The pathogenic potential of these bacteria in humans and animals and their widespread distribution has meant significant interest remains in understanding the biology and epidemiology of *Chlamydiae*. While studies to characterize the genetic diversity were initially very limited, the advent of genome sequencing technology has now meant that significant new information is available for chlamydial pathogens of humans and animals [4]. As with many other bacterial species, to understand their intraspecies and interspecies genetic diversity, population structure and distribution of strains, multilocus sequence typing (MLST) has proven to be a successful tool for fine-detailed chlamydial epidemiology and diversity studies. Developed two decades ago as a molecular typing method to provide definite characterization of a bacterial isolate in straightforward and reproducible manner [16], MLST has been applied to numerous pathogenic and nonpathogenic bacterial as well eukaryotic organisms and is globally adopted as a universal epidemiological tool. Even in the genomic analyses and epidemiology era, MLST is still used as the first point for strain identification [17–19].

MLST is a typing method that uses DNA sequence from several (four-12) conserved housekeeping (HK) genes. HK genes targeted in MLST schemes must fulfil the following criteria: they are (1) evolutionary conserved, (2) widely separated on the chromosome, and

(3) not adjacent to putative outer membrane, secreted, or hypothetical proteins that might be under diversifying selection. In addition, each MLST locus has a similar extent of nucleotide diversity [18]. MLST can based on two generic techniques: (1) PCR amplification of the approximately 450–500 bp HK gene fragments, followed by Sanger DNA sequencing; and (2) extracting sequences of MLST loci from whole-genome sequencing (WGS) data sets. The retrieved DNA sequences are then assigned an allelic number which, when combined with the allelic numbers assigned to the other HK gene sequences, collectively constitute an allelic profile or sequence type (ST), a "molecular barcode" of the strain which can be documented and/or referred to in a well curated reference database (19). MLST data are fully portable and widely shared among researchers via open access online MLST databases such as the PubMLST (https://pubmlst.org/) [20]. With the advent of a variety of sequence and phylogenetic analyses software tools, MLST data can be optimally analyzed for molecular epidemiology investigations. In recent years, the PubMLST database has been updated with other two other MLST schemes for *C. trachomatis*. In addition, the database has been upgraded to the Bacterial Isolate Genome Sequence Database (BIGSdb) platform that enables users to perform characterization of strains at the whole genome level [18, 21].

Here, we describe the principles and molecular approaches for generating MLST data for an analysis of a bacterial species in the order *Chlamydiales*. The example workflow described will be for the use of a *C. pecorum*-specific MLST scheme; however, this approach is anticipated to be applicable to all chlamydial species. Next, we also describe the general analytical approaches used for basic analyses of MLST data, utilizing a *Chlamydiales* MLST scheme as an example of how to assess the population structure, genetic diversity and molecular epidemiology of chlamydiae [22–27]. Finally, this chapter concludes with an overview of the use of whole genome data in MLST analyses.

2 Materials

Standard molecular biology equipment (pipettes, heating block, thermocycler, gel electrophoresis equipment, UV transilluminator, benchtop centrifuge, spectrophotometer, microwave, vortex, autoclave, magnetic stir bars, heating plate), consumables (sterile pipette (filter) tips, plastic sample racks, Parafilm, DNAse/RNAse-free Eppendorf microfuge tubes (0.6–2 mL), commercial DNA purification kits and molecular grade chemicals and reagents are required. Unless stated otherwise, all solutions are prepared with MilliQ (mQ) water.

2.1 Clinical Swab Processing

1. *C. pecorum*-positive DNA extracts from a range of samples, respective of the population to be investigated using MLST (*see* **Note 1**).
2. Clinical swab (*see* **Notes 1–3**).
3. Sterile 1× Tris–EDTA (TE) buffer (10 mM Tris–HCl—1 mM EDTA pH 7.0) (*see* **Note 4**). To prepare 100 mL 1× TE buffer pH 7.0: Add 1 mL of 1 M Tris–HCl in MilliQ water (pH 7.0) and 0.2 mL 0.5 M EDTA in MilliQ water (pH 8.0) in a glass beaker and make up to 100 mL with MilliQ water. Autoclave or filter-sterilize using 0.2 μm filters (Millipore) and store at room temperature.
4. Sterile 1.5–2 mL Eppendorf centrifuge tubes.
5. Vortex.
6. Heating block.
7. Benchtop centrifuge.

2.2 Preparation of Killed Chlamydial Culture

1. Cryovial of purified chlamydial elementary bodies (EBs).
2. Heating block.
3. Benchtop centrifuge.

2.3 DNA Extraction

1. Commercially available DNA extraction kit (*see* **Note 5**).
2. Heating block.
3. Benchtop centrifuge.
4. 1.5–2 mL DNAse/RNAse-free Eppendorf tubes.

2.4 PCR Amplification of the MLST Gene Fragments

1. DNA template.
2. Forward and reverse primers (*see* **Note 6**).
3. PCR Master mix (*see* **Note 7**).
4. 0.2 mL (or other) PCR tubes.
5. Thermocycler.

2.5 Gel Electrophoresis

1. Agarose.
2. Running buffer: Make 1 L of 1× Tris–Borate–EDTA (TBE) electrophoresis buffer by adding 100 mL of 10× TBE buffer to 900 mL MilliQ. 10× TBE buffer: Measure 108 g of Tris base, 55 g of boric acid, 7.5 g of EDTA (disodium salt) and dissolve in 800 mL MilliQ Water. Top up to 1 L with MilliQ Water. A magnetic stir bar and hot plate can aid the process.
3. Ethidium bromide (EtBr): 10 μg/mL stock solution (*see* **Note 8**).
4. 6× DNA loading dye (Thermo Scientific).
5. DNA Molecular Weight Marker VIII (Sigma Aldrich).

6. Gel electrophoresis apparatus, including gel casts, combs, and power source.
7. 1.5% (w/v) agarose gel: Measure 1.5 g of agarose and add to 100 mL 1× TBE buffer (aliquoted in a 500-mL bottle). Heat in microwave until the agarose is melted and the solution is liquid. Leave it to cool for 1–2 min on room temperature and add 5 μL of EtBr. Pour the gel in the cast, insert the well comb and wait until it solidifies.
8. UV transilluminator.

2.6 PCR Amplicon Purification

1. Commercial PCR amplicon purification kit (*see* **Note 9**).
2. 1.5–2 mL DNAse/RNAse-free Eppendorf tubes.
3. Benchtop centrifuge.

2.7 Sequencing and Analysis

1. Purified PCR amplicons (*see* **Note 10**).
2. Forward and reverse primers.

3 Methods

All procedures are carried out at room temperature, unless otherwise specified.

3.1 Clinical Swab Processing (See Note 2)

1. Dispense 0.5 mL 1× TE buffer (or substitute, *see* **Note 4**) in each Eppendorf tube. Prepare replicate empty tubes as well and set them aside. Label the tubes with the swab/sample ID.
2. Place the dry swab in the Eppendorf tube and cut off the handle so that swab fits into the tube. Discard the cut-out swab handle. Firmly close the tube.
3. Vortex on high setting up to 3 min.
4. Briefly centrifuge to combine the swab suspension and to minimize aerosol production.
5. Transfer 0.25 mL of the swab suspension into each corresponding replicate tube. The processed swab (with 0.25 mL residue swab suspension) can now be stored at −20 °C.
6. Place the tubes with swab suspension on a (preheated) heating block at 95 °C for 20 min. This step kills any organisms present in the swab suspension, so the samples can be handled as noninfectious material.

3.2 Preparation of Killed Chlamydial Culture (See Note 2)

1. Preheat the heating block at 95 °C.
2. Thaw chlamydial cultures on room temperature.
3. Place the tubes containing the chlamydial cultures in a centrifuge and briefly centrifuge to combine the culture suspension and to minimize aerosol production.

4. Place the tubes in the heating block for 20 min. Following this step the cultures are killed and considered as noninfectious material.
5. Briefly centrifuge to combine the culture suspension and to minimize aerosol production.

3.3 DNA Extraction

1. Extract DNA from the samples according to the manufacturer's instructions (*see* **Note 5**).
2. It is important to elute the extracted DNA into 1.5–2 mL DNAse/RNAse-free Eppendorf tubes to minimize DNA degradation in the samples. Total elution volumes are 100–200 μL. It is also recommended to assess the DNA for purity and concentration using a spectrophotometer (e.g., NanoDrop) (*see* **Note 5**). Store the samples at −20 °C until ready to PCR the MLST gene fragments.

3.4 PCR Amplification of the MLST Gene Fragments

1. Prepare the PCR mix using the species-specific MLST primers (*see* **Note 6**) (e.g., *C. pecorum* primers outlined in Table 1), dispense into PCR tubes and add 5 μL DNA template. A total reaction volume for PCR reactions may vary between 25 and 50 μL (*see* **Note 7**).
2. Include negative (e.g., MilliQ H_2O) and positive (DNA from cultured chlamydial isolate) controls in each amplification assay.
3. Run the assay in the thermocycler.

   Briefly, standard cycling conditions are as follows:

   (a) An initial denaturation (may vary between 1 and 10 min dependent on the type of polymerase used, 95 °C).

   (b) Followed by 30–40 cycles of denaturation (30s, 95 °C), annealing (30 s, 50 °C–60 °C dependent on the melting temperatures of the primers) and extension (1 min, 72 °C).

   (c) A final extension (7 min, 72 °C).

   (d) Hold on 4 °C until removed from the thermocycler.

3.5 Gel Electrophoresis (See Note 8)

1. Place the agarose gel into the electrophoresis tank filled with 1× TBE and remove the comb.
2. In a separate PCR tube and/or on a Parafilm, mix 5 μL of the PCR product as well as the 5 μL of the DNA marker with 2 μL of the 6× loading dye and load the gel wells.
3. Connect the tank to the power source and run on 100 V for 20–30 min.
4. Visualize the amplicons on a gel with a UV transilluminator and compare with the DNA marker for size. Expected band sizes are also listed in Table 1.

Table 1
***Chlamydiales* MLST primers sets for seven loci, with *C. pecorum* (Cpec), *C. trachomatis/C. pneumoniae* (Ct/Cpn), and *C. psittaci* (Cpsi) primer sets used as an example**

Allele	Name: Forward 5′-3′ sequence spp	Name: Reverse 5′-3′ sequence spp	Amplicon size (bp)	MLST fragment size (bp)
gatA	MJgatF: GCTTTAGAGTTGAGAGAAGCT Cpec gatA_fw: GCTTTAGAATTARSARAWGCT Ct/Cpn YPgatA3: GCCTTAGAGTTAAGAAATGCCG Cpsi	MJgatR: GATCCTCCTGTATCTGATCC Cpec gatA_rp: GATCCTCCGGTATCYGATCC Ct/Cpn YPgatA4: CCCCCTGTATCGGAACCTAACGC Cpsi	500–600	425
*oppA*_3	MJoppF: ATGTGCAAGATCCCAGTGGG Cpec oppA_fw: ATGCGCAAGATATCAGTGGG Ct/Cpn YPoppA3: ATGCGCAAGATATCAATGGG Cpsi	MJoppR: GGCGCTACTTGTTATGGG Cpec oppA_rp: AAAGCTCCRSTWGMTATMGGWAG Ct/Cpn YPoppA4: GGCAAGGTTTGGTGTAACTCGC Cpsi	500–610	483 Ct 473/ Cpn 494 483
hflX	MJhflF: TGAGGAGATCTCTGCATCG Cpec hflX_fw: GCTTCTARAGTACTTTTAAATG Ct/Cpn YPhflX3: GAGATTTTTGCTAATCGAGCG Cpsi	MJhflR: ATCTTCATGCAAAGCAGCC Cpec hflX_rp: TATTTRGAAATYTTTKCSAGYCG Ct/Cpn YPhflX4: GTAAAACATCTTCATGTAACGC Cpsi	500–610	435
gidA	MJgidF: GCGTCACAACAAAAGAAGGC Cpec gidA_fw: GGAGTCWCTACWAAAGAAGG Ct/Cpn YPgidA1: GCTTATTAGAGAGCTGTCCTGGC Cpsi	MJgidR: TGACGCTGTATATCACACGG Cpec gidA_rp: TCGTAYTGYACATCRAAAGG Ct/Cpn YPgidA2: CGCGTTTTCTAACCCACGG Cpsi	500–670	474
enoA	MJenoF: TGATGGTAGATGCTGACGG Cpec enoA_fw: CCTATGATGAATCTKATCAATGG Ct/Cpn YPenoA3: CCTATGATGAATCTCATTAATGG Cpsi	MJenoR: GATTTTTACTGATCCCTTCG Cpec enoA_rp: TCTTCTTCRGCWAGMCCATCT Ct/Cpn YPenoA4: CCCAACCATCAAAATCTTCTTCCG Cpsi	450–500	381

(continued)

Table 1 (continued)

Allele	Name: Forward 5′-3′ sequence spp	Name: Reverse 5′-3′ sequence spp	Amplicon size (bp)	MLST fragment size (bp)
*hem*N	MJhemF: GATCGCGATAGAGATAGACCC Cpec hemN_fw: AGATCTTCTTCWGGRGGWAGAGA Ct/Cpn YPhemN1: GGATCCATTTCGGAGGAGGC Cpsi	MJhemR: ATCTTCTCCTGATAGATATCG Cpec hemN_rp: TTCYTTCAKAACSTAGGTTTT Ct/Cpn YPhemN2: CCTGAAAGGATTTTCTCATGG Cpsi	500–630	432
fumC	MJfumF: TGATTAAGAAATGTGCAGC Cpec fumC_fw: ATTAAAAAATGTGCTGCT Ct/Cpn YPfumC1: GGGCTCCTGAGGTTATGCC Cpsi	MJfumR: CCTTCAGGTACATTAAGCC Cpec fumC_rp: CCTTCAGGAACATTYAACCC Ct/Cpn YPfumC2: CGCAAATAATGAATCACCTTATC Cpsi	500–600	465

R = A or G; S = G or C; W = A or T; Y = C or T; M = A or C; K = G or T

3.6 PCR Amplicon Purification

1. Once it has been confirmed that only a single band of the expected size has been amplified, purify PCR amplicons using the PCR purification kit according to the manufacturer's instructions (*see* **Note 9**). This step may involve: incubation at specified temperature, washing with specified volume of buffers, and centrifuging to discard the buffers.
2. It is important to elute the purified amplicons into 1.5–2 mL DNAse/RNAse-free Eppendorf tubes to minimize DNA degradation in the samples. Usually, purified amplicons are eluted in volumes of 30–50 μL. Store the samples at −20 °C until ready to sequence each MLST gene fragment.
3. Repeat steps in Subheading 3.5 to ensure that the purification has been successful.

3.7 Sequencing and Analysis (See Note 10)

1. Prepare the amplicon for Sanger sequencing according to the manufacturer's instructions. Make sure that amplicon sequencing is performed in both directions. This will include that the same template is provided separately with forward and reverse primers.
2. Sequence analyses (*see* **Note 11**): Check the quality of the ABI format output sequence files for each gene using sequence analyses software.
3. Align forward and reverse trace sequencing files of each sequenced gene and obtain a consensus sequence by using a sequence analysis software.
4. Align the consensus sequence (with or without trace sequencing files) to the reference allele (*see* **Note 12**) and trim to an appropriate size to correspond with established MLST gene sequence fragments (Fig. 1). Export this region into a basic text editor and save as a FASTA sequence file (∗.fas/∗.fasta). Repeat this for all genes of the MLST scheme.
5. Allele and Sequence Type designation. The allele designation for each aligned sequence can be obtained at the *Chlamydiales*

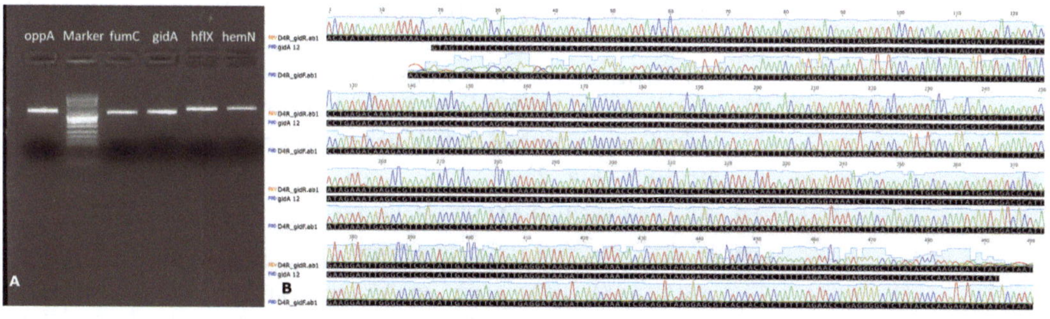

Fig. 1 MLST based on PCR and sequencing. (**a**) Visualization of MLST PCR amplicons on an agarose gel; and (**b**) Alignment of the forward and reverse trace sequencing files of the PCR amplicon and reference allele

PubMLST database using the "Sequence query" option at https://pubmlst.org/bigsdb?db=pubmlst_chlamydiales_seqdef &page=sequenceQuery by submitting the FASTA file. The database will output a result that either indicates the allele number or, if the allele is novel, the output will indicate that the FASTA file and both forward and reverse trace sequencing files will have to be submitted to the database curator that will assign the novel allele identifier. Repeat this for all MLST loci and obtain the allelic profile. Interrogate the allelic profile using "Search by combinations of alleles" option: https://pubmlst.org/bigsdb?db=pubmlst_chlamydiales_seqdef&page=profiles.

6. If the ST is novel, the database curator will assign the novel STs identifier (*see* **Note 13**).
7. Determination of the MLST alleles and ST using WGS reads (*see* **Note 14**): Extract the MLST genes from the sequencing reads and/or de novo assembled contigs.
8. Obtain the allele designation for each aligned sequence from *Chlamydiales* PubMLST database using the "Sequence query" option at https://pubmlst.org/bigsdb?db=pubmlst_chlamydiales_seqdef&page=sequenceQuery by submitting the FASTA file. The database will output a result that either indicates the allele number or, if the allele is novel, the output will indicate that the FASTA file with a notification that this sequence has been generated by WGS will have to be submitted to the database curator that will assign the novel allele identifier. Repeat this for all MLST loci and obtain the allelic profile.
9. Interrogate the allelic profile using "Search by combinations of alleles" option: https://pubmlst.org/bigsdb?db=pubmlst_chlamydiales_seqdef&page=profiles
10. If the ST is novel, the database curator will assign the novel STs identifier.
11. MLST sequence analyses (*see* **Note 15**). To understand the level of genetic diversity present in the interrogated population, sequences analyses such as: (a) calculating the number of polymorphic (segregating) sites and haplotypes in each MLST locus; (b) haplotype diversity; (c) determining the number of synonymous (d_s) and nonsynonymous (d_n) substitutions per site in each MLST loci; and (d) estimating ST diversity by calculating number of STs per number of analyzed strains can be done (Fig. 2).
12. Align individual MLST alleles from all analyzed strains.
13. Concatenate individual MLST allele in specific order (e.g., *gat*A, *opp*A, *hfl*X, *gid*A, *eno*A, *hem*N and *fum*C) for each analyzed strain. In any of the analyses, keep the order of the alleles the same. Align the concatenated sequence.

14. Perform desired analysis using alignment of each allele or the allele concatenated sequence and record the output.
15. Cluster analyses: One of the most common analysis using MLST data is to define clonal complexes (clusters) using allelic profiles of the investigated strains (Fig. 2). This analysis also determines "the putative progenitor (central genotype) of a complex" (*see* **Note 16**). For this analysis, the input is the allelic (numeric) MLST profile.
16. Collate allelic profiles for all seven loci, STs and sample names for all analyzed strains in an excel format.
17. Input into the eBurst software, define the threshold and analyzed. The output is a graphical representation of the clonal clusters.
18. Phylogenetic and evolutionary MLST analyses (*see* **Notes 17** and **18**): Concatenated MLST sequences were previously used to evaluate phylogenetic relationships among chlamydial strains on the basis that analysis of seven HK fragments used in *Chlamydiales* MLST are likely representative of the whole-genome phylogeny [10, 23, 27–29].
19. Align concatenated MLST allele sequences for each strain and perform phylogenetic analyses.
20. Phylogenetic analyses provide graphical output (a tree) as well as distance matrix in a table format.
21. Further annotation of the generated tree can be performed to highlight certain phylogenetic feature(s) of the population.
22. Whole-genome MLST (see **Notes 19** and **20**).

4 Notes

1. The strain population should be carefully determined considering several important criteria: the research question itself; are they representative of the investigated genetic diversity, host(s), geographic locations, outbreak(s), different clinical manifestations, etc.
2. Due to the use of PCR to generate MLST allele amplicons, and availability of next-gen polymerases, we can use a variety of samples for DNA extractions, such as semipurified chlamydial cultures, dry swabs collected from different (e.g., ocular, nasal, vaginal, and/or rectal) anatomical sites, and/or clinical samples where live chlamydia is difficult to isolate (e.g., joint fluid, different tissue (e.g., brain, lung, placenta) samples, urine, and/or fecal samples).

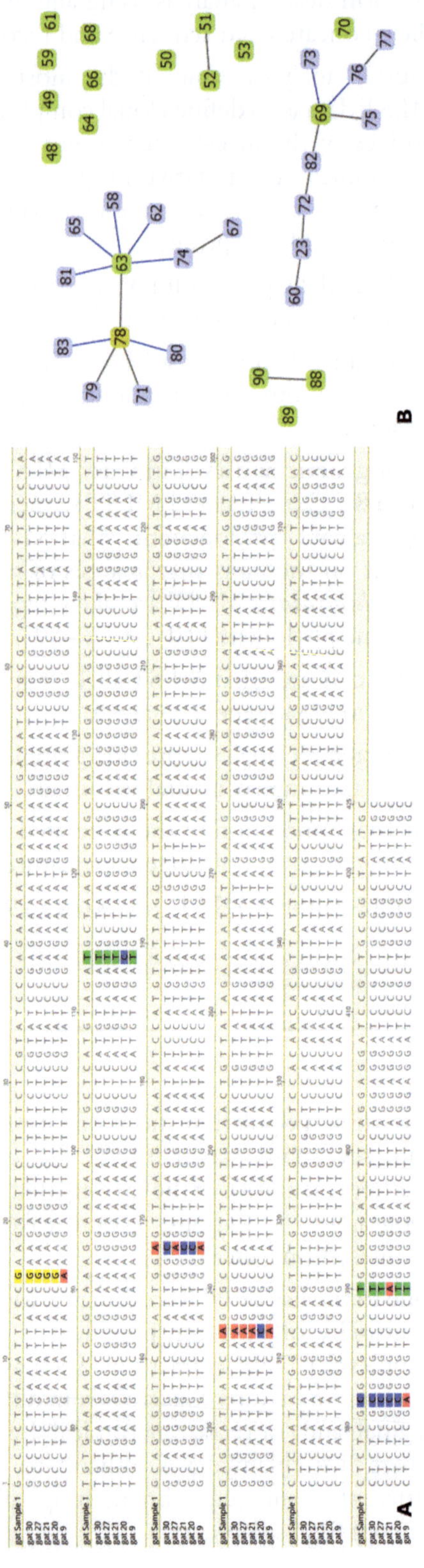

Fig. 2 MLST data analyses. (**a**) Alignment of *gatA* alleles, where SNPs are highlighted in different colors; (**b**) Example of cluster analyses of *C. pecorum* STs, where STs highlighted in bright green represent putative progenitors in a clonal complex

3. Prior to applying MLST, make sure that the interrogated samples are positive for the targeted species (e.g., *C. pecorum*-, *C. psittaci*-, *C. abortus*-positive) by screening with the species-specific qPCR assay.
4. For diagnostics and/or prevalence studies swab samples are often the most commonly taken for detection mucosal shedding TE buffer can be substituted with sterile phosphate buffered saline (PBS) and/or MilliQ water (Ultrapure water). It is advised that the swabs taken from both diseased as well as asymptomatic hosts and/or chlamydial cultures are handled with care in a biosafety cabinet in case that carry infectious material. Before taking out the swab samples from the cabinet for vortexing and/or centrifugation, spray the outside of the tubes with 70% Ethanol.
5. Any commercially available DNA extraction kit can be used in this step (e.g., QIAamp DNA Mini Kit (Qiagen), or GenElute™ genomic DNA purification kits (Sigma-Aldrich)), following the DNA extraction protocol as per the manufacturer's instructions. For use in PCR (and other), the extracted DNA should have a 260/280 ratio of ~1.8 and concentration at least 10 ng/μL. RNaseA may be used to digest RNA during the procedure.
6. For certain *Chlamydia* species, the MLST primers sequences have been already optimized to ensure specificity and sensitivity. In MLST, primers are usually designed to have similar annealing temperatures and to produce amplicons of similar sizes, so that PCR conditions are the same for all or at least several loci. For use of novel primers, it is recommended to perform in silico sequence analyses: such blast against the nucleotide collection nr/nt database (https://blast.ncbi.nlm.nih.gov/Blast.cgi) to assess primers sequence specificity and potential unspecific binding; and using tools such as OligoAnalyser 3.1 (available from http://sg.idtdna.com/calc/analyzer) to evaluate potential primer dimerization, hairpins, and melting temperatures. It is also likely that the PCR assay optimization will be needed.
7. A commercial PCR Master Mix can be substituted for individual PCR Mix components including DNA polymerase, a buffer for the specific DNA polymerase, and deoxynucleotides (dNTPs), in addition to magnesium chloride ($MgCl_2$) if not supplied with the buffer. For a detailed PCR protocol refer to [30].
8. EtBr as a DNA intercalating agent is a suspected carcinogen, thereby gloves should always be worn when handling gels containing EtBr. Other alternative dyes (e.g., SYBR dyes) for

the staining of DNA are available. For the detailed protocol of how to prepare agarose gels and set the apparatus refer to [31].

9. A variety of commercial DNA amplicon purification kits can be used. After the purification, it is highly recommended to visualize purified products on an electrophoresis gel.

10. It is an imperative that the MLST amplicon sequencing is performed in both directions in order to accurately determine the allele sequence. In our studies, we have used DNA sequencing service providers; however, in-house sequencing reactions are also commonly used.

11. For trace sequencing files and sequence analyses a variety of commercial as well as open source software platforms are available (e.g., Geneious Pro software package [32], MEGA [33], DNAAlignEditor [34], BioEdit v7 (available from http://www.mbio.ncsu.edu/BioEdit/page2.htmL)). In order to denote novel alleles, it is imperative that trace sequencing files are of a good quality and that they cover the length of reference allele. To achieve so, *Chlamydiales* MLST primers are designed to amplify longer fragment that can be trimmed as often the 5′ and 3′ ends may produce a low quality read (Table 1, Fig. 1). It is strongly recommended that all interrogated including both previously denoted as well novel alleles and STs must be deposited at the PubMLST *Chlamydiales* database to maintain data standardization and availability.

12. The reference allele (e.g., *C. psittaci gid*A allele 13) can be downloaded from the *Chlamydiales* pubMLST database: https://pubmlst.org/bigsdb?db=pubmlst_chlamydiales_seqdef&page=downloadAlleles&tree=1

13. Submission of novel alleles and Sequence Types: Novel alleles or STs, as well as details of *Chlamydia* strains, can be submitted any time for inclusion to the database. E-mail all required information (*see* for details https://pubmlst.org/chlamydiales/submission.shtml). Be sure to submit the trace sequencing files as well as the FASTA files.

14. We have included this step (where applicable) to describe a workflow if WGS reads are used for MLST and where the user has access to WGS analyses platforms and/or software. Prior to MLST gene and/or fragment sequences extraction, the sequenced paired reads must be assessed for quality by trimming and filtering reads below the specified criteria, and that they cover chromosomal regions which contain the MLST genes. MLST data can be extracted from WGS by direct mapping [35] and/or de novo assembly/BLAST approaches [36] using the referenced tools. If the sequence analyses platform permits, MLST data can be extracted by the user itself using several approaches such as: (a) mapping the reads onto

MLST concatenated species-specific sequence; (b) de novo assembling reads into contigs and BLAST searching for the MLST gene by homology; and (c) mapping the reads onto referenced annotated genome and extracting the MLST fragments.

15. Prior to any sequence analyses, align the sequences using any of the sequence analyses software, as most of the analyses require the sequence alignment. Alignment can be generated using Clustal [37] and/or MUSCLE [38] or other sequence alignment tool. In our MLST sequence analyses we have used the open source software DNASp v.5 [39] to evaluate the polymorphisms in individual allele as well as concatenated allele sequences, however other software with similar functions can be used. Besides the basic analyses outlined in **step 11** of Subheading 3.7, estimation of average number of nucleotide substitutions per site between different populations (D_{xy}) and average number of nucleotide differences between populations and estimation of the minimum number of recombination events (R_m) can also be done (dependable on the research question asked) using the MLST set.
16. The commonly used goeBURST program and instructions are available from http://goeburst.phyloviz.net/ [40]. Briefly, MLST profiles are divided into clonal complexes (CCs) under a user-defined threshold level of identity. In our studies, we used a stringent single locus variant (SLV) level; where CCs were defined as groups of STs, which share six out of seven loci (alleles) with at least one other ST in the group, while singletons (satellites) were defined as STs that differ by at least two loci (alleles) from all other STs. A "putative progenitor" in CCs is a ST that has the most SLV links to other STs. In larger CCs, a ST (besides the predicted primary progenitor) that has several SLV links of their own is considered a "putative progenitor" of a subgroup. Where population is clonal, a more "relaxed" double or triple locus variant threshold can be used.
17. Prior to phylogenetic tree reconstruction, we recommend using a software jModelTest v.2 [41] to estimate best-fit models of nucleotide substitution for constructing phylogenies. Several phylogenetic tree-building methods can be used for reconstructing phylogenetic trees using MLST concatenated sequences, with neighbor-joining, maximum likelihood (ML), and Bayesian methods most commonly used. In our studies we have used maximum likelihood and Bayesian methods as these methods apply a model of sequence evolution and provide robust trees with node support values [19]. The phylogenetic software can be used separately or as implemented into a sequence and phylogenetic analyses platform such as MEGA, Geneious, and/or CLC Genomics.

18. Further annotation of the phylogenetic tree can include: (a) highlighting clades: by hosts, species, and/or geographical location; (b) adding legend or a key; and (c) rerouting. These functions can be implemented in the phylogenetic software itself, or web-based tools such as iTol Interactive Tree Of Life (https://itol.embl.de/) can be used.
19. Whilst MLST has proven its value as a "gold standard" molecular epidemiology tool, whole genome sequencing (WGS) is the "ultimate molecular barcode" of a bacterium which facilitates new understandings of within-host evolution, transmission, and population structure with an increased sensitivity for the detection of rare genetic variants [18]. Due to the success of *Chlamydiae* in infecting a wide range of hosts across the globe, the documented emergence of new genetically variant strains indicates the need for more comprehensive studies such as using wgMLST approaches to better understand the genetic population structure of these organisms [36].
20. For such analyses, it is recommended to use the Bacterial Isolate Genome Sequence Database (BIGSdb) [20] which extends the MLST by gene-by-gene approach, described in detail in [18, 26, 36]. Briefly, *Chlamydiales* wgMLST uses any combination of >900 defined chromosomal as well as plasmid loci to evaluate allelic variants between the strains with The Genome Comparator tool implemented in BIGSdb.

References

1. Taylor-Brown A, Vaughan L, Greub G, Timms P, Polkinghorne A (2015) Twenty years of research into *Chlamydia*-like organisms: a revolution in our understanding of the biology and pathogenicity of members of the phylum *Chlamydiae*. Path Dis 73:1–15
2. Pillonel T, Bertelli C, Greub G (2018) Environmental metagenomic assemblies reveal seven new highly divergent chlamydial lineages and hallmarks of a conserved intracellular lifestyle. Front Microbiol 9:79
3. Staub E, Marti H, Biondi R, Levi A, Donati M, Leonard CA et al (2018) Novel *Chlamydia* species isolated from snakes are temperature-sensitive and exhibit decreased susceptibility to azithromycin. Sci Rep 8:5660
4. Bachmann NL, Polkinghorne A, Timms P (2014) *Chlamydia* genomics: providing novel insights into chlamydial biology. Trends Microbiol 22:464–472
5. Ziklo N, Huston WM, Hocking JS, Timms P (2016) *Chlamydia trachomatis* genital tract infections: when host immune response and the microbiome collide. Trends Microbiol 24:750–765
6. Taylor HR, Burton MJ, Haddad D, West S, Wright H (2014) Trachoma. Lancet 384:2142–2152
7. Roulis E, Polkinghorne A, Timms P (2013) *Chlamydia pneumoniae*: modern insights into an ancient pathogen. Trends Microbiol 21:120–128
8. Rodolakis A, Laroucau K (2015) *Chlamydiaceae* and chlamydial infections in sheep or goats. Vet Microbiol 181:107–118
9. Knittler MR, Sachse K (2015) *Chlamydia psittaci*: update on an underestimated zoonotic agent. Path Dis 73:1–15
10. Jelocnik M, Branley J, Heller J, Raidal S, Alderson S, Galea F et al (2017) Multilocus sequence typing identifies an avian-like *Chlamydia psittaci* strain involved in equine placentitis and associated with subsequent human psittacosis. Emerg Microbes Infect 6:e7
11. Polkinghorne A, Hanger J, Timms P (2013) Recent advances in understanding the biology,

epidemiology and control of chlamydial infections in koalas. Vet Microbiol 165:214–223

12. Walker E, Lee EJ, Timms P, Polkinghorne A (2015) *Chlamydia pecorum* infections in sheep and cattle: A common and under-recognised infectious disease with significant impact on animal health. Vet J 206:252–260
13. Sachse K, Laroucau K, Riege K, Wehner S, Dilcher M, Creasy HH et al (2014) Evidence for the existence of two new members of the family *Chlamydiaceae* and proposal of *Chlamydia avium* sp. nov. and *Chlamydia gallinacea* sp. nov. Syst Appl Microbiol 37:79–88
14. Taylor-Brown A, Polkinghorne A (2017) New and emerging chlamydial infections of creatures great and small. New Microbes New Infect 18:28–33
15. Taylor-Brown A, Spang L, Borel N, Polkinghorne A (2017) Culture-independent metagenomics supports discovery of uncultivable bacteria within the genus *Chlamydia*. Sci Rep 7:10661
16. Maiden MCJ, Bygraves JA, Feil E, Morelli G, Russell JE, Urwin R et al (1998) Multilocus sequence typing: a portable approach to the identification of clones within populations of pathogenic microorganisms. Proc Natl Acad Sci U S A 95:3140–3145
17. Maiden MCJ (2006) Multilocus sequence typing of bacteria. Annu Rev Microbiol 60:561–588
18. Maiden MCJ, van Rensburg MJJ, Bray JE, Earle SG, Ford SA, Jolley KA et al (2013) MLST revisited: the gene-by-gene approach to bacterial genomics. Nat Rev Microbiol 11:728–736
19. Pérez-Losada M, Cabezas P, Castro-Nallar E, Crandall KA (2013) Pathogen typing in the genomics era: MLST and the future of molecular epidemiology. Infect Genet Evol 16:38–53
20. Jolley K, Maiden M (2010) BIGSdb: scalable analysis of bacterial genome variation at the population level. BMC Bioinformatics 11:595
21. Maiden MCJ, Harrison OB (2016) Population and functional genomics of the *Neisseria* revealed with gene-by-gene approaches. J Clin Microbiol 54:1949–1955
22. Pannekoek Y, Morelli G, Kusecek B, Morre S, Ossewaarde J, Langerak A et al (2008) Multi locus sequence typing of *Chlamydiales*: clonal groupings within the obligate intracellular bacteria *Chlamydia trachomatis*. BMC Microbiol 8:42
23. Pannekoek Y, Dickx V, Beeckman DSA, Jolley KA, Keijzers WC, Vretou E et al (2010) Multi Locus sequence typing of *Chlamydia* reveals an association between *Chlamydia psittaci* genotypes and host species. PLoS One 5:e14179
24. Jelocnik M, Frentiu FD, Timms P, Polkinghorne A (2013) Multi-locus sequence analysis provides insights into the molecular epidemiology of *Chlamydia pecorum* infections in Australian sheep, cattle and koalas. J Clin Microbiol 51:2625–2632
25. Guo W, Jelocnik M, Li J, Sachse K, Polkinghorne A, Pannekoek Y et al (2017) From genomes to genotypes: molecular epidemiological analysis of *Chlamydia gallinacea* reveals a high level of genetic diversity for this newly emerging chlamydial pathogen. BMC Genomics 18:949
26. Versteeg B, Bruisten SM, Pannekoek Y, Jolley KA, Maiden MCJ, van der Ende A, Harrison OB (2018) Genomic analyses of the *Chlamydia trachomatis* core genome show an association between chromosomal genome, plasmid type and disease. BMC Genomics 19(1):130
27. Li M, Jelocnik M, Yang F, Gong J, Kaltenboeck B, Polkinghorne A et al (2017) Asymptomatic infections with highly polymorphic *Chlamydia suis* are ubiquitous in pigs. BMC Vet Res 13:370
28. Jelocnik M, Walker E, Pannekoek Y, Ellem J, Timms P, Polkinghorne A (2014) Evaluation of the relationship between *Chlamydia pecorum* sequence types and disease using a species-specific multi-locus sequence typing scheme (MLST). Vet Microbiol 174:214–222
29. Jelocnik M, Self R, Timms P, Borel N, Polkinghorne A (2015) Novel sequence types of *Chlamydia pecorum* infect free-ranging Alpine ibex (*Capra ibex*) and red deer (*Cervus elaphus*) in Switzerland. J Wildl Dis 51:479–483
30. Lorenz TC (2012) Polymerase chain reaction: basic protocol plus troubleshooting and optimization strategies. J Vis Exp 63:3998
31. Lee PY, Costumbrado J, Hsu C-Y, Kim YH (2012) Agarose gel electrophoresis for the separation of DNA fragments. J Vis Exp 62:3923
32. Kearse M, Moir R, Wilson A, Stones-Havas S, Cheung M, Sturrock S et al (2012) Geneious basic: an integrated and extendable desktop software platform for the organization and analysis of sequence data. Bioinformatics 28:1647–1649
33. Tamura K, Peterson D, Peterson N, Stecher G, Nei M, Kumar S (2011) MEGA5: molecular evolutionary genetics analysis using maximum likelihood, evolutionary distance, and maximum parsimony methods. Mol Biol Evol 28(10):2731–2739

34. Sanchez-Villeda H, Schroeder S, Flint-Garcia S, Guill KE, Yamasaki M, McMullen MD (2008) DNAAlignEditor: DNA alignment editor tool. BMC Bioinformatics 9:154–154
35. Inouye M, Dashnow H, Raven L-A, Schultz MB, Pope BJ, Tomita T et al (2014) SRST2: rapid genomic surveillance for public health and hospital microbiology labs. Genome Med 6:90
36. Jolley KA, Maiden MCJ (2013) Automated extraction of typing information for bacterial pathogens from whole genome sequence data: *Neisseria meningitidis* as an exemplar. Euro Surveill 18:20379–20379
37. Larkin MA, Blackshields G, Brown NP, Chenna R, McGettigan PA, McWilliam H et al (2007) Clustal W and Clustal X version 2.0. Bioinformatics 23:2947–2948
38. Edgar RC (2004) MUSCLE: a multiple sequence alignment method with reduced time and space complexity. BMC Bioinformatics 5:113
39. Librado P, Rozas J (2009) Dna SP v5: a software for comprehensive analysis of DNA polymorphism data. Bioinformatics 25:1451–1452
40. Francisco A, Bugalho M, Ramirez M, Carrico J (2009) Global optimal eBURST analysis of multilocus typing data using a graphic matroid approach. BMC Bioinformatics 10:152
41. Darriba D, Taboada GL, Doallo R, Posada D (2012) jModelTest 2: more models, new heuristics and parallel computing. Nat Methods 9:772–772

Chapter 8

Identification and Discrimination of *Chlamydia trachomatis* Ocular and Urogenital Strains and Major Phylogenetic Lineages by *Ct*GEM Typing, A Double-Locus Genotyping Method

Deborah C. Holt, Patiyan Andersson, Cameron Buckley, David M. Whiley, and Philip M. Giffard

Abstract

*Ct*GEM typing was developed to subdivide the bacterial species *Chlamydia trachomatis* on the basis of genome phylogeny and anatomical tropism. The rationale was facilitation of surveillance for ocular strains, although the method is applicable to essentially any *C. trachomatis* surveillance application that does not require high resolution. *Ct*GEM is a double-locus genotyping method. The loci included in the assay were identified by computerized analysis of 65 complete genomes for resolution optimized sets of single nucleotide polymorphisms (SNPs). From this, two PCR amplifiable fragments were defined. One, *rgl*, is within a hypothetical gene annotated as Jali-1891 within the *C. trachomatis* B_Jali20 genome. The other, *ofr*, is within the *ompA* gene which encodes the major outer membrane protein. Variation in *rgl* is conferred by two SNPs defining four haplotypes that exhibit concordance with genome phylogeny. Variation within *ofr* is more complex and allows for inference of *ompA* genotype, either to the level of single genotype, or group of closely related genotypes. Two *Ct*GEM formats were developed. One is based on interrogation of the two loci by high resolution melting analysis (HRMA), and the other based on analysis of the loci by Sanger sequencing. The genotypes defined identify known ocular genotypes, discriminate known ocular genotypes from each other, discriminate the major phylogenetic lineages of the species, and discriminate all *ompA* genotypes with the exception of closely related variants within the genotypes H, I, J cluster. The Sanger sequencing format provides slightly more resolution that the HRMA format with respect to *ompA* genotype. An unusual aspect of this method is that all possible combinations of *rgl* haplotype, and inferred *ompA* genotype(s) have been given *Ct*GEM typing numbers. This includes types that at this time have not been shown to exist.

Key words Genotyping, *Chlamydia trachomatis*, Phylogeny, Population structure

1 Introduction

1.1 Background

Chlamydia trachomatis is a bacterial pathogen exclusive to humans, which affects mucosal surfaces. In common with other members of the *Chlamydiaceae*, *C. trachomatis* is an obligate intracellular

Amanda Claire Brown (ed.), *Chlamydia trachomatis: Methods and Protocols*, Methods in Molecular Biology, vol. 2042, https://doi.org/10.1007/978-1-4939-9694-0_8,

parasite with a distinctive life cycle that encompasses infective non-growing elementary bodies, and reticulate bodies which grow inside a host cell vacuole. *C. trachomatis* can only be propagated in vitro in mammalian cell culture.

C. trachomatis has long been regarded as a urogenital tract (UGT) and ocular pathogen [1, 2], although other sites such as the gastrointestinal tract can support *C. trachomatis* growth [3]. *C. trachomatis* urogenital tract infections represent the most common bacterial sexually transmitted infection (STI) of humans in the world [2]. *C. trachomatis* ocular infections include ophthalmia neonatorum [4] which is acquired by infants during childbirth; mild self-limiting conjunctivitis in adults acquired by transmission from an STI [5], and trachoma, which is a potentially blinding chronic condition [6]. Trachoma has a distinctive ecology and natural history [7]. The syndrome is defined by conjunctivitis acquired during childhood by transfer of *C. trachomatis* infected facial secretions. Repeated infections lead to a hypersensitivity response that in turn leads to anatomical changes, including turned in eyelashes that abrade the cornea. If untreated for many years, irreversible blindness due to corneal scarring can result. Although previously widespread, trachoma has been eliminated from much of the world, with significant prevalence remaining in some regions of Africa and South West Pacific [8]. Australia is the only developed country in the world where trachoma is yet to be eliminated. Trachoma in Australia is almost entirely confined to Aboriginal people living in very remote areas [9]. Ongoing efforts at elimination in Australia mean that trachoma is now typically seen in the form of localized outbreaks in children in remote Aboriginal communities. An important defining feature of trachoma is that it is not regarded as an STI.

In the last decade, comparative genomics has provided a comprehensive understanding of *C. trachomatis* evolutionary history, diversity, and population structure [10–13]. A phylogeny based on genome-wide orthologous SNPs of representative strains is shown in Fig. 1 [11, 14]. Three major lineages are evident. The earliest branching “LGV” lineage encompasses strains associated with a rare and potentially invasive form of STI, termed “lymphogranuloma venereum” (LGV) [15]. The T1 and T2 lineages both encompass strains associated with STIs, and ocular disease consistent with trachoma.

For many years, variation in the *C. trachomatis* major outer membrane protein of *C. trachomatis*, (termed OmpA (used here) or MOMP) was used to define strains [16, 17]. The historical reasons were twofold: (1) OmpA is strongly immunogenic, and so formed the basis of a serological classification scheme that predated DNA-based methods, and (2) This serological classification correlated with natural history, specifically whether the strain was associated with LGV, STIs-plus-nontrachoma ocular infections, or trachoma. The serological "serovar" terminology has been retained to refer to *ompA* genotypes based on *ompA* sequence variation. Types A, B, Ba, and C are trachoma associated, types D, E, F, G, H, I, Ia, J, Ja, and K are associated with the common noninvasive form of STI, and types L1, L2, L2b, and L3 are LGV associated.

Variation in *ompA* is not completely consistent with phylogeny as defined by genome-wide SNPs [10–13]. Thus, the phylogenetic position of a strain cannot be reliably inferred from an *ompA* sequence. A distinctive feature of the T2 lineage (Fig. 1) is an early branching sublineage that encompasses strains with "ocular" *ompA* genotypes. These are primarily of African and East Asian origin, and for some time, a model was that this "classical ocular lineage" encompassed all ocular strains, and thus adaptation to this niche was of relatively ancient origin. However, recent genome-based studies of Australian strains associated with ocular disease in noninfant Indigenous children have revealed strains that are not within the classical ocular lineage but do carry "ocular" B, Ba, and C variants of *ompA* [11]. These strains are termed AusB, AusBa, and AusC, in accordance with Australian origin and *ompA* genotype. They are all very closely related to unremarkable STI strains, and in essence differ from these STI strains only at *ompA*, and, in the case of AusBa and AusC, at the surface proteins encoding gene cluster *pmpEFGH*. All of these strains were strongly associated with ocular infections in children. AusBa and AusC were not detected in adult UGT specimens, even though "conventional" STI strains were detected with high prevalence in the same study. This suggests recent origins of ocular strains by three independent recombination events, and also that *ompA* and *pmpEFGH* directly confer anatomical tropism. The AusBa and AusC strains are quite closely related to each other within lineage T1, and phylogenetically remote from the classical ocular lineage. AusB is within lineage T2, remote from AusBa and AusC, and also not within the classical ocular lineage. This emphasizes the strength of evidence for convergent evolution of ocular strains. More recent studies indicate that AusB is representative of a globally distributed lineage that may be associated with STIs as well as ocular infections, while AusBa and AusC appear to be unique to Australia.

The *Ct*GEM genotyping method [14] was designed to meet the following performance specifications as cheaply and efficiently as possible:

- Discriminate known ocular strains from nonocular strains.
- Discriminate known ocular strains from each other.
- Discriminate LGV strains from non-LGV strains.
- Resolve the major phylogenetic lineages as defined by genome-wide orthologous SNPs.
- Usable with primary clinical specimens.

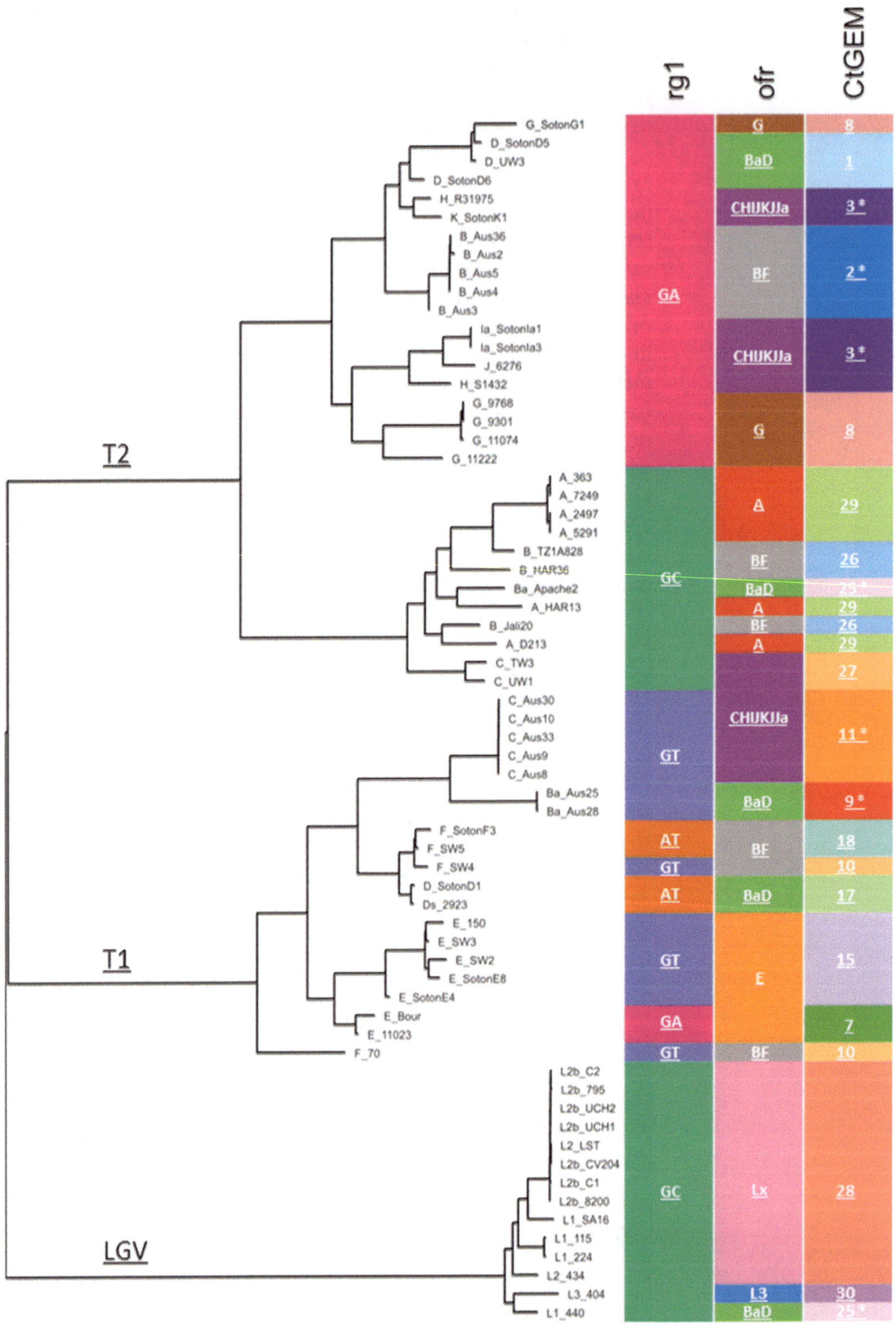

Fig. 1 *Ct*GEM typing compared to *C. trachomatis* genome diversity. *Ct*GEM typing is consistent with the major branches of a phylogenetic tree based on genome-wide orthologous SNPs from 65 *C. trachomatis* genome sequences. *Isolates within *Ct*GEM groups 2, 3, 9, 11, and 25 can be further distinguished by sequencing of

An inevitable corollary of meeting these specifications is that the method will reveal, by elimination, whether or not a strain is one associated with "conventional" STIs.

The rationale for the development of the *Ct*GEM method was to facilitate *C. trachomatis* genotype surveillance in the Australian Northern Territory. A proximate reason for such surveillance is enhancement of the evidence base informing service providers responding to instances of positive *C. trachomatis* diagnostic tests in young children. As has been detailed elsewhere [14, 18], this creates a very difficult medicolegal environment in which the imperatives of child protection can conflict with imperatives to not act on/reinforce stereotypes of Indigenous populations, by instigating baseless investigations of child abuse. The potential role of *Ct*GEM typing is to provide low-resolution comparator genotypes of strains identified in young children who are potentially at risk. A number of mechanisms that could give rise to positive *C. trachomatis* in pediatric UGT specimens in the absence of activities that would be of concern to child protection authorities have been suggested [18–20]. The mechanism that *Ct*GEM typing is designed to address is autoinoculation of an ocular strain from the eye to the UGT site. It may be reasoned that simply genotyping a strain from a potentially at-risk child to determine if it is a "ocular" would be sufficient to answer this question. However, a complication is that it was documented that in the 1980s and 1990s in the Northern Territory, the AusB strain was transmitted extensively as an STI agent through adult sexual networks, as well as being detected in eye swabs from noninfant children [11, 21, 22]. This was not the case in another survey in 2012–2013 [18]. Therefore, the presence or absence of nominally ocular strains circulating in sexual networks at any time point cannot be assumed, and can only be determined empirically by surveillance. We reasoned that the informative power of a *C. trachomatis* genotype from a potentially at-risk child is a function of strains currently circulating in adult sexual networks—with the key point being that if a UGT specimen from a child contained a *C. trachomatis* genotype that is (1) ocular and (2) not detectable in adult sexual networks, then there is no reason to regard the child as having acquired an STI [23, 24]. This can have a large impact on mandated responses of child protection service providers; for example, it could make a critical difference as to whether or not an investigation is triggered.

Fig. 1 (continued) the *ofr* amplicon (as shown in Table 3). Of the isolates represented in the phylogenetic tree, sequencing would further separate L1–440 (25-D) from Ba-Apache2 (25-Ba), and isolate K-SotonK1 (3-K) from the other CtGEM 3 isolates (3-HIIaJ)

We stress that *Ct*GEM typing is not designed to provide the resolution inferring particular transmission events, and so is completely unsuitable for use in any forensic or forensic-like approach to the identification of a particular individual who may be the source of a strain [14]. Notwithstanding the proximate reason for the development of *Ct*GEM typing, we believe this method has considerable potential utility as a general purpose, cost-effective method for *C. trachomatis* genotype surveillance. For example, it may be particularly useful in regions in which trachoma elimination is being pursued, to identify cryptic reservoirs of ocular strains.

While a detailed review of other *C. trachomatis* genotyping methods is beyond the scope of this document, we acknowledge that there have been such methods developed [25]. Examples include methods based purely upon *ompA* variation [26, 27], multilocus sequence typing [28, 29], or, as with *Ct*GEM, based on *ompA* and one other gene [30].

1.2 Method Design and Implementation

The *Ct*GEM method [14] is similar to Minim typing [31, 32]. This approach entails the derivation of a resolution optimized SNP set from an alignment of known DNA sequence variation, and then the development of high resolution melting analysis (HRMA) methods to interrogate the SNPs in the set. The amplified fragments for HRMA potentially include other "bystander" SNPs that in a properly designed method will increase resolving power in a manner that is fully understood with respect to the population structure. HRMA is a well-established methodology that is based upon carefully controlled melting of a PCR product in the presence of a dye for which fluorescence is dependent upon binding to double stranded nucleic acid [31, 33]. The relationship between temperature and fluorescence defines a "melting curve." This indicates the melting temperature (Tm), but also the curve shape is, in effect, a spectrum that encompasses additional information.

A detailed account of the design of the *Ct*GEM method has been published elsewhere [14]. In brief, method development made use of the SNP-mining software Minimum SNPs. This derives SNP sets from DNA sequence alignments on the basis of optimized resolving power [34, 35]. For *Ct*GEM design, the alignment was a genome wide orthologous SNP matrix from 65 *C. trachomatis* genomes encompassing much of the known *C. trachomatis* diversity [11]. The measure of resolving power used by Minimum SNPs was the Simpsons index of Diversity (D) [36], which in this context is the probability that any two of the 65 genomes will be discriminated by the SNP set. Experimentation using Minimum SNPs showed that it was feasible for an HRMA of two fragments to meet the method specifications. This is elaborated below.

Minimum SNPs analysis revealed a particularly promising triallelic SNP at the position corresponding to position 222,834 in the *C. trachomatis* B_Jali20 genome (accession number NC012686, the reference genome of choice in our descriptions of the method). This is within hypothetical gene Jali-1891. It was determined in silico that a genotyping method based on this SNP, and any of a number of SNPs in *ompA*, may meet specifications, so HRMA assays were devised. Development of the final form of the position 222,812/Jali-1891 HRMA method was not straightforward. First, there is another SNP at position 222,810 (i.e., two bases from triallelic SNP). Not only will this SNP affect the outcomes of any HRMA interrogation of SNP 222812, but we determined that interrogation of SNP 222810 would markedly improve the resolving power of the method, in a manner concordant with large scale genome phylogeny. Accordingly, we developed an HRMA method to resolve all four haplotypes defined by the two SNPs. Initially, a PCR to amplify the region encompassing the two SNPs and encompassing no other SNPs was devised. The amplimer was termed "region 1" (*rg1*). At this point we found it necessary to implement two elaborations of conventional HRMA. First, we developed an unlabeled probe plus asymmetric PCR-based method to discriminate *rg1* alleles with the same G + C content. This reflected a previously reported assay format [37]. Second, a nested approach that involved a first round PCR amplification prior to a second amplification and HRMA procedure was developed. The two-stage approach was found to be necessary to ensure sufficient yield of the PCR product that is subject to HRMA. In reactions with low yield, primer depletion is insufficient to support production of single stranded DNA from the asymmetric PCR, with consequent failure of probe binding and allele discrimination.

The development of the *ompA* HRMA assay was also not initially straightforward, but we were able to successfully repurpose an assay previously described by this research group [24]. Attempts to design HRMA assays to interrogate *ompA* SNPs identified using Minimum SNPs were unsuccessful, because of too great a number of bystander SNPs to make HRMA interpretable, and/or an absence of sufficiently conserved PCR primer sites. However, this was circumvented by using an HRMA method developed to discriminate *ompA* genotypes B and Ba [24]. We now term the fragment amplified in this assay "*ompA* fragment" (*ofr*). We developed a semi-nested PCR embodiment of the *ofr* HRMA. We regard this as optimal, although a single step procedure (i.e., second PCR only) is often successful. Although *ofr* does not encompass any of the *ompA* SNPs that we identified using Minimum SNPs, it was determined that HRMA of *rg1* and *ofr* is effective in meeting the performance specifications for this typing method.

Detailed testing of the HRMA embodiment of *Ct*GEM, both with specimens and in silico with known genome variation [10] did identify a low but nonzero frequency of classification of nonocular strains as ocular. We determined that these instances can be resolved by Sanger sequencing of *ofr*, which can reveal sequence variation that is cryptic in *ofr* HRMA. We regard the Sanger sequencing as "optional but integral" so include this in the method description.

In summary, *Ct*GEM is a double-locus *C. trachomatis* typing method that yields biologically meaningful genotypes throughout the diversity of *C. trachomatis*. We regard the default mode as HRMA because of its low cost and simplicity, particularly for large numbers of specimens. However, depending on user requirements, Sanger sequencing of these fragments can also be used, either, for example, as a confirmation step in a screen for ocular strains using the HRMA mode, or as a primary method. The primary use of Sanger sequencing may be appropriate if small numbers of specimens are to be analyzed; an advantage of the sequencing approach is that control strains are not required.

2 Materials

2.1 CtGEM Typing, HRMA Embodiment

1. Materials for DNA extraction of control strain or test samples (*see* **Note 1**).
2. Platinum® SYBR® Green qPCR SuperMix-UDG (Invitrogen).
3. Primers and probes diluted as indicated in sterile double distilled H_2O.

10 μM CT_11508_F_outer	5′ CATCCACTTTACCAGCGATTC 3′
10 μM CT_11508_R_outer	5′ CCATCTTTGGCTCGTGCTAT 3′
1 μM CT_11508_F	5′ GGAAAGGTCAGCACTCTTGAA 3′ (*see* **Note 2**)
10 μM CT_11508_R	5′ GGGGTTCTTGTTGTTTTAGGG 3′
10 μM CT_11508_probe	5′ TAGCTAGATAGGCTATCCGAA AACAAATCC 3′ (phos)
10 μM ofr_F_outer	5′ TCCTACTGCAATACCGCAAG 3′
10 μM ofr_R_nested	5′ GCTTCTTTCCAATATGCTCAATC 3′
10 μM ofr_F_nested	5′ GCCATTCATGGTAATCAATAGAG 3′

4. Q-solution (5 M betaine) (Invitrogen).
5. Sterile double distilled H_2O.
6. Qiagen 0.1 mL Strip tubes and caps.

7. Qiagen Rotor-Gene Q device and Rotor-Gene Q series software.
   - Set software to acquire to channel Cycling A green.
   - Use the default gain optimization settings.

2.2 Analysis of HRMA Data for CtGEM Typing

1. Data from melting of PCR products in the Qiagen Rotorgene device.

2.3 Supplementary HRMA to Discriminate AusC Strains from H/I/Ia/J/Ja/K Strains

1. Reaction products from first round *ofr* amplification (Subheading 3.1.2, First Round PCR).
2. Platinum® SYBR® Green qPCR SuperMix-UDG (Invitrogen).
3. Primers and probes diluted as indicated in sterile double distilled H_2O.

10 μM ompA_CK_HRM_F	5′ TAGTCCCTGTCGCAGCTTCT 3′
10 μM ompA_CK_HRM_R	5′ GATATGTTGGGGCGGAATTT 3′

4. Q-solution (5 M betaine) (Invitrogen).
5. Sterile double distilled H_2O.
6. Qiagen 0.1 mL Strip tubes and caps.
7. Qiagen Rotor-Gene Q device and Rotor-Gene Q series software.
   - Set software to acquire to channel Cycling A green.
   - Use the default gain optimization settings.

2.4 CtGEM Typing, Sanger Sequencing Embodiment

1. Materials for DNA extraction of test samples (*see* **Note 1**).
2. SensiFAST™ SYBR® No-ROX Mastermix kit (Bioline, Australia).
3. Sterile double distilled H_2O.
4. Qiagen 0.1 mL Strip tubes and caps.
5. Qiagen Rotor-Gene Q device.
6. The following PCR/sequencing primers:

10 μM CT_11508_F_seq	5′ CCCATTGCCGAGAGATAAAA 3′
10 μM CT_11508_R_seq	5′ CTCCTGCGGAGGTTAGATTG 3′
10 μM ofr_F_outer	5′ TCCTACTGCAATACCGCAAG 3′
10 μM ofr_R_seq	5′ TGAACCAAGCCTTATGATCG 3′

3 Methods

3.1 CtGEM Typing, HRMA Embodiment

3.1.1 Amplification and Melting of rg1

First Round PCR

1. For each sample to be genotyped and each DNA extraction of known genotype to be used as a control, combine the following into a Qiagen 0.1 mL strip tube:

   5.0 μL of 2× Platinum® SYBR® Green qPCR SuperMix-UDG.

   0.5 μL of 10 μM CT_11508_F_outer primer.

   0.5 μL of 10 μM CT_11508_R_outer primer.

   0.5 μL of Q-solution.

   1.0 μL of DNA extraction (or H_2O for a no template negative control) (*see* **Note 1**).

   2.5 μL of sterile double distilled H_2O.
2. Attach a lid to each tube and label appropriately.
3. Place tubes into a Rotor-Gene Q device, commencing at position 1.
4. Perform PCR amplification as follows:

   50 °C 2 min;

   95 °C 2 min;

   50 cycles of 95 °C 10 s, 61–56 °C (1 °C per cycle) 10 s, 72 °C 20 s;

   95 °C 1 min;

   40 °C 1 min
5. Remove tubes from Rotor-Gene Q device and store at 4 °C.

Second Round Asymmetric Nested PCR and Amplimer Melting

1. For each of the first round PCR reactions, combine the following into a Qiagen 0.1 mL strip tube:

   5.0 μL of 2× Platinum® SYBR® Green qPCR SuperMix-UDG.

   0.4 μL of 1 μM CT_11508_F primer (*see* **Note 2**).

   0.4 μL of 10 μM CT_11508_R primer.

   1.0 μL of 10 μM CT_11508 probe.

   0.5 μL of Q-solution.

   1.0 μL of first round PCR amplification (or H_2O for second round no template negative control).

   1.7 μL of sterile double distilled H_2O.
2. Attach a lid to each tube and label appropriately.
3. Place tubes into a Rotor-Gene Q device, commencing at position 1.
4. Perform PCR amplification and HRMA using the following conditions:

50 °C 2 min;

95 °C 2 min;

50 cycles of 95 °C 10 s, 61–56 °C (1 °C per cycle) 10 s, 72 °C 20 s;

95 °C 1 min;

40 °C 1 min;

50–95 °C in 0.1 °C increments for 2 s each.

3.1.2 *Amplification and Melting of* ofr

First Round PCR

1. For each sample to be genotyped and each control DNA extraction of known genotype to be used as a control, combine the following into a Qiagen 0.1 mL strip tube:

   5.0 μL of 2× Platinum® SYBR® Green qPCR SuperMix-UDG.

   0.4 μL of 10 μM ofr_F_outer primer.

   0.4 μL of 10 μM ofr_R_nested primer.

   0.5 μL of Q-solution.

   1.0 μL of DNA extraction (or H_2O for a no template negative control) (*see* **Note 1**).

   2.7 μL of sterile double distilled H_2O.
2. Attach a lid to each tube and label appropriately.
3. Place tubes into a Rotor-Gene Q device, commencing at position 1.
4. Perform PCR amplification using the following conditions:

   50 °C 2 min;

   95 °C 2 min;

   50 cycles of 95 °C 10 s, 59–54 °C (1 °C per cycle) 10 s, 72 °C 40 s;

   95 °C 1 min;

   40 °C 1 min.
5. Remove tubes from Rotor-Gene Q device and store at 4 °C.

Second-Round Semi-nested PCR and Amplimer Melting (*See* **Note 3**)

1. For each of the first round PCR reactions, combine the following into a Qiagen 0.1 mL strip tube:

   5.0 μL of 2× Platinum® SYBR® Green qPCR SuperMix-UDG.

   0.4 μL of 10 μM ofr_F_nested primer.

   0.4 μL of 10 μM ofr_R_nested primer.

   0.5 μL of Q-solution.

   1.0 μL of first round PCR amplification (or H_2O for second round no template negative control).

   2.7 μL of sterile double distilled H_2O.
2. Attach a lid to each tube and label appropriately.

3. Place tubes into a Rotor-Gene Q device, commencing at position 1.
4. Perform PCR amplification and HRMA using the following conditions:

   50 °C 2 min;

   95 °C 2 min;

   50 cycles of 95 °C 10 s, 59–54 °C (1 °C per cycle) 10 s, 72 °C 40 s;

   95 °C 1 min;

   40 °C 1 min;

   65–95 °C in 0.1 °C increments for 2 s each.

3.2 Analysis of HRMA Data for CtGEM Typing

1. Open the second round PCR in the Rotor-Gene Q software.
2. Select Analysis → Quantitation → Cycling A Green to view the amplification curves.
   - Inspect the amplification curves and exclude any samples which do not reach the default cycle threshold setting.

3.2.1 Assignment of Melting Curves to HRMA Alleles

rg1 (*See* **Note 4**)

3. Select Analysis → HRM → HRM A HRM to view the melt curves (the HRM tab may be found under the "Other" tab when the file is initially opened).
   - In the HRM analysis window, normalize the data at 53–54 °C on the left, and 80–81 °C on the right.
   - View the normalized HRM curves by selecting "Normalised Graph" at the top of the HRM analysis window.
   - Select all the controls of known genotype from the sample names panel on the far right.
   - Sequentially compare each of the unknown genotypes to the control samples.
4. Initially compare each sample of unknown genotype between ~50 and 70 °C which encompasses the probe melting event (Fig. 2 upper inset). As the second round PCR is asymmetric, there is an excess of the strand complementary in sequence to the probe (*see* **Note 2**). The dissociation of the probe from its complementary strand occurs in a manner dependent on the sequence mismatches with the probe.
   - The AT genotype has two mismatches to the probe sequence, and consequently the probe dissociates from this genotype at the lowest temperature.
   - The GA genotype is an exact match to the probe sequence, and consequently the probe dissociates from this genotype at the highest temperature.
   - The GC and GT genotypes have a single mismatch with the probe sequence, and consequently the probe dissociates with these genotypes at a temperature between the other two genotypes.

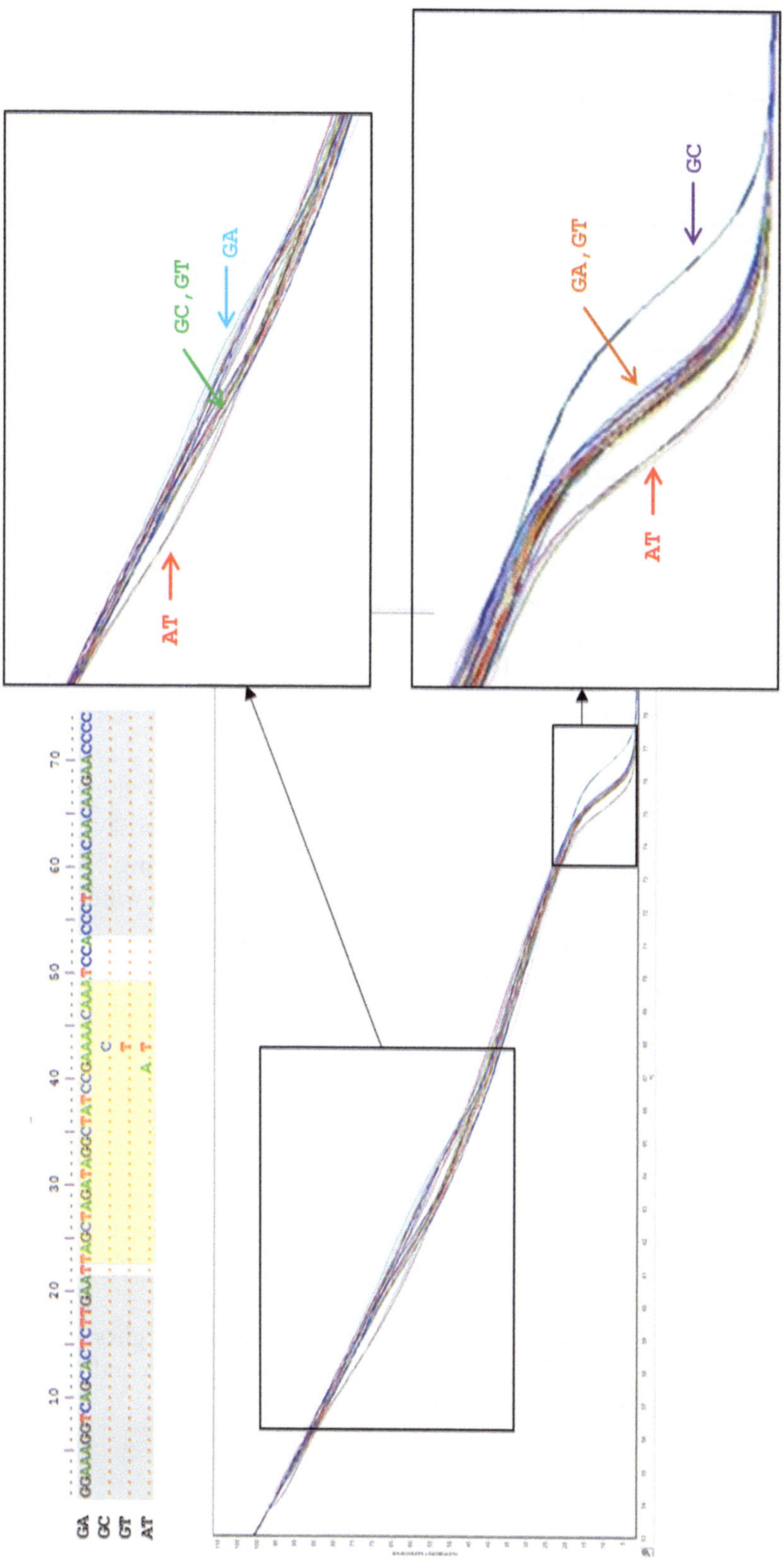

Fig. 2 HRMA of the second-round *rg1* PCR product. The asymmetric nature of the second round PCR (*see* **Note 2**) enables the melting of the probe from its complementary strand to be visualized and separates the genotypes into AT, GC/GT, and GA (upper inset). The melting of the remaining double stranded PCR product separates the genotypes into AT, GA/GT, and GC (lower inset). The combination of the two melting events allows for all four genotypes to be distinguished. Gray shading indicates the second round PCR primer binding sites and the yellow shading indicates the probe binding site

5. Then compare the melt curves of each sample of unknown genotype after 70 °C, which is the melting of the full length double stranded PCR amplicon (Fig. 2 lower inset).
   - The AT genotype amplicon has the lowest G + C content, and consequently it melts at the lowest temperature.
   - The GA and GT genotypes have one additional G + C compared with the AT genotype, and consequently these amplicons melt ~0.5 °C later than the AT genotype.
   - The GC genotype has two additional G + Cs compared with the AT genotype, and consequently the amplicons melts ~1.2 °C later than the AT genotype amplicons (and ~0.7 °C later than the GA and GT genotype amplicons).

ofr (*See* **Note 5**)

1. Open the second round PCR in the Rotor-Gene Q software.
2. Select Analysis → Quantitation → Cycling A Green to view the amplification curves.
   (a) Inspect the amplification curves and exclude any samples which do not reach the default cycle threshold setting.
3. Select Analysis → HRM → HRM A HRM to view the melt curves (the HRM tab may be found under the "Other" tab when the file is initially opened).
   (a) In the HRM analysis window, normalize the data at 73–74 °C on the left, and 83–84 °C on the right.
   (b) View the normalized HRM curves by selecting "Normalised Graph" at the top of the HRM analysis window.
   (c) Select all the controls of known genotype, as well as all of the samples of unknown genotype to be identified, from the sample names panel on the far right.
   (d) Sequentially identify each of the known genotypes and assign the unknowns to a genotype by sequentially identifying each of the genotypes and removing those samples from the analysis, as described below.
4. Genotypes G, E, and L1/L2.
   (a) Genotype G amplicons have the highest G + C content, and consequently melt at the highest temperature (Fig. 3a).
   (b) Genotype E amplicons have the second highest G + C content, and consequently melt at the second highest temperature (Fig. 3a).

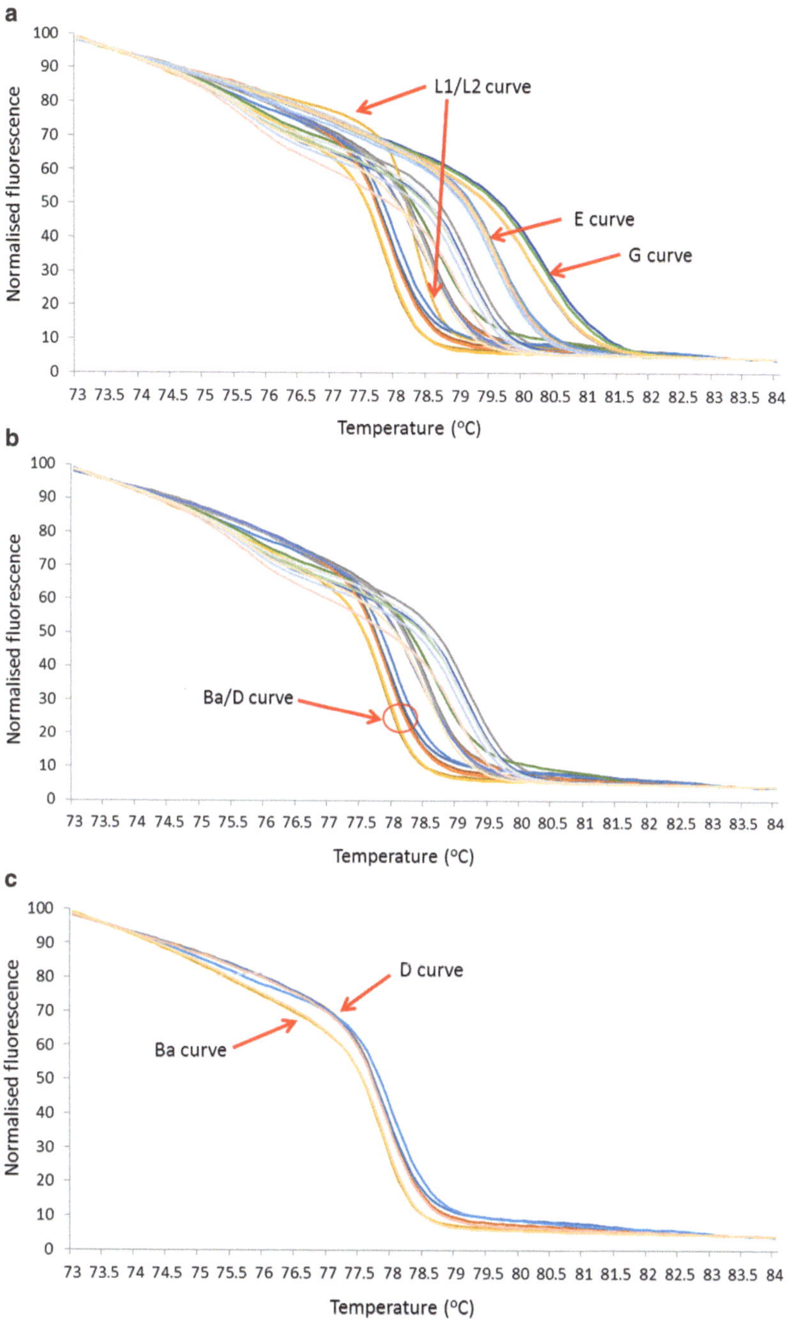

Fig. 3 Detailed method for analysis of *ofr* curves. (**a**) HRMA of *ofr* curves showing seven of the eight predicted curves (L3 curve not represented). The E curve, G curve, and L1/L2 curve can be identified first as indicated. (**b**) With the E, G, and L1/L2 curves removed, the Ba/D curve can be identified as the curve with the lowest Tm. (**c**) Although we conservatively group the Ba and D curves together in our scheme, it can be seen that the Ba curve starts to melt slightly earlier than the D curve, likely due to a G-A SNP at the 3′ end of the fragment. This feature, or sequencing of the fragment, can be used to distinguish between Ba and D samples if required (*see* **Note 5**). (**d**) The remaining curves can be split into two groups- those with a clear early melting domain (C/H/I/Ia/J/Ja/K), and those without (A, B/F). (**e**) The remaining curves (C/H/I/Ia/J/Ja/K) have a range of Tms which reflects the sequence differences shown in Fig. 2. (**f**) The A curve and the B/F curves can be easily distinguished by their Tm

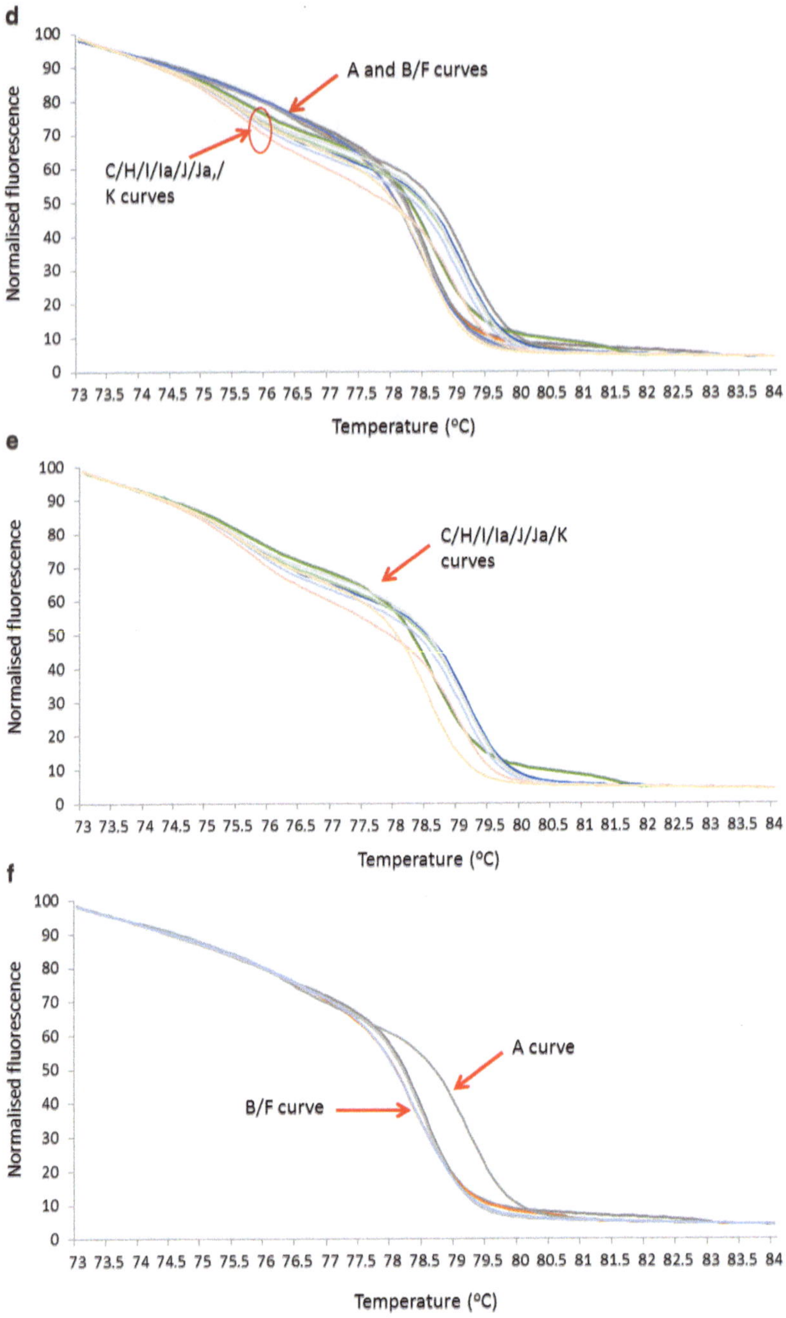

Fig. 3 (continued)

(c) Genotype L1 and L2 amplicons have a distinct shape which is discernible from the shape of the melt curves of all other genotype (Fig. 3a).

(d) Although we have not experimentally tested a L3 strain, in silico prediction of the L3 genotype melt curve using uMELT analysis indicates it will have the same distinctive shape as the L1 and L2 melt curves.

(e) Remove all of the controls and samples of these genotypes by deselecting from the sample name panel at the far right.

5. Genotypes Ba and D.

(a) Genotype Ba and D amplicons have the lowest G + C content, and consequently can be identified as the curve with the lowest Tm (Fig. 3b).

(b) Although our genotyping scheme conservatively groups the Ba and D curves together as they have the same G + C content, we have found that the Ba curve starts to melt at a slightly lower temperature than the D curve (Fig. 3c), likely due to a G → A SNP at the 3′ end of the fragment (Fig. 3c, *see* **Note 5**).

(c) Remove all of the controls and samples of these genotypes by deselecting from the sample name panel at the far right.

6. Genotypes C, H, I, Ia, J, Ja, and K.

(a) Of the remaining genotypes, amplicons of genotypes C, H, I, Ia, J, Ja, and K all have an early melting domain which distinguishes them from amplicons of genotype A, B, and F (Fig. 3d).

(b) Amplicons of genotype C, H, I, Ia, J, Ja, and K have a range of Tms and cannot be reliably distinguished from each other (Fig. 3e).

(c) Remove all of the controls and samples of genotypes C, H, I, Ia, J, Ja, and K by deselecting from the sample name panel at the far right.

7. Genotypes A, B, and F.

(a) Amplicons of genotype A can clearly be distinguished by a higher melting temperature than the other remaining amplicons of genotype B and F (which cannot be resolved from each other) (Fig. 3f).

3.2.2 Inference of CtGEM Type from HRMA Alleles

1. Use Table 1 to determine which *Ct*GEM type corresponds to the combination of the determined *ofr* and *rgl* HRMA alleles. All possible combinations of alleles are given a type number, including those that have not been shown to exist to date.

Table 1
***Ct*GEM HRMA types**

	HRMA allele of *ofr*							
***Rg1* haplotype**	**Ba D**	**BF**	**CHIaJJJa K**	**Lx**	**A**	**L3**	**E**	**G**
GA	1	2	3	4	5	6	7	8
GT	9	10	11	12	13	14	15	16
AT	17	18	19	20	21	22	23	24
GC	25	26	27	28	29	30	31	32

Types in shaded cells have not been demonstrated to exist to date, either in analysis of specimens, or in silico analyses of known *C. trachomatis* genome variation

3.3 Supplementary HRMA to Discriminate AusC Strains from H/I/Ia/J/Ja/K Strains (See Note 6)

3.3.1 Amplification and Melting of ofr Subfragment

1. For each of the first round *ofr* PCR reactions to be tested, combine the following into a Qiagen 0.1 mL strip tube:

   5.0 μL of 2× Platinum® SYBR® Green qPCR SuperMix-UDG.

   0.4 μL of 10 μM ompA_CK_HRM_F primer.

   0.4 μL of 10 μM ompA_CK_HRM_R primer.

   0.5 μL of Q-solution.

   1.0 μL of first round *ofr* PCR amplification (or H_2O for second round no template negative control).

   2.7 μL of sterile double distilled H_2O.
2. Attach a lid to each tube and label appropriately.
3. Place tubes into a Rotor-Gene Q device, commencing at position 1.
4. Perform PCR amplification and HRMA using the following conditions:

   50 °C 2 min;

   95 °C 2 min;

   50 cycles of 95 °C 10 s, 59–54 °C (1 °C per cycle) 10 s, 72 °C 40 s;

   95 °C 1 min;

   40 °C 1 min;

   65–95 °C in 0.1 °C increments for 2 s each.

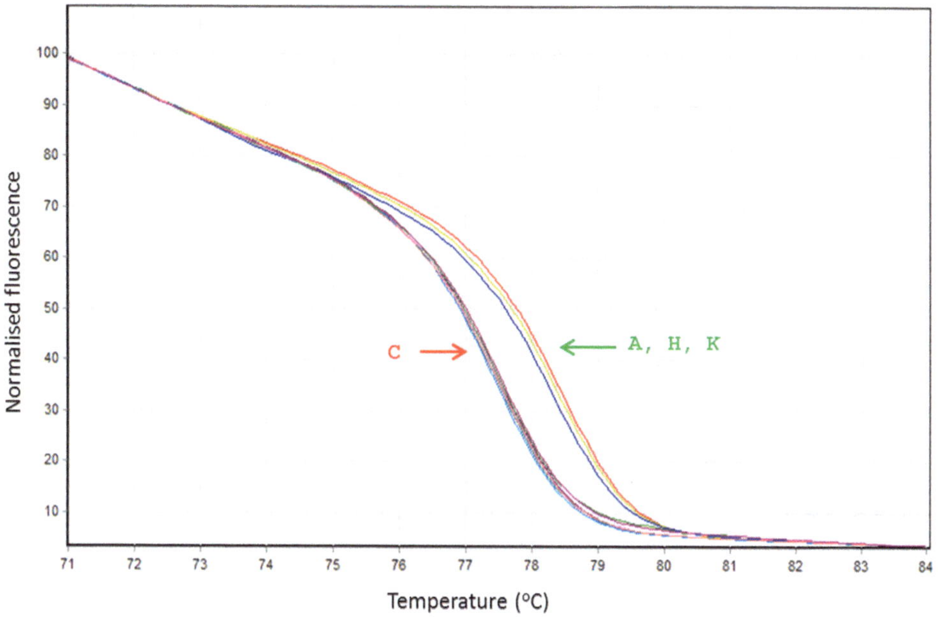

Fig. 4 Supplementary HRMA to discriminate AusC strains from H, I, Ia, J, Ja, and K strains. Amplification of a subfragment of *ofr* allows for AusC strains to be discriminated from H, I, Ia, J, Ja, and K strains by HRMA

3.3.2 Assignment of Melting Curves to HRMA Alleles

1. Open the second round PCR in the Rotor-Gene Q software.
2. Select Analysis → Quantitation → Cycling A Green to view the amplification curves.
3. Inspect the amplification curves and exclude any samples which do not reach the default cycle threshold setting,
4. Select Analysis → HRM → HRM A HRM to view the melt curves (the HRM tab may be found under the "Other" tab when the file is initially opened).
5. In the HRM analysis window, normalize the data at 69–70 °C on the left, and 84–85 °C on the right,
6. View the normalized HRM curves by selecting "Normalized Graph" at the top of the HRM analysis window. *Ofr* C genotype strains will melt at a lower temperature than the H/I/Ia/J/Ja/K curves (Fig. 4).

3.4 CtGEM Typing, Sanger Sequencing Embodiment (See Note 7)

3.4.1 Amplification of Region Containing rg1*, Suitable for Sanger Sequence Analysis*

1. For each clinical sample to be amplified, combine the following into a Qiagen 0.1 mL or 0.2 mL tube:

   10.0 μL of SensiFast SYBR

   1.0 μL of 10 μM CT_11508_F_seq Primer

   1.0 μL of 10 μM CT_11508_R_seq Primer

   2.0 μL of DNA Extraction

   6.0 μL of Sterile Double Distilled H_2O

2. Place a lid on each tube and label appropriately.
3. Place tubes into a Rotor-Gene Q device, commencing at position 1.
4. Perform PCR amplification using the following conditions:

   95 °C 10 min;

   40 cycles of 95 °C 20 s, 60 °C 30 s, 72 °C 60 s;

   65–95 °C in 1 °C increments
5. Remove tubes from Rotor-Gene Q device and store at −20 °C until amplicons are ready to be sent off for sequencing.
6. Perform sequence determination using the PCR primers. This was carried out using a DNA sequencing service provider (*see* **Note 8**).

3.4.2 Amplification of Region Containing ofr*, Suitable for Sanger Sequence Analysis*

1. For each clinical sample to be amplified, combine the following into a Qiagen 0.1 mL or 0.2 mL tube:

   10 μL of SensiFast SYBR.

   1.0 μL of 10 μM ofr_F_outer Primer.

   1.0 μL of 10 μM ofr_R_seq Primer.

   2.0 μL of DNA Extraction.

   6.0 μL of Sterile Double Distilled H_2O.
2. Attach a lid to each tube and label appropriately.
3. Place tubes into a Rotor-Gene Q device, commencing at position 1.
4. Perform PCR amplification using the following conditions:
5. 95 °C 10 min;

   40 cycles of 95 °C 20 s, 60 °C 30 s, 72 °C 60 s;

   65–95 °C in 1 °C increments.
6. Remove tubes from Rotor-Gene Q device and store at −20 °C until amplicons are ready to be sent off for sequencing.
7. Perform sequence analysis using the PCR primers. This was carried out using a DNA sequencing service provider (*see* **Note 8**).

3.4.3 rg1 *Haplotype Determination from Sequence Data (See* ***Note 7****)*

1. Obtain *rgl* sequences (both directions) from sequencing service provider.
2. Open Web BLASTn [38] at https://blast.ncbi.nlm.nih.gov/Blast.cgi?PROGRAM=blastn&PAGE_TYPE=BlastSearch&LINK_LOC=blasthome1.
3. Select the "Align two or more sequences" box.
4. Copy and paste the *rgl* sequence you have obtained into the "Enter Query Sequence" box. The sequence does not have to be trimmed—it only needs to encompass the *rgl* sequence.

5. Copy and paste the known *rgl* variants in Fasta format into the "Enter Subject Sequence" box. This alignment is here, with polymorphic sites underlined:

   >GC haplotype

   GGAAAGGTCAGCACTCTTGAATTAGCTAGATAGGCTATCCGACAACAAATCCACCCTAAAACAACAAGAACCCC

   >GA haplotype

   GGAAAGGTCAGCACTCTTGAATTAGCTAGATAGGCTATCCGAAAACAAATCCACCCTAAAACAACAAGAACCCC

   >GT haplotype

   GGAAAGGTCAGCACTCTTGAATTAGCTAGATAGGCTATCCGATAACAAATCCACCCTAAAACAACAAGAACCCC

   >AT haplotype

   GGAAAGGTCAGCACTCTTGAATTAGCTAGATAGGCTATCCAATAACAAATCCACCCTAAAACAACAAGAACCCC

   Alternatively, the alignment can be uploaded using the "choose file" button. This requires saving the alignment as a file first.
6. Press the "BLAST" button.
7. Scroll down to observe result when analysis complete.
8. The *rgl* sequence(s) in the Subject Sequence library with a 100% match over the full length of the subject sequence represents the *rgl* haplotype defined by your sequence.
9. This procedure should be done with both forward and reverse sequences (*see* **Note 9**).

3.4.4 *ompA* Genotype Inference from ofr Sequence Data

1. Obtain *ofr* sequences (both directions) from sequencing service provider.
2. Open Web BLASTn [38] at https://blast.ncbi.nlm.nih.gov/Blast.cgi?PROGRAM=blastn&PAGE_TYPE=BlastSearch&LINK_LOC=blasthome1.
3. Select the "Align two or more sequences" box.
4. Copy and paste the *ofr* sequence you have obtained into the "Enter Query Sequence" box. The sequence does not have to be trimmed – it only needs to encompass the *ofr* sequence.

5. Copy and paste the known *ofr* variants in Fasta format into the "Enter Subject Sequence" box. The alignment is given here:

```
>ompA genotype A
GCCACTCATGGTAGTCAATAGAGGCATCCTTAGTCCCTGTCGCAGCTTCTGTTCCTGCGG
TAATATCAAGTGGAAATTCCGCCCCAACATATCCTTTCGGCTTATTAATAGTAAATTCGG
ATGCATTACAAAGAACATTCAACTCTTCTACTTTAGGTTTAGATTGAGCATATTGGAAAG
AAGC
>ompA genotype B
GCCATTCATGGTAATCAATAGAGGCATCCTTAGTTCCTGTCGCAGCATCTGTTCCTGCTG
TAAGATCAAGAGGCAACTCCTTACCTACATACCCTTTAGGTTTATTAATAGTAAACTCTG
CTGCATTGCAGAGAACGTTTAATTCTTCTACTTTAGGTTTAGATTGAGCATATTGGAAAG
AAGC
>ompA genotype Ba,ofr variant in AusBa
GCCATTCATGGTAATCAATAGAGGCATCCTTAGTTCCTGTCGCAGCATCTGTTCCTGATG
TAAGATCAAGAGGCAACTCCTTACCTACATACCCTTTAGGTTTATTAATAGTAAACTCTG
CTGCATTGCAGAGAACGTTTAATTCTTCTACTTTAAGTTTAGATTGAGCATATTGGAAAG
AAGC
>ompA genotype Ba,ofr variant in Apache-2 and others
GCCATTCATGGTAATCAATAGAGGCATCCTTAGTTCCTGTCGCAGCATCTGTTCCTGATG
TAAGATCAAGAGGCAACTCCTTACCTACATACCCTTTAGGTTTATTAATAGTAAACTCTG
CTGCATTGCAGAGAACGTTTAATTCTTCTACTTTAGGTTTAGATTGAGCATATTGGAAAG
AAGC
>ompA genotype Ba,ofr variant in IU-FW0305 only
GCCATTCATGGTAATCAATAGAGGCATCCTTAGTTCCTGTCGCAGCATCTGTTCCTGATG
TAAGATCAAGAGGCAACTCCTTACCTCCATACCCTTTAGGTTTATTAATAGTAAACTCTG
CTGCATTGCAGAGAACGTTTAATTCTTCTACTTTAAGTTTAGATTGAGCATATTGGAAAG
AAGC
>ompA genotype C
GCCACTCATGGTAGTCAATAGAGGCATCCTTAGTCCCTGTCGCAGCTTCTGTTCCTGCGG
TAATATTAAGTGGAAATTCCGCCCCAACATATCCTTTCGGCTTATTAATAGTAAATTCGG
ATGCATTACAAAGAACATTTAACTCTTCTACTTTAGGTTTAGATTGAGCATATTGGAAAG
AAGC
>ompA genotype D variant 1
GCCATTCATGGTAATCAATAGAGGCATCCTTAGTTCCTGTCGCAGCATCTGTTCCTGCTG
TAAGATCAAGAGGAAACTCCTTACCTACATACCCTTTAGGTTTATTAATAGTAAACTCTG
CTGCATTGCAGAGAACGTTTAATTCTTCTACTTTAGGTTTAGATTGAGCATATTGGAAAG
AAGC
```

>*ompA* genotype D variant 2
GCCATTCATGGTAATCAATAGAGGCATCCTTAGTTCCTGTCGCAGCATCTGTTCCTGCTG
TAAGATCAAGAGGAAACTCCTTACCTACATACCCTTTAGGTTTATTAATAGTAAACTCTG
CTGCATTGCAGAGAACGTTTAATTCTTCTACTTTAGGTTTAGATTGAGCATATTGGAATG
AAGC

>*ompA* genotype E
GCCACTCATGGTAATCAATAGAGGCATCTTTAGTGCCCGTCGCTGCATCAGTTCCTGCTA
TGAGTGCAAGAGGGAATTCTTGCCCTACATATCCTTTAGGCTTATTGATAGTAAACTCAG
CTGCGTTACAGAGAACGTTTAATTCTTCGACTTTAGGTTTAGATTGAGCGTATTGGAAAG
AAGC

>*ompA* genotype E rare variant, strain E-C58 only
GCCACTCATGGTAATCAATAGAGGCGTCTTTAGTGCCCGTCGCTGCATCAGTTCCTGCTA
TGAGTGCAAGAGGGAATTCTTGCCCTACATATCCTTTAGGCTTATTGATAGTAAACTCAG
CTGCGTTACAGAGAACGTTTAATTCTTCGACTTTAGGTTTAGATTGAGCGTATTGGAAAG
AAGC

>*ompA* genotype F
GCCACTCATGGTAATCAATAGAGGCATCTTTAGTGCCCGTCGCTGCATCTGTTCCTGCTG
TAAGATCAAGAGGAAACTCCTTACCTACATACCCTTTAGGTTTATTAATAGTAAACTCTG
CTGCGTTACAGAGAACGTTTAATTCTTCGATTTTAGGTTTAGATTGAGCATATTGGAAAG
AAGC

>*ompA* genotype G most common variant
GCCACTCATGGTAATCAATAGAGGCATCTTTAGTGCCCGTCGCTGCATCAGTTCCTGCTG
TGAGTGCAAGAGGGAATTCTTGCCCTACATATCCTTTAGGCTTATTGATAGTAAACTCAG
CTGCGTTACAGAGAACGTTTAATTCTTCGACTTTAGCTTTAGATTGAGCGTATTGGAAAG
AAGC

>*ompA* genotype G less common variant e.g. strain SotonG1
GCCACTCATGGTAATCAATAGAGGCATCTTTAGTGCCCGTCGCTGCATCAGTTCCTGCTG
TGAGTGCAAGAGGGAATTCTTGCCCTACATATCCTTTAGGCTTATTGATAGTAAACTGAG
CTGCGTTACAGAGAACGTTTAATTCTTCGACTTTAGGTTTAGATTGAGCGTATTGGAAAG
AAGC

>*ompA* genotypes H, I, Ia, J
GCCACTCATGGTAGTCAATAGAGGCATCCTTAGTCCCTGTCGCAGCTTCTGTTCCTGCGG
TAATATCAAGTGGAAATTCCGCCCCAACATATCCTTTCGGCTTATTAATAGTAAATTCGG
ATGCATTACAAAGAACATTTAACTCTTCTACTTTAGGTTTAGATTGAGCATATTGGAAAG
AAGC

>*ompA* genotype I rare variant strain I-UW12 only
GCCACTCATGGTAGTCAATAGAGGCATCCTTAGTCCCTGTCGCAGCTTCTGTTCCTGCGA
TAATATCAAGTGGAAATTCCGCCCCAACATATCCTTTCGGCTTATTAATAGTAAATTCGG
ATGCATTACAAAGAACATTTAACTCTTCTACTTTAGGTTTAGATTGAGCATATTGGAAAG
AAGC

>*ompA* genotype J rare variant
GCCACTCATGGTAGTCAATAGAGGCATCCTTAGTCCCTGTCGCAGCTTCTGTTCCTGCGG
TAATATCAAGTGGAAATTCCGCCCCAACATATCCTTTCGGCTTATTAATAGTAAATTCGG
ATGCATTACAAAGAACATTTAATTCTTCTACTTTAGGTTTAGATTGAGCATATTGGAAAG
AAGC

>*ompA* genotype Ja
GCCACTCATGATAGTCAATAGAGGCATCCTTAGTCCCTGTCGCAGCTTCTGTTCCTGCGG
TAATATCAAGTGGAAATTCCGCCCCAACATATCCTTTCGGCTTATTAATAGTAAATTCGG
ATGCATTACAAAGAACATTCAATTCTTCTACTTTAGGTTTAGATTGAGCATATTGGAAAG
AAGC

>*ompA* genotype K
GCCACTCATGGTAGTCAATAGAGGCATCCTTAGTCCCTGTCGCAGCTTCTGTTCCTGCGG
TAATATCAAGTGGAAATTCCACCCCAACATATCCTTTCGGCTTATTAATAGTAAATTCGG
ATGCATTACAAAGAACATTTAACTCTTCTACTTTAGGTTTAGATTGAGCATATTGGAAAG
AAGC

>*ompA* genotypes L1, L2 common variant
GCCATTCATGGTAATCAATAGAGGCATCCTTAGTTCCTGTCACACCATCTGTTCCTGCTT
TAAGATCAAGAGGGAATTCTTGCCCTACATATCCTTTAGGCTTATTGATAGTAAACTCAG
CTGCGTTACAGAGAACGTTTAATTCTTCGACTTTAGGCTTGGATTGAGCGTATTGGAAAG
AAGC

>*ompA* genotype L1 rare variant, strain 1322 only
GCCATTCATGGTAATCAATAGAGGCATCCTTAGTTCCTGTCGCAGCATCTGTTCCTGCTG
TAAGATCAAGAGGAAACTCCTTACCTACATACCCTTTAGGTTTATTAATAGTAAACTCTG
CTGCATTGCAGAGAACGTTTAATTCTTCTACTTTAGGCTTGGATTGAGCGTATTGGAAAG
AAGC

>*ompA* genotype L1 rare *ofr* variant strain 115 only
GCCATTCATGGTAATCAATAGAGGCATCCTTAGTTCCTGTCACATCATCTGTTCCTGCTT
TAAGATCAAGAGGGAATTCTTGCCCTACATATCCTTTAGGCTTATTGATAGTAAACTCAG
CTGCGTTACAGAGAACGTTTAATTCTTCGACTTTAGGCTTGGATTGAGCGTATTGGAAAG
AAGC

>*ompA* genotype L1 rare *ofr* variant, strain 440 only
GCCATTCATGGTAATCAATAGAGGCATCCTTAGTTCCTGTCGCAGCATCTGTTCCTGCTG
TAAGATCAAGAGGAAACTCCTTACCTACATACCCTTTAGGTTTATTAATAGTAAACTCTG
CTGCATTGCAGAGAACGTTTAATTCTTCTACTTTAGGTTTAGATTGAGCATATTGGAAAG
AAGC

>*ompA* genotype L3
GCCACTCATGGTAGTCAATAGAGGCATCCTTAGTCCCTGTCGCAGCTTCTGTTCCTGCGG
TAATATCAAGTGGAAATTCCGCCCCAACATATCCTTTCGGCTTATTAATAGTAAATTCGG
ATGCATCACAAAGAACATTCAACTCTTCTACTTTAGGTTTAGATTGAGCATATTGGAAAG
AAGC

Table 2
Ct*GEM types resolved by sequencing of *rg1* and *ofr

	Sequence-based allele of *ofr*												
rg1 haplotype	Ba	D	B	F	C	H, I, Ia, J	Ja	K	L1, L2	A	L3	E	G
GA	1-Ba	1-D	2-B	2-F	3-C	3-HIaJ	3-Ja	3-K	4-L1L2	5-A	6-L3	7-E	8-G
GT	9-Ba	9-D	10-B	10-F	11-C	11-HIaJ	11-Ja	11-K	12-L1L2	13-A	14-L3	15-E	16-G
AT	17-Ba	17-D	18-B	18-F	19-C	19-HIaJ	19-Ja	19-K	20-L1L2	21-A	22-L3	23-E	24-G
GC	25-Ba	25-D	26-B	26-F	27-C	27-HIaJ	27-Ja	27-K	28-L1L2	29 - A	30-L3	31-E	32-G

We term these "STypes" to designate sequence-based types. In the interests of clarity, the numbering is concordant with the numbering for the HRMA-based method in Table 1. Because some HRMA-based types are further resolved by sequencing, more than one SType can have the same number. To discriminate these, the STypes are given suffixes to designate the *ompA* genotype(s) defined by the *ofr* sequence. For example, SType 2-B represents HRMA type 2 for which the HRMA-defined BF allele of *ofr* has been resolved by sequencing to *ompA* genotype B. For consistency, these suffixes are retained, even for types where *ofr* sequencing provides no additional resolution. For example, Type 7 (HRMA) and SType 7-E (Sanger sequencing) are exactly equivalent on the basis of currently known *C. trachomatis* diversity. For *rg1*, sequencing provides no additional resolution as compared to HRMA, so the numbers of rows in this table is the same as for Table 1. STypes in shaded cells have not been demonstrated to exist, either in analysis of specimens, or from in silico analyses of known *C. trachomatis* genome variation

6. Press the "BLAST" button.
7. Scroll down to observe result when analysis complete.
8. The *ofr* sequence(s) in the Subject Sequence library with a 100% match over the full length with the query sequence represents the inferred *ompA* genotype(s).
9. It is prudent to compare forward and reverse reads of *ofr* sequence (*see* **Note 9**).

3.4.5 Determination of CtGEM Sequence Type (SType) from rg1 *Haplotypes and Inferred* ompA *Genotypes*

1. Identify the combination of *rg1* haplotype and inferred *ompA* genotype in Table 2.
2. Read off the *Ct*GEM SType matching the combination from **step 1**. Table 2 encompasses all possible combination combinations of *rg1* haplotype and inferred *ompA* genotype, including those that have not been shown to exist.
3. If more information regarding *Ct*GEM types and/or STypes is needed, *see* Table 3. This presents short free-form descriptions of the types and STypes.

4 Notes

1. The control strains utilized in the development of *Ct*GEM are shown in Table 4. DNA extraction can be performed by any standard method. We have performed *Ct*GEM typing using:

Table 3
Detailed descriptions of types defined by both HRMA and sequencing

*Ct*GEM type (HRMA)	*Ct*GEM type (sequencing)	Description and comments
Type 1	SType 1-D	*ompA genotype D strains in lineage T2. ompA* genotype D strains are of moderate–high prevalence in both lineage T1 and T2, and this type is specific for T2
Type 2	SType 2-B	*AusB: ompA genotype B strain that in northern Australia was associated with both ocular and UGT specimens in 1980s–1990s.* This T2 strain appears to be internationally distributed, and be weakly tropic, so can manifest as either ocular or UGT associated
Type 2	SType 2-F	*Rare European ompA genotype F strains with the rgl GA haplotype.* This lineage T2 strain is not discriminated from SType 2-B (AusB) by HRMA so *ofr* sequencing is suggested for AusB confirmation
Type 3	SType 3-HIIaJ	*Typical ompA genotype H, I, Ia, and J strains.* These are in the T2 lineage
Type 3	SType 3-Ja	*Typical ompA genotype Ja strains.* These are in the T2 lineage
Type 3	SType 3-K	*Typical ompA genotype K strains.* These are in the T2 lineage
Type 7	SType 7-E	*OmpA genotype E strains with the rgl GA haplotype.* All known *ompA* genotype E strains are in lineage T1, but they vary in *rgl*
Type 8	SType 8-G	*Typical OmpA genotype G strains, in T2 lineage.* The very great majority of *ompA* genotype G strains are this type
Type 9	SType 9_Ba	*AusBa: OmpA genotype Ba strain strongly associated with ocular infections in remote Australia.* This lineage T1 strain appears to be unique to Australia
Type 9	SType 9-D	*Rare European ompA genotype D strains with the rgl GT haplotype.* This is not discriminated from SType 9-Ba (AusB) by HRMA, so *ofr* sequencing is advised for AusBa confirmation
Type 10	SType10-F	*Typical ompA genotype F strains.* These are in the T1 lineage
Type 11	SType 11-C	*AusC: OmpA genotype C strain strongly associated with ocular infections in remote Australia.* This lineage T1 strain appears to be unique to Australia
Type 11	SType 11-J	*Rare ompA genotype J strains with the rgl GT haplotype.* This is a European strain. This can be discriminated from SType 11-C (AusC) by a supplementary HRMA method (*see* Subheading 3.3) as well as by *ofr* sequencing. Unlike typical *ompA* genotype J strains, this is in the T1 lineage
Type 11	SType 11-Ja	*Rare ompA genotype Ja strains with the rgl GT haplotype.* This genotype was identified in Queensland, Australia. They can be discriminated from SType 11-C (AusC) by a supplementary HRMA method (*see* Subheading 3.3) as well as by *ofr* sequencing. The phylogenetic lineage is not currently known

(continued)

Table 3 (continued)

*Ct*GEM type (HRMA)	*Ct*GEM type (sequencing)	Description and comments
Type 15	SType 15-E	*OmpA genotype E strains with the rgl GT haplotype.* All known *ompA* genotype E strains are in lineage T1, but they vary in *rgl*. These "GT E's" appear to be the most prevalent – And may be the most prevalent of all STI genotypes
Type 17	SType 17_D	*OmpA genotype D strains in lineage T1 (cf type 1).* These have the *rgl* AT haplotype
Type 18	SType 18-F	*OmpA genotype F strains, in the T1 lineage.* These have the *rgl* AT haplotype (cf type 3)
Type 19	SType 19-J	*Rare ompA genotype J strains with the rgl AT haplotype.* Unlike typical *ompA* genotype J strains, this is in the T1 lineage
Type 23	SType 23-E	*OmpA genotype E strains with the rgl AT haplotype.* All known *ompA* genotype E strains are in lineage T1, but they vary in *rgl*
Type 24	SType 24-G	*Rare ompA genotype G strain with the rgl AT haplotype.* As with other *ompA* genotype G strains, this is in the T2 lineage. However, it has the *rgl* AT haplotype, which is highly atypical for the T2 lineage
Type 25	SType 25-Ba	*OmpA genotype Ba in the classical ocular lineage*
Type 26	SType 26-B	*OmpA genotype B in the classical ocular lineage*
Type 27	SType 27-C	*OmpA genotype C in the classical ocular lineage*
Type 28	SType 28-L1 L2	*LGV strains, ompA genotypes L1, L2, L2a, L2b*
Type 29	SType 29-A	*OmpA genotype A in the classical ocular lineage*
Type 30	SType 30_L3	*LGV strains, ompA genotype L3*

Only types that have been shown to exist either in specimens, or in silico analysis of publicly available genetic data have been included. Information informing the notes concerning each genotype is from [10, 11, 14]

cultured strains extracted using a QIAamp DNA mini kit (Qiagen) or a GenElute Bacterial Genomic DNA kit (Sigma-Aldrich); human urine samples extracted using a QIAamp Viral RNA mini kit; remnant human urine samples from a Roche 4800 CT/NG test (Roche Diagnostics); and human swab samples extracted using the Siemens Versant system (Siemens Healthcare Australia).

2. The concentration of primer "CT_11508_F" is 10% the concentration of primer "CT_11508_R". This is because this is an asymmetric PCR reaction that yields a product for which an appreciable proportion is in single stranded form. This is essential because it provides a target for probe binding. For the great majority of the PCR, the reaction is not asymmetric, and the amplimer is double stranded. It is only near the end of the reaction that depletion of primer "CT_11508_F" causes the

Table 4
Control isolates used during development of *Ct*GEM typing

ofr controls	Genome accession	*ompA* genotype	*ofr* curve
A2497	FM872306	A	A
Aus3	ERS153020	B	BF
Aus25	ERS351392	Ba	BaD
Aus9	ERS153045	C	CHIJK
Aus11	ERS153024	D	BaD
Aus15[a]	–	E	E
Aus20	ERS153031	F	BF
Aus19	ERS153037	G	G
UCH1	ERS001407	L2b	L

rg1 controls	Genome accession	*rg1* genotype	*rg1* curve
ATCC VR-902B[b]	–	GC	GC
Aus3	ERS153020	GA	GA
Aus9	ERS153045	GT	GT
PS297[c]	–	AT	AT

[a]Aus15 was serotyped as a serovar E [11, 21] and shown to have an identical *ofr* curve to clinical samples analyzed in this study which had an *ofr* E sequence
[b]Amplirun® *C. trachomatis* DNA from reference strain ATCC VR-902B (Vircell)
[c]PS297 is a clinical sample for which *rg1* was sequenced and confirmed as an AT genotype

PCR to become asymmetric and single stranded DNA to accumulate. We have not performed a quantitative determination of the kinetics of single stranded amplimer production. However, we have determined that PCR yield needs to be close to optimal. We consistently observed that a low yield was correlated with failure of probe binding as determined by examination of the HRMA data. This is the reason that a two-stage PCR is performed. In our hands we were unable to achieve acceptable HRMA reproducibility using a single stage embodiment, and failure appeared to be closely correlated with low PCR yield. We designed two "outer" primers for the first stage amplification in the interests of specificity; any spurious PCR products generated in the first round reaction will not be complementary to the second round primers. As has been previously noted [14], the nested PCR format does not appear to increase the sensitivity of the PCR; an absence of detectable product from the first round PCR is a strong predictor of an absence of detectable PCR product from the second round PCR. When *Ct*GEM was tested with *C. trachomatis* positive clinical specimens, 46% were typed successfully by HRMA, with PCR failure being essentially the only failure mode [14]. We have not systematically explored strategies to increase PCR sensitivity. However, we have observed on occasion that simply repeating the reaction can lead to PCR positivity, indicating a stochastic element in PCR failure. Also, our protocol makes use of just 1 μL of purified nucleic acid from the specimen. A strategy that is likely to increase PCR sensitivity is concentrating the nucleic preparation by, for example, ethanol precipitation followed by redissolving in a smaller volume.

3. We adopted a two-stage PCR for *ofr*, even though, unlike *rgl*, this method does not encompass asymmetric PCR and probe binding. The justification was that this was still likely to give some improvement to the HRMA performance, and it is technically straightforward to perform the two-stage reactions for *rgl* and *ofr* simultaneously. Unlike the *rgl* reaction, the second stage PCR is semi-nested (one of the primers is the same as the first round PCR). This was a driven by primer design constraints. We considered this appropriate, as a spurious product from the first round will not be exponentially amplified in the second round.
4. To the knowledge of the authors, among the second-round PCR primers, there are no SNPs or SNP alleles in *rgl* apart from those that define the four haplotypes encompassed in our data analysis protocols. This conclusion was reached from analysis of data from 563 diverse *C. trachomatis* strains reported by Hadfield and co-workers [10], and has been previously reported [14]. We have also searched GenBank and found no further SNPs or SNP alleles. Known *rgl* sequences and associated haplotypes are shown in Subheading 3.4.3.
5. There are instances of variation within *ofr* sequences from the same *ompA* genotype. This was determined by the same searches for variation as for *rgl* (*see* **Note 4**). These have been carefully defined and their potential to confound HRMA analysis considered—*see* the list below. The outcome of this analysis is that there is negligible potential for sequence variation in the amplified regions to confound the *Ct*GEM method. Known *ofr* sequences and inferred *ompA* genotypes are shown in Subheading 3.4.4.
   (a) *Ofr variants associated with ompA genotype Ba and B:*
   The most complex of these instances relates to *ompA* genotype Ba, and, by extension, to *ompA* genotype B. We have now identified three *ofr* variants that are annotated as being from *ompA* genotype Ba. We also identified an

Table 5
Haplotypes with G + C content SNPs at *ofr* positions 58, 87, 156, and 161

ofr position	58	87	156	161	G + C
AusBa and Ba Ap-2	A	A	A	A	0
Ba/Apache-2 and others	A	A	A	G	1
Ba IU-FW03035	A	C	A	A	1
B	C	A	G	A	2

ambiguity in strain naming that led to a sequence label in Fig. 2a in [14] being potentially misleading. The top reference sequence in that figure is labeled "Ba_apache". We have now determined that this sequence is identical to *ofr* of the AusBa strain [11] and to *ofr* of a strain named "Ap-2" (GenBank: AB915581.1). We identified two additional *ofr* variants from *ompA* gene sequences that have been classed as genotype Ba. One of these was identified in strains Ba/Apache 2 (GenBank DQ064282.1, (classed as genotype Ba), ATCC VR-347 (GenBank KP120856.1, genotype not stated in GenBank entry), strain HIMPH28 (GenBank JF437559.1, classed as genotype B in GenBank entry), and strain 3,969,853 (GenBank AY378285.1, genotype not stated in GenBank entry). The other has been identified in strain U-FW0305 only (GenBank FJ261930, identified as genotype Ba in GenBank entry). We did not identify any variation in the *ofr* from *ompA* genotype B, unless the sequence from strain HIMPH28, (likely misclassified as genotype B) is taken into account. Therefore, on the basis of these data, there are four *ofr* variants from genotypes B/Ba. These are defined by SNPs at *ofr* positions 58, 87, 156, and 161. The haplotypes with G + C contents are as in Table 5.

At this time the HRMA data interpretation protocol has been validated only with the AusBa, Ap-2 sequence, and the genotype B sequence. However, it can be seen that the other genotype Ba *ofr* variants have different G + C contents with respect to the *ompA* genotype B *ofr*, making it highly likely that the different sequences can be discriminated by HRMA. Analyses using the HRMA curve prediction software uMELT [39] indicated that the A allele at position 161 confers a distinctive early melting domain while the other SNPs affect the position with respect to temperature of the remainder of the melting curve. It is noteworthy that a distinctive early melting domain for the *ofr* from AusBa has been observed and noted (Fig. 3c), consistent with the uMELT prediction. In our HRMA curve interpretation protocol, we take the conservative approach of not using the presence/absence of this early melting domain to discriminate *ofr* sequences from *ompA* genotypes Ba and D, but note that this is likely feasible. On the basis of the uMELT analysis, we thus predict that the melting curve from the Apache-2 variant of *ofr* will be even more similar to the melting curve from *ompA* genotype D, than the *ofr* melting curve from AusBa. Conversely, the *ofr* melting curve from strain U-FW0305 is predicted to possess the distinctive early melting domain, but the "late melting part" of the curve to be shifted upward in temperature by ~0.4 °C with respect to AusBa.

Thus, all four variants are predicted to be discriminated from each other on the basis of combination of the presence of an early melting domain (AusBa and IU-FW03035), and overall position of melting curve, with genotype B melting at a higher temperature than the other three. Furthermore, given that the effect of the SNP at position 161 is predicted to be confined to the early part of the melting curve, the position of the remainder of the melting curves will be conferred by the other three SNPs only. Thus, if SNP 161 is ignored, sequences from AusBa and Ba/Apache-2 have a difference of two in G + C content with respect to genotype B. This is predicted to confer a Tm difference of ~0.7 °C, which is large enough to underpin a very robust HRMA method.

(b) *Variants for ofr in ompA genotype D*:

The two variants of genotype D differ within the reverse primer site. This cannot affect the melting curve because the sequence of the primer regions is imposed by the primer sequences, not the template.

(c) *Variants of ofr in ompA genotype E:*

Analysis of the data reported by Hadfield and co-workers [10] revealed a single *ompA* genotype E strain (E-C58) with a variant *ofr* sequence. This varies from the common *ofr* variant at a single A/G SNP. uMELT analysis [39] indicates that this will yield a melting curve similar to the *ompA* genotype G *ofr*. Thus there is a potential for a miscall. However, uMELT analysis does indicate that the melting curve from the rare variant would be concordant with the common variant of the genotype E *ofr* in the early part of the melting curve, and concordant with *ompA* genotype G *ofr* in the later part of the melting curve. However, we feel this would need to be validated. We note that the rare variant of the genotype E *ofr* is very rare indeed, being identified in one of the 146 (0.6%) genotype E strains reported by Hadfield and co-workers [10]. Also, the rare variant can be unambiguously identified by *ofr* sequencing.

(d) *Variants of ofr in ompA genotype G*:

uMELT analysis [39] indicates that the melting curves of the two *ofr* G variants will be essentially identical with no discernible difference. The difference between the variants is a single G/C SNP at *ofr* position 118. This does not affect G + C content, so would not be expected to have a significant effect on the melting curve. An example of a strain with the less common variant is SotonG1 (GenBank HE601807.1).

(e) *Variants of ofr in ompA genotype I*:

The common variant of *ofr* in *ompA* genotype I is identical to *ofr* in *ompA* genotypes H, J, and Ia Analysis of large scale comparative genomic data reported by Hadfield and coworkers [10] revealed a single *ompA* genotype I strain (I-UW12) with a variant *ofr* sequence. A prediction from uMELT analysis [39] is that the melting curve from this sequence will be well within the range of variation of the *ompA* genotype H, I, Ia, J, Ja, K HRMA allele. It shares the very marked early melting domain with these similar *ofr* variants.

(f) *Variants of ofr in ompA genotype J:*

The common variant of *ofr* in *ompA* genotype J is identical to *ofr* in *ompA* genotypes H, I, and Ia. Analysis generated by Hadfield and coworkers, [10] yielded evidence of a rare variant *ofr* sequence from three genotype J strains [14]. The strains are J-C114, J-NL55, and J-Sou106. This differs from the ompA genotype H, I, Ia, J, sequence at a single G/T SNP. A prediction from uMELT analysis [39] is that the melting curve from this sequence will be well within the range of variation of the *ompA* genotype H, I, Ia, J, Ja, K HRMA allele. It shares the very marked early melting domain with these similar *ofr* variants.

(g) *Variants of ofr in ompA genotypes L1 and L2:*

There are four variants, three of which have only been recorded in one or very few isolate(s). On the basis of uMELT analysis [39], the common variant, and the variants identified in strains 1322 (GenBank HE601951.1) and 115 (GenBank HE601952.1) are predicted to yield very similar melting curves. In particular, they all have very similar predicted Tm values, and all are predicted to display the very steep reduction of fluorescence with temperature that we have demonstrated experimentally with the common L1/L2 *ofr* variant (Fig. 3a). This appears to be a robust diagnostic feature for genotype L1/L2. As we have previously noted [14], the *ofr* sequence from strain 440 (GenBank AB915592.1) will not be identified as such by our HRMA method. This *ofr* variant is identical to *ofr* from *ompA* Genotype D, variant 1. This means that the HRMA embodiment of *Ct*GEM will not discriminate *ompA* genotype Ba strains in the classical ocular clade, from LGV strains with this aberrant *ofr* sequence. However, they will be discriminated by the Sanger sequencing embodiment.

6. In the initial design of *Ct*GEM, which involved SNP mining from 65 genome sequences, the combination of *rgl* genotype GT and *ofr* genotype C/H/I/Ia/J/Ja/K, was diagnostic for the Australian ocular C strains (AusC, *Ct*GEM type 11). However, analysis of a further 494 genomes identified two J strains with a GT *rgl* genotype, which would falsely type as AusC under this assumption (J_NL55, J_Sou106). In order to definitively distinguish between these possibilities, an additional HRMA assay was designed to discriminate *ofr* C curves from *ofr* H/I/Ia/J/Ja/K curves. This is conducted as a nested PCR using the using the first round *ofr* PCR product as template. The amplimer is from *ofr* positions 31–92, the informative SNP is at *ofr* position 67. On the basis of extensive searches of available sequence data, a "T" at this position is completely sensitive and specific for genotype C, with all other *C. trachomatis ompA* sequences having a "C" at that position. In other words, the T allele is informative for *ompA* genotype C beyond the context of discriminating genotype C from genotypes H, I, Ia, J, Ja, and K. However, because of other SNPs in the amplified fragment in genotypes other than H, I, Ia, J, and C, this HRMA is not validated and likely ineffective for identification of genotype C where there is no prior knowledge of *ompA* genotype.
7. This is a conventional PCR and Sanger sequencing procedure. The primers amplify regions that are larger than, and encompass, *rgl* and *ofr*. The *rgl* amplimer is 353 bp, and the *ofr* amplimer is 1011 bp.
8. Sanger sequencing of the PCR products can be performed by essentially any DNA sequencing service provider. We obtained sequencing service provision from the Australian Genome Research Facility (University of Queensland, Brisbane, Australia). We commissioned the PCR product purification and Sanger sequencing service (the "CS+" option, as specified here: http://www.agrf.org.au/docs/sanger-sequencing-sample-preparation-guide.pdf.

9. It is prudent to compare forward and reverse reads of the PCR product. BLASTn searches sequences in both directions, so a rapid and convenient approach is to perform the BLAST search with forward and reverse sequences. If they do not provide the same result, then there is an inconsistency between the forward and reverse sequences that needs to be resolved. This may require manual examination of the sequence traces. It is also possible that even if the forward and reverse sequences define identical variants of *ofr*, but this is not identical to any member of the library. If this occurs, a new variant of *rgl* or *ofr* has been identified.

References

1. Mariotti SP, Pascolini D, Rose-Nussbaumer J (2009) Trachoma: global magnitude of a preventable cause of blindness. Br J Ophthalmol 93(5):563–568. https://doi.org/10.1136/bjo.2008.148494
2. WHO (2011) Prevalence and incidence of selected sexually transmitted infections (trans: research DoRHa). World Health Organization, Geneva, Switzerland
3. Chan PA, Robinette A, Montgomery M, Almonte A, Cu-Uvin S, Lonks JR, Chapin KC, Kojic EM, Hardy EJ (2016) Extragenital infections caused by *Chlamydia trachomatis* and *Neisseria gonorrhoeae*: a review of the literature. Infect Dis Obstet Gynecol 2016:5758387. https://doi.org/10.1155/2016/5758387
4. Gupta NK, Bowman CA (2012) Managing sexually transmitted infections in pregnant women. Womens Health 8(3):313–321. https://doi.org/10.2217/whe.12.16
5. Garland SM, Malatt A, Tabrizi S, Grando D, Lees MI, Andrew JH, Taylor HR (1995) *Chlamydia trachomatis* conjunctivitis. Prevalence and association with genital tract infection. Med J Aust 162(7):363–366
6. Derrick T, Roberts C, Last AR, Burr SE, Holland MJ (2015) Trachoma and ocular chlamydial infection in the era of genomics. Mediat Inflamm 2015:791847. https://doi.org/10.1155/2015/791847
7. Taylor HR (ed) (2008) Trachoma: a blinding scourge from the bronze age to the twenty-first century. Centre for Eye Research Australia, Victoria, Australia
8. Diab MM, Allen RC, Gawdat TI, Saif AS (2018) Trachoma elimination, approaching 2020. Curr Opin Ophthalmol 29 (5):451–457. https://doi.org/10.1097/ICU.0000000000000504
9. Taylor HR, Anjou MD (2013) Trachoma in Australia: an update. Clin Exp Ophthalmol 41 (5):508–512. https://doi.org/10.1111/ceo.12023
10. Hadfield J, Harris SR, Seth-Smith HMB, Parmar S, Andersson P, Giffard PM, Schachter J, Moncada J, Ellison L, Vaulet MLG, Fermepin MR, Radebe F, Mendoza S, Ouburg S, Morre SA, Sachse K, Puolakkainen M, Korhonen SJ, Sonnex C, Wiggins R, Jalal H, Brunelli T, Casprini P, Pitt R, Ison C, Savicheva A, Shipitsyna E, Hadad R, Kari L, Burton MJ, Mabey D, Solomon AW, Lewis D, Marsh P, Unemo M, Clarke IN, Parkhill J, Thomson NR (2017) Comprehensive global genome dynamics of *Chlamydia trachomatis* show ancient diversification followed by contemporary mixing and recent lineage expansion. Genome Res 27 (7):1220–1229. https://doi.org/10.1101/gr.212647.116
11. Andersson P, Harris SR, Smith HM, Hadfield J, O'Neill C, Cutcliffe LT, Douglas FP, Asche LV, Mathews JD, Hutton SI, Sarovich DS, Tong SY, Clarke IN, Thomson NR, Giffard PM (2016) *Chlamydia trachomatis* from Australian aboriginal people with trachoma are polyphyletic composed of multiple distinctive lineages. Nat Commun 7:10688. https://doi.org/10.1038/ncomms10688
12. Harris SR, Clarke IN, Seth-Smith HM, Solomon AW, Cutcliffe LT, Marsh P, Skilton RJ, Holland MJ, Mabey D, Peeling RW, Lewis DA, Spratt BG, Unemo M, Persson K, Bjartling C, Brunham R, de Vries HJ, Morre SA, Speksnijder A, Bebear CM, Clerc M, de Barbeyrac B, Parkhill J, Thomson NR (2012)

Whole-genome analysis of diverse *Chlamydia trachomatis* strains identifies phylogenetic relationships masked by current clinical typing. Nat Genet 44 (4):413–419, https://doi.org/10.1038/ng.2214

13. Jeffrey BM, Suchland RJ, Quinn KL, Davidson JR, Stamm WE, Rockey DD (2010) Genome sequencing of recent clinical *Chlamydia trachomatis* strains identifies loci associated with tissue tropism and regions of apparent recombination. Infect Immun 78 (6):2544–2553. https://doi.org/10.1128/IAI.01324-09
14. Giffard PM, Andersson P, Wilson J, Buckley C, Lilliebridge R, Harris TM, Kleinecke M, O'Grady KF, Huston WM, Lambert SB, Whiley DM, Holt DC (2018) *Ct*GEM typing: discrimination of *Chlamydia trachomatis* ocular and urogenital strains and major evolutionary lineages by high resolution melting analysis of two amplified DNA fragments. PLoS One 13 (4):e0195454. https://doi.org/10.1371/journal.pone.0195454
15. Dal Conte I, Mistrangelo M, Cariti C, Chiriotto M, Lucchini A, Vigna M, Morino M, Di Perri G (2014) Lymphogranuloma venereum: an old, forgotten re-emerging systemic disease. Panminerva Med 56 (1):73–83
16. Caldwell HD, Schachter J (1982) Antigenic analysis of the major outer membrane protein of *Chlamydia* spp. Infect Immun 35 (3):1024–1031
17. Brunham RC, Plummer FA, Stephens RS (1993) Bacterial antigenic variation, host immune response, and pathogen-host coevolution. Infect Immun 61(6):2273–2276
18. Giffard PM, Singh G, Garland SM (2017) What does *Chlamydia trachomatis* detection in a urogenital specimen from a young child mean? Sex Transm Infect 93(4):236–237. https://doi.org/10.1136/sextrans-2015-052473
19. Andersson P, Tong SY, Lilliebridge RA, Brenner NC, Martin LM, Spencer E, Delima J, Singh G, McCann F, Hudson C, Johns T, Giffard PM (2014) Multisite direct determination of the potential for environmental contamination of urine samples used for diagnosis of sexually transmitted infections. J Pediatr Infect Dis Soc 3(3):189–196. https://doi.org/10.1093/jpids/pit085
20. Giffard PM, Lilliebridge RA, Wilson J, Murray G, Phillips S, Tabrizi SN, Garland SM, Martin L, Singh G, Tong SYC, Holt DC, Andersson P (2017) Contaminated fingers: a potential cause of *Chlamydia trachomatis*-positive urine specimens. Sex Transm Infect. https://doi.org/10.1136/sextrans-2016-053081
21. Asche VL, Hutton SI (1990) Proceedings of the seventh international symposium on human chlamydial infections. In: Bowie WR, Caldwell HD, Jones RP et al (eds) International symposium on human chlamydial infections. Press Syndicate of the University of Cambridge, Harrison Hot Springs
22. Douglas FP, Bidawid-Woodroffe S, McDonnell J, Hyne SG, Mathews JD (1986) Proceedings of the sixth international symposium on human chlamydial infections. In: Oriel D, Ridgway GL, Schachter J, Taylor-Robinson D, Ward J (eds) International symposium on human chlamydial infections. Press Syndicate of the University of Cambridge, Surrey
23. Giffard PM, Singh G, Garland SM (2016) What does *Chlamydia trachomatis* detection in a urogenital specimen from a young child mean? Sex Transm Infect. https://doi.org/10.1136/sextrans-2015-052473
24. Giffard PM, Brenner NC, Tabrizi SN, Garland SM, Holt DC, Andersson P, Lilliebridge RA, Tong SY, Karimi M, Boylan P, Ryder N, Johns T, Singh G (2016) *Chlamydia trachomatis* genotypes in a cross-sectional study of urogenital samples from remote northern and Central Australia. BMJ Open 6(1):e009624. https://doi.org/10.1136/bmjopen-2015-009624
25. Pedersen LN, Herrmann B, Moller JK (2009) Typing *Chlamydia trachomatis*: from egg yolk to nanotechnology. FEMS Immunol Med Microbiol 55(2):120–130
26. Lysen M, Osterlund A, Rubin CJ, Persson T, Persson I, Herrmann B (2004) Characterization of *ompA* genotypes by sequence analysis of DNA from all detected cases of *Chlamydia trachomatis* infections during 1 year of contact tracing in a Swedish County. J Clin Microbiol 42(4):1641–1647
27. Stevens MP, Twin J, Fairley CK, Donovan B, Tan SE, Yu J, Garland SM, Tabrizi SN (2010) Development and evaluation of an *ompA* quantitative real-time PCR assay for *Chlamydia trachomatis* serovar determination. J Clin Microbiol 48(6):2060–2065. https://doi.org/10.1128/JCM.02308-09
28. Herrmann B, Isaksson J, Ryberg M, Tangrot J, Saleh I, Versteeg B, Gravningen K, Bruisten S (2015) Global multilocus sequence type analysis of *Chlamydia trachomatis* strains from 16 countries. J Clin Microbiol 53

(7):2172–2179. https://doi.org/10.1128/JCM.00249-15

29. Dean D, Bruno WJ, Wan R, Gomes JP, Devignot S, Mehari T, de Vries HJ, Morre SA, Myers G, Read TD, Spratt BG (2009) Predicting phenotype and emerging strains among *Chlamydia trachomatis* infections. Emerg Infect Dis 15(9):1385–1394
30. Schaeffer A, Henrich B (2008) Rapid detection of *Chlamydia trachomatis* and typing of the Lymphogranuloma venereum associated L-Serovars by TaqMan PCR. BMC Infect Dis 8:56. https://doi.org/10.1186/1471-2334-8-56
31. Tong SY, Giffard PM (2012) Microbiological applications of high-resolution melting analysis. J Clin Microbiol 50(11):3418–3421. https://doi.org/10.1128/JCM.01709-12
32. Lilliebridge RA, Tong SY, Giffard PM, Holt DC (2011) The utility of high-resolution melting analysis of SNP nucleated PCR amplicons--an MLST based *Staphylococcus aureus* typing scheme. PLoS One 6(6):e19749. https://doi.org/10.1371/journal.pone.0019749
33. Erali M, Voelkerding KV, Wittwer CT (2008) High resolution melting applications for clinical laboratory medicine. Exp Mol Pathol 85 (1):50–58. https://doi.org/10.1016/j.yexmp.2008.03.012
34. Robertson GA, Thiruvenkataswamy V, Shilling H, Price EP, Huygens F, Henskens FA, Giffard PM (2004) Identification and interrogation of highly informative single nucleotide polymorphism sets defined by bacterial multilocus sequence typing databases. J Med Microbiol 53(Pt 1):35–45. https://doi.org/10.1099/jmm.0.05365-0
35. Price EP, Inman-Bamber J, Thiruvenkataswamy V, Huygens F, Giffard PM (2007) Computer-aided identification of polymorphism sets diagnostic for groups of bacterial and viral genetic variants. BMC Bioinformatics 8:278. https://doi.org/10.1186/1471-2105-8-278
36. Hunter PR, Gaston MA (1988) Numerical index of the discriminatory ability of typing systems: an application of Simpson's index of diversity. J Clin Microbiol 26(11):2465–2466
37. Bell JM, Turnidge JD, Andersson P (2010) Aac (6′)-Ib-cr genotyping by simultaneous high-resolution melting analyses of an unlabeled probe and full-length amplicon. Antimicrob Agents Chemother 54(3):1378–1380. https://doi.org/10.1128/AAC.01476-09
38. Altschul SF, Gish W, Miller W, Myers EW, Lipman DJ (1990) Basic local alignment search tool. J Mol Biol 215(3):403–410. https://doi.org/10.1016/S0022-2836(05)80360-2
39. Dwight Z, Palais R, Wittwer CT (2011) uMELT: prediction of high-resolution melting curves and dynamic melting profiles of PCR products in a rich web application. Bioinformatics 27(7):1019–1020. https://doi.org/10.1093/bioinformatics/btr065

Chapter 9

Dual RNA-Seq of Chlamydia and Host Cells

James W. Marsh, Regan J. Hayward, Amol Shetty, Anup Mahurkar, Michael S. Humphrys, and Garry S. A. Myers

Abstract

During the infection of a host cell by a bacterial pathogen, a cascading series of gene expression changes occurs as each organism manipulates or responds to the other via defense or survival strategies. Unraveling this complex interplay is key for our understanding of bacterial virulence and host response pathways for the development of novel therapeutics. Dual RNA sequencing (dual RNA-Seq) has recently been developed to simultaneously capture host and bacterial transcriptomes from an infected cell. Leveraging the sensitivity and resolution allowed by RNA-seq, dual RNA-Seq can be applied to any bacteria–eukaryotic host interaction. We pioneered dual RNA-Seq to simultaneously capture *Chlamydia* and host expression profiles during an in vitro infection as proof of principle. Here we provide a detailed laboratory protocol and bioinformatics analysis guidelines for dual RNA-seq experiments focusing on *Chlamydia* as the organism of interest.

Key words *Chlamydia*, Dual RNA-Seq, Transcriptomics, Protocol, Method, Host–pathogen

1 Introduction

The interaction between host and pathogen during infection is defined by cascading changes in gene expression as each organism deploys mechanisms of defense and survival [1]. Dual RNA-Sequencing (dual RNA-Seq) leverages this interaction to measure the total host–pathogen transcriptome during an infection, providing insight into the coordinated regulatory interactions, virulence mechanisms, and innate responses that are key to therapeutic development [2]. In a typical experiment, total mRNA from a sample is subjected to high-throughput next-generation sequencing and mapped to a reference genome to deduce the structure and/or expression state of each transcript. Gene expression changes can be accurately measured between samples with high coverage and sensitivity, while alternative splicing analyses can be applied to identify novel isoforms and transcripts, RNA editing, and allele-specific expression [3]. The high sensitivity and

Amanda Claire Brown (ed.), *Chlamydia trachomatis: Methods and Protocols*, Methods in Molecular Biology, vol. 2042, https://doi.org/10.1007/978-1-4939-9694-0_9,

dynamic range of RNA-Seq has expanded our capability for whole transcriptome analysis and enabled new insight into the functional elements of the genome [4].

As an obligate intracellular bacterial pathogen that relies heavily on its host for nutrients and survival, *Chlamydia trachomatis* is an ideal bacterium for the application of dual RNA-Seq. We applied dRNA-Seq to map host and bacteria transcriptomes from *Chlamydia*-infected host epithelial cells, which highlighted a dramatic, early response to infection and numerous altered pathways within the host cell [1]. Here, we present a detailed dual RNA-Seq protocol for the in vitro infection of human epithelial cells with *Chlamydia trachomatis*. First, *Chlamydia*-infected HeLa cells are collected, which are then subjected to total cellular lysis and RNA extraction. DNase I is used for the removal of contaminating DNA and the remaining RNA is subjected to rRNA depletion to remove uninformative ribosomal species and minimize the required sequencing depth. RT-PCR is used to confirm DNA removal and validate transcriptional data. From here, the user may pursue cDNA library construction according to the desired high-throughput sequencing method and the requirements of the sequencing facility.

A subsequent bioinformatic analysis is performed to calculate [1] significant differences between infected host RNA abundance compared to uninfected host RNA, and [2] the relative abundances of chlamydial RNA throughout the course of the experiment. This process usually comprises the trimming and quality control of raw sequence files, the quantification of RNA transcripts according to the feature of interest (usually genes), and a statistical estimation of differential expression between uninfected and infected treatment groups (host). These dynamics can then be correlated with the relative abundance of the bacterial transcriptional profile during infection.

From here, several downstream applications are possible, including but not limited to time-series analysis, pathogen-induced alternative isoform analysis, gene set enrichment and pathway analysis, coexpression and biclustering analysis, for the purpose of addressing the specific biological question of the researcher [5]. While the breadth of possible applications is wide and thus out of the scope of this protocol, we provide informed guidelines to assist the researcher in their pursuit of further analysis. Furthermore, dual RNA-Seq data may be integrated with other data sources to establish a more complete picture of gene regulation, including genotyping data to identify genetic loci responsible for gene expression variation, epigenetic information to highlight the influence of transcription factor binding, histone modification, and methylation, miRNA-Seq data to identify the regulatory mechanisms of gene expression changes via noncoding RNA [6, 7].

2 Materials

General laboratory equipment and consumables, such as micropipettes and sterile pipette tips, as well as specific PPE is assumed and are therefore not listed here.

2.1 Cell Culture

1. HeLa 229 epithelial cells (ATCC® CCL-2.1™) **CRITICAL** All experiments that use human or animal tissues must comply with governmental and institutional guidelines and regulations.
2. *Chlamydia trachomatis* serovar E **CAUTION** *Chlamydia* is a human bacterium that poses a risk of infection. All work with this organism should be conducted in a class II biosafety cabinet while wearing appropriate personal protective equipment (PPE).
3. Six-well plates.
4. Serological pipets.
5. Class II biosafety cabinet.
6. Flask rocker.
7. Centrifuge.
8. 37 °C/5% CO_2 Incubator.
9. Dulbecco's modified Eagle's medium with high glucose and pyruvate, with gentamicin, streptomycin, 10% v/v heat-inactivated fetal bovine serum.
10. Sucrose phosphate glutamate (SPG) media.
11. 5% trypsin–EDTA.
12. Hemocytometer.
13. Cell scrapers.
14. 15 mL centrifuge tubes.
15. Dulbecco's phosphate buffered saline, no calcium, no magnesium.
16. Ethanol **CAUTION** Highly flammable. Causes skin and serious eye irritation. Handle using appropriate safety equipment.
17. RNase-free water.
18. Heating block.

2.2 RNA Extraction, Purification, and rRNA Depletion

1. Heating block.
2. MasterPure™ RNA Purification Kit (Epicenter, cat. no. MCR85102).
3. 1.5 mL Eppendorf tubes
4. Microcentrifuge.

5. Vortex.
6. Ice.
7. 70% ethanol (diluted with RNase-free water).
8. TURBO DNA-*free*™ kit (Thermo Fisher, cat. no. AM1907).
9. AMPure XT beads (Beckman Coulter, cat. no. A63880).
10. Ribo-Zero™ rRNA Removal Kit (Human/Mouse/Rat) (Epicenter, cat. no. RZH1046).
11. RNA Clean & Concentrator™-5 (Zymo Research, cat. no. R1015).
12. Ribo-Zero™ rRNA Removal Kit (gram-negative bacteria) (Epicenter, cat. no. RZNB1056).
13. TaqMan® Gene Expression Assay (Thermo Fisher, cat. no. 4453320).
14. TaqMan® Universal Master Mix (Thermo Fisher, cat. no. 4352042).
15. MicroAmp® optical 96-well reaction plates (Thermo Fisher, cat. no. 4306737).
16. MicroAmp® optical adhesive film (Thermo Fisher, cat. no. 4311971).
17. MicroAmp® optical film compression pad (Thermo Fisher, cat. no. 4312639).
18. RT-PCR machine.

2.3 Bioinformatic Analysis

1. QC, trimming, and filtering tools for FASTQ files (e.g., FASTQC, Cutadapt, trimmomatic, and FASTX toolkit).
2. Transcript abundance quantifier (i.e., Salmon, Kallisto).
3. Statistical analysis of gene-level count matrices (e.g., Limma, edgeR, DESeq2).

3 Methods

3.1 Cell Culture (See Notes 1–3)

1. One day prior to infection, detach precultured confluent HeLa cells from the base of T75 flask(s) by adding 4 mL trypsin–EDTA and incubate for 5 min at 37 °C, 5% CO_2.
2. Use hemocytometer to determine HeLa cell concentration and seed 8×10^5 HeLa cells per well in all wells of a six-well plate. Ensure there are three plates per time-point (one infected plate and two noninfected control plates) (*see* **Note 4**). Incubate plates overnight at 37 °C, 5% CO_2.
3. Following seeding, sit plates on a bench at room temperature for 15 min to allow cells to settle, ensuring an even distribution of cells.

4. The next day, use light microscope to visualize confluent cell monolayers in each well.
5. Infect plates at a MOI of 1.5 to ensure that 100% of the host cells will be infected. Mock-infect additional wells with SPG buffer only (*see* **Note 5**).
6. Centrifuge plates at 500 × *g* for 30 min at room temperature. Incubate plates at 37 °C, 5% CO_2 (*see* **Note 6**).
7. Wash cells twice with DPBS and overlay with warm, fresh DMEM media (containing streptomycin, gentamicin, 10% FBS). Incubate at 37 °C, 5% CO_2.
8. At each time-point, wash cells twice with DPBS and add 1 mL DPBS to each well. Harvest cells with a cell scraper and dispense solution into a 15 mL centrifuge tube. Store tubes at −80 °C until all time-points are complete. Cells can be stored for up to 6 weeks at −80 °C.

3.2 Cell Lysis, RNA Extraction, Purification, and rRNA Depletion

1. Cell lysis (45 min) (*see* **Note 7**). Remove centrifuge tubes from −80 °C freezer and thaw at room temperature.
2. Preset heating block to 65 °C.
3. Add 1 μL of 50 μg/mL Proteinase K (MasterPure™ RNA Purification Kit; Epicenter) to 300 μL of Tissue and Cell Lysis Buffer (MasterPure™ RNA Purification Kit; Epicenter) for each sample.
4. Pellet cells by centrifugation at 5000 × *g* for 30 min. Discard the supernatant, leaving ~25 μL of liquid.
5. Vortex for 10 s to resuspend the pellet.
6. Add 300 μL of Tissue and Cell Lysis Solution (containing Proteinase K) to each 25 μL sample and mix thoroughly by vortexing.
7. Incubate tubes in a heating block at 65 °C for 15 min, vortexing briefly every 5 min.
8. Place samples on ice for 3–5 min (*see* **Note 8**).
9. Total nucleic acid precipitation (1 h). Add 175 μL of MPC Protein Precipitation Reagent (MasterPure™ RNA Purification Kit; Epicenter) to each 300 μL of lysed sample and vortex vigorously for 10 s.
10. Pellet the debris by centrifugation at 10,000 × *g* for 10 min at 4 °C.
11. Transfer the supernatant (containing total nucleic acid) to a clean 1.5 mL microcentrifuge tube and discard the pellet.
12. Add 500 μL of isopropanol to the recovered supernatant and invert the tube 30–40 times. Do not vortex.

13. Pellet the total nucleic acid by centrifugation at 10,000 × *g* for 10 min at 4 °C.
14. Carefully pour off the isopropanol without dislodging the pellet.
15. Rinse the pellet twice with 70% ethanol, being careful not to dislodge the pellet. Centrifuge briefly if pellet is dislodged.
16. Remove residual ethanol with a pipette and resuspend the pellet in 35 μL of TE Buffer (MasterPure™ RNA Purification Kit; Epicenter). Samples can be stored in TE Buffer overnight at 4 °C.
17. DNA digestion (45 min). Preset heating block to 37 °C.
18. Add 4 μL of 10× TURBO DNase Buffer (TURBO™ DNA--*free*™ Kit; Thermo Fisher) to each 35 μL sample (*see* **Note 9**).
19. Add 1 μL of TURBO™ DNase to each sample to digest contaminating DNA. Gently flick the tubes to mix and pulse-spin to distribute liquid to the bottom of the tube.
20. Increase DNase volume to 2–3 μL if digesting a higher amount of DNA (*see* **Note 10**).
21. Incubate tubes in a heating block at 37 °C for 30 min.
22. After incubation, add 8 μL of DNase Inactivation Solution (TURBO™ DNA-*free*™ Kit; Thermo Fisher) and incubate tubes at room temperature for 5 min, mixing occasionally (*see* **Note 11**).
23. Centrifuge tubes at 10,000 × *g* for 1.5 min.
24. Transfer supernatant (containing the RNA) to a fresh 1.5 mL microcentrifuge tube.
25. Validation of DNA removal by RT qPCR (2 h) (*see* **Note 12**). Remove 5 μL from each RNA sample and place in a clean 1.5 mL microcentrifuge tube.
26. Add 95 μL of nuclease-free water (1:20 dilution).
27. Prepare enough RT qPCR master mix to assay each sample in triplicate, as well as nontemplate controls: 1 μL 20× TaqMan® Gene Expression Assay, 10 μL 2× TaqMan® Gene Expression Master Mix, 5 μL RNase-free water, *see* Table 1.

Table 1
qRT-PCR mastermix

Assay name	Gene target	ABI assay ID
ACTB	Beta-actin	Hs99999903_ml
MT-ATP6	Mitochondrially encoded ATP synthase 6	Hs02596862_gl
18S	Eukaryotic 18 s rRNA	Hs99999901_sl

28. Cap the tube and invert the tube several times to mix the reaction components. Pulse vortex.
29. Aliquot 16 μL of master mix into individual wells of a 96-well reaction plate (Thermo Fisher). Include a triplicate reaction for each sample and nontemplate controls.
30. Add 5 μL of each diluted template to appropriate wells and gently tap plate on benchtop to distribute contents to the base of the well.
31. Place adhesive film (Thermo Fisher) over the plate and seal with compression pad (Thermo Fisher) (*see* **Note 13**).
32. Place plate in RT-PCR machine and run assay according to the following cycling conditions:

Hold:	95 °C, 10 min
Cycle (40×):	95 °C, 15 s
	60 °C, 1 min

33. rRNA depletion (90 min) (*see* **Note 14**). Preset one heating block to 68 °C and one at 50 °C.
34. Remove Ribo-Zero™ rRNA Removal Magnetic Core Kit from 4 °C and allow to warm to room temperature (*see* **Notes 15** and **16**). Remove Human/Mouse/Rat and gram-negative bacteria components of the Ribo-Zero™ rRNA Removal kits from −80 °C and thaw on ice.
35. Vigorously mix magnetic beads (Ribo-Zero™ rRNA Removal kit; Illumina) for 20 s by vortexing.
36. Carefully pipet 65 μL of magnetic beads into 2 mL microsphere wash tubes (Ribo-Zero™ rRNA Removal kit; Illumina); two tubes per sample (*see* **Note 17**).
37. Centrifuge microspheres at 12,000 × *g* for 3 min. Carefully remove supernatant without dislodging the pellet (*see* **Note 18**).
38. Wash the microsphere wash tubes by adding 130 μL of microsphere wash solution (Ribo-Zero™ rRNA Removal kit; Illumina) to each tube. Vortex vigorously.
39. Centrifuge microsphere wash tubes at 12,000 × *g* for 3 min. Carefully remove supernatant without dislodging the pellet.
40. Add 65 μL of microsphere resuspension solution (Ribo-Zero™ rRNA Removal kit; Illumina) to each tube and vortex vigorously until a homogenous suspension is produced.
41. Add 1 μL of RiboGuard RNase inhibitor (Ribo-Zero™ rRNA Removal kit; Illumina) to each tube. Mix by vortexing for 10 s and set aside (at room temperature) (*see* **Note 19**).

42. Treat two aliquots of each sample with Ribo-Zero rRNA removal solution (Ribo-Zero™ rRNA Removal kit; Illumina) according to the following preparation (two removal preps per sample): 1 μL RNase-free water (Ribo-Zero™ rRNA Removal Kit), 4 μL Ribo-Zero Reaction Buffer, 25 μL RNA sample, 5 μL Ribo-Zero rRNA Removal Solution (gram negative bacteria kit), and 5 μL Ribo-Zero rRNA Removal Solution (Human/Mouse/Rat kit).
43. Fully mix the samples by pipette-mixing 10–15 times.
44. Gently mix the reactions and incubate at 68 °C for 10 min in heating block. Return the Ribo-Zero reaction buffer to −80 °C.
45. Remove the microsphere wash tubes from the heating block and incubate at room temperature for 15 min.
46. Vortex the microsphere wash tubes at medium speed for 20 s to ensure a homogenous slurry.
47. Pipet hybridized RNA sample to the resuspended microsphere wash tubes, pipette-mixing 10–15 times to mix. Immediately vortex the microsphere wash tubes at medium speed for 5 s.
48. Incubate microsphere wash tubes at room temperature for 10 min. Vortex at medium speed for 5 s, every 3–4 min.
49. Following incubation, mix samples again by vortexing at medium speed for 5 s.
50. Incubate samples in heating block at 50 °C for 10 min.
51. Transfer the RNA-microsphere suspension to a Microsphere Removal Unit (Ribo-Zero™ rRNA Removal kit; Illumina) and centrifuge at 12,000 × *g* for 1 min at room temperature. Save the eluate (~100 μL) and discard the removal unit.
52. Purification of rRNA-depleted samples (45 min). Add two volumes of RNA Binding Buffer (RNA Clean & Concentrator™-5; Zymo Research) to each volume of RNA sample and mix well (*see* **Note 20**).
53. Add one volume of 100% ethanol to the mixture and mix well.
54. Transfer the mixture to a Zymo-Spin IC column (RNA Clean & Concentrator™-5; Zymo Research) in a collection tube and centrifuge at 12,000 × *g* for 1 min. Discard the flow-through.
55. Combine two reactions of the same sample to one column and spin multiple times until the entire mixture passes through the column. The column capacity is 5 μg of RNA.
56. Add 400 μL of RNA Prep Buffer (RNA Clean & Concentrator™-5; Zymo Research) to the column and centrifuge at 12,000 × *g* for 1 min. Discard the flow-through.

57. Add 800 μL of RNA Wash Buffer (RNA Clean & Concentrator™-5; Zymo Research) to the column and centrifuge at 12,000 × *g* for 1 min. Discard the flow-through.
58. Add 400 μL of RNA wash buffer to the column and centrifuge at 12,000 × *g* for 1 min. Discard the flow-through.
59. Centrifuge the column in an emptied collection tube at 12,000 × *g* for 2 min. Carefully remove the column from the collection tube and transfer to a new RNase-free microcentrifuge tube.
60. Add 20 μL of DNase/RNase-free water directly to one column matrix and let stand for 1 min at room temperature. Centrifuge at 10,000 × *g* for 30 s (*see* **Note 21**).
61. Eluted RNA is now ready for RNA-Seq library prep appropriate to the sequencing platform being utilized (*see* **Notes 22** and **23**).

3.3 Bioinformatic Analyses

1. Raw FASTQ files are trimmed and filtered for quality using one of several available tools, include Cutadapt, trimmomatic, and FASTX toolkit.
2. Quantify counts per transcript with alignment-independent tools such as Kallisto [8] or Salmon [9].
3. The Bioconductor package *tximport* [10] is then used to aggregate transcript counts to gene-level counts (*see* **Notes 24** and **25**).
4. Statistical analysis of differentially expressed genes is performed using Limma [11], DESeq2 [12], or edgeR [13] (*see* **Note 26**). The reader is encouraged to investigate the advantages of each method. A suitable starting-point for defining differential expression is genes with a log2-fold change significantly different from 0.5 below a *p*-value of 0.05. However, this is highly dependent on the experimental design and question of interest.
5. The bacterial counts are converted to transcripts-per-million (TPM), which is a useful measure of relative transcript abundance (*see* **Notes 27** and **28**).

4 Notes

1. Prior to commencing, the RNA species to be investigated (i.e., mRNA, miRNA, snRNA, etc.), number of replicates, time-points, and multiplicity of infection (MOI) need to be decided to address to experimental question of interest. This protocol describes an experiment that collects total mRNA from *Chlamydia*-infected host cells at an MOI of 1.5 from three biological replicates, at four time-points of infection: 1-, 12-, 24-, and 48-hpi, with time-matched mock-infected controls included.

2. To calculate the amount of RNA required from organisms, an estimation of both the desired fold coverage and the ratio of bacteria to host genome size is an appropriate starting point. Further considerations include the number of replicates, the expected influence of housekeeping and structural RNA (rRNA and tRNA), the possibility of host–bacteria sequence overlap, the number of time-points, and the multiplicity of infection (MOI). The *Chlamydia* to HeLa genome size is ~1:3200 MB, indicating that *Chlamydia*'s RNA accounts for ~0.03% of total host–bacteria RNA. As ~95% of this will be uninformative rRNA and tRNA [2], 0.0015 and 4.9985% of total RNA will be informative bacteria and host mRNA, respectively. Given this ratio, 1×10^{10} host reads and $\sim 3.33 \times 10^{9}$ bacterial reads would be required to capture sufficient RNA from both organisms. Thus, to achieve sufficient coverage overall, $>1 \times 10^{10}$ reads would be required for dual RNA-Seq of *Chlamydia* and host.
3. Approximately 250 ng of total *Chlamydia*–host RNA is required per replicate prior to rRNA depletion, which can be extracted from two confluent wells of a six-well plate (per replicate).
4. The initiation and period of transcriptional response is likely to differ between host and bacteria [14]. It is recommended that multiple time-points be measured to robustly capture the host–bacteria transcriptional landscape throughout the course of infection.
5. The MOI should be selected to maximize the transcriptional signal from both host and pathogen without resulting in a heightened and/or distorted response with little biological relevance.
6. Centrifugation is important to synchronize the infections and time-points, with free, dead, and/or nonviable bacterial cells are removed by washing monolayers twice with DPBS.
7. Extra attention should be paid to preventing DNA contamination and RNA degradation. To achieve this, adhere to protocol time and temperature recommendations, and use only RNase-free reagents and consumables. It is advised to carry out all RNA work in a clean environment that is separated from any non-RNA work.
8. To avoid RNA degradation, minimize the delay between the host cell lysis and RNA extraction steps.
9. To remove contaminating DNA, two treatments with TURBO DNA-*free*™ DNase (Thermo Fisher) is most effective.

10. Alternatively, add half the DNase to each reaction, incubate for 30 minutes, then add the remainder of the enzyme and incubate for another 30 min.
11. Environments colder than 22 °C can reduce the inactivation of TURBO™ DNase. Move tubes to a heating block to control the temperature if necessary.
12. To confirm DNA removal, three real-time qPCR assays for human targets and one endpoint PCR assay for *C. trachomatis* is performed. The qPCR assays are based on TaqMan® Gene Expression assays (Applied Biosystems) with primer and probe sets targeting beta-actin, mitochondrially encoded ATP synthase 6, and eukaryotic 18S rRNA [1]. The endpoint PCR is based on custom-designed primer sets that are specific for *C. trachomatis*, which were designed using PrimerExpress software (Applied Biosystems). Each sample should be tested in triplicate in RT-minus and RT-plus format.
13. If any bubbles are visible in the wells or liquid is present on the sides of the wells, centrifuge plate at 500 × *g* for 2 min. Do not touch the film with bare hands at any point.
14. This protocol describes the use of both a hybridization-based rRNA depletion and poly(A)-depletion step to maximize host–bacteria mRNA species, while additionally enriching bacterial mRNA. For hybridization, cDNA oligonucleotides attach to complementary rRNA that is immobilized on magnetic beads; ensure compatibility between oligonucleotide and organism. An equal volume of Ribo-Zero beads from both a Human/Mouse/Rat-specific and gram-negative bacteria-specific Ribo-Zero™ rRNA Removal Kit (Epicenter) are combined to remove both host and bacterial rRNA. This method will not enrich immature mRNAs and noncoding RNAs and additional targeted enrichment techniques should be considered for this purpose if required. Poly(A)-depleted and rRNA-depleted eluates are then combined for library construction.
15. Do not place the Ribo-Zero™ Magnetic Core Kit on ice.
16. The washed magnetic beads must be at room temperature for use in this step. The order of the addition (hybridized RNA *to* the magnetic beads) is critical for rRNA removal efficiency.
17. Store unused beads at 4 °C. Do not place the magnetic beads on ice.
18. The supernatant contains sodium azide.
19. Avoid creating air bubbles when adding RNase inhibitor.
20. The minimum recommended sample volume for use with this kit is 50 μL.
21. The eluted RNA can be used immediately or stored at −80 °C.

22. There are a number of sequencing platforms currently available, including Illumina, SOLID, Ion Torrent, Roche 454, Nanopore, and Pacific Biosciences, and each are suited to specific purposes and should be investigated by the user according to their desired outcome.
23. While we suggest that paired-end reads >50 nucleotides as a good starting-point, longer reads will enable greater coverage, reduced multimapping, and improved transcript identification [15].
24. Alignment-independent quantification of reads is generally preferred over alignment-dependent methods due to its higher sensitivity, ability to accommodate differences in gene lengths between samples, and computational speed.
25. IGV is used to visualize genomic regions of high/low coverage that could indicate technical or biological errors, as well as host exon–intron boundaries, splice sites, exon junction read counts, and read strand. The alignment files are then sorted by read name to allow feature counting.
26. For Limma and edgeR-based analysis, we recommend Trimmed Mean of M-Values (TMM) normalization, which corrects for differences in RNA composition and sample outliers, while providing better across-sample comparability [16].
27. TPM is the most appropriate measure of relative transcript abundance for the bacterial counts. Alternatively, absolute abundance may be calculated with the use of spike-in controls.
28. For a detailed description and methodology for dual RNA-Seq bioinformatic analyses *see* ref. 5.

References

1. Humphrys MS, Creasy T, Sun Y, Shetty AC, Chibucos MC, Drabek EF, Fraser CM, Farooq U, Sengamalay N, Ott S, Shou H, Bavoil PM, Mahurkar A, Myers GSA (2013) Simultaneous transcriptional profiling of bacteria and their host cells. PLoS One 8(12): e80597. https://doi.org/10.1371/journal.pone.0080597
2. Westermann AJ, Gorski SA, Vogel J (2012) Dual RNA-seq of pathogen and host. Nat Rev Microbiol 10(9):618–630. https://doi.org/10.1038/nrmicro2852
3. Wu D-D, Ye L-Q, Li Y, Sun Y-B, Shao Y, Chen C, Zhu Z, Zhong L, Wang L, Irwin DM, Zhang YE, Zhang Y-P (2015) Integrative analyses of RNA editing, alternative splicing, and expression of young genes in human brain transcriptome by deep RNA sequencing. J Mol Cell Biol 7(4):314–325. https://doi.org/10.1093/jmcb/mjv043
4. Nagalakshmi U, Wang Z, Waern K, Shou C, Raha D, Gerstein M, Snyder M (2008) The transcriptional landscape of the yeast genome defined by RNA sequencing. Science 320 (5881):1344–1349. https://doi.org/10.1126/science.1158441
5. Marsh JW, Hayward RJ, Shetty AC, Mahurkar A, Humphrys MS, Myers GSA (2017) Bioinformatic analysis of bacteria and host cell dual RNA-sequencing experiments. Brief Bioinform 19(6):1115–1129. https://doi.org/10.1093/bib/bbx043
6. Angelini C, Costa V (2014) Understanding gene regulatory mechanisms by integrating ChIP-seq and RNA-seq data: statistical solutions to biological problems. Front Cell Dev

Biol 2:51. https://doi.org/10.3389/fcell.2014.00051

7. Klein H-U, Schäfer M (2016) Integrative analysis of histone ChIP-seq and RNA-seq data. Curr Protoc Hum Genet 90:20.23.21–20.23.16. https://doi.org/10.1002/cphg.17

8. Bray NL, Pimentel H, Melsted P, Pachter L (2016) Near-optimal probabilistic RNA-seq quantification. Nat Biotechnol 34 (5):525–527. https://doi.org/10.1038/nbt.3519

9. Patro R, Duggal G, Love MI, Irizarry RA, Kingsford C (2017) Salmon provides fast and bias-aware quantification of transcript expression. Nat Methods 14(4):417–419. https://doi.org/10.1038/nmeth.4197

10. Soneson C, Love MI, Robinson MD (2015) Differential analyses for RNA-seq: transcript-level estimates improve gene-level inferences. F1000Res 4:1521. https://doi.org/10.12688/f1000research.7563.2

11. Ritchie ME, Phipson B, Wu D, Hu Y, Law CW, Shi W, Smyth GK (2015) Limma powers differential expression analyses for RNA-sequencing and microarray studies. Nucleic Acids Res 43(7):e47. https://doi.org/10.1093/nar/gkv007

12. Love MI, Huber W, Anders S (2014) Moderated estimation of fold change and dispersion for RNA-seq data with DESeq2. Genome Biol 15(12):550. https://doi.org/10.1186/s13059-014-0550-8

13. Anders S, McCarthy DJ, Chen Y, Okoniewski M, Smyth GK, Huber W, Robinson MD (2013) Count-based differential expression analysis of RNA sequencing data using R and bioconductor. Nat Protoc 8 (9):1765–1786. https://doi.org/10.1038/nprot.2013.099

14. Schulze S, Henkel SG, Driesch D, Guthke R, Linde J (2015) Computational prediction of molecular pathogen-host interactions based on dual transcriptome data. Front Microbiol 6:65. https://doi.org/10.3389/fmicb.2015.00065

15. Cho H, Davis J, Li X, Smith KS, Battle A, Montgomery SB (2014) High-resolution transcriptome analysis with long-read RNA sequencing. PLoS One 9(9):e108095. https://doi.org/10.1371/journal.pone.0108095

16. Robinson MD, Oshlack A (2010) A scaling normalization method for differential expression analysis of RNA-seq data. Genome Biol 11 (3):R25. https://doi.org/10.1186/gb-2010-11-3-r25

Chapter 10

Isolation and Propagation of Single Inclusion-Derived Chlamydia Using Laser Microdissection

Oleg V. Podgorny, Nadezhda F. Polina, and Vassili N. Lazarev

Abstract

Other than its routine application for capturing pure cell populations from fixed tissue sections for diverse downstream molecular assays, laser microdissection enables isolation of single live cells. Here we describe a method for the isolation of single *Chlamydia trachomatis*-infected cells using a laser microdissection system, in which the dissected samples are captured via gravity. Cells infected by *C. trachomatis* at low multiplicity of infection are marked with the fluorescent Golgi-specific probe BODIPY® FL C5-ceramide, to facilitate identification of the cells with chlamydial inclusions under the microscope. Individual *C. trachomatis*-infected cells are harvested into separate wells with a pregrown host cell monolayer. Inclusions in harvested cells maturate, and the released elementary bodies infect the host cell monolayer, thus initiating propagation of single inclusion-derived *Chlamydia*. The method can be used for generation of microbiological clones of *C. trachomatis* and recovery of transformants and mutants. Isolated single *Chlamydia*-infected cells can also be examined by diverse downstream molecular assays to reveal unknown features of the *Chlamydia* replication at a single inclusion level.

Key words Laser microdissection, *Chlamydia*, Elementary bodies, BODIPY® FL C5-ceramide

1 Introduction

The obligate intracellular lifestyle of *Chlamydia trachomatis* has hindered the use of common molecular tools to genetically manipulate this bacterium. A limited genetic toolbox for the elucidation of molecular mechanisms underlying extracellular survival, intracellular multiplication, tissue tropism, the niche-specific adaptation, and persistence of *C. trachomatis* hampers developing efficient treatments of *C. trachomatis*-associated diseases and, therefore, restricts a progress in public health. A lack of methods for the fast and efficient selection and cloning of transformants or mutants is one of the major reasons why *Chlamydia* species remain intractable organisms for genetic manipulation. *Chlamydia* resides in the inclusion within an infected cell and, therefore, impedes routine antibiotic selection, because antibiotics must penetrate both a

Amanda Claire Brown (ed.), *Chlamydia trachomatis: Methods and Protocols*, Methods in Molecular Biology, vol. 2042,
https://doi.org/10.1007/978-1-4939-9694-0_10,

plasma membrane of the host cell and an inclusion membrane to reach the bacterium. This makes some antibiotics useless for selection. Additionally, higher concentrations of antibiotics are required to reach their appropriate levels in an inclusion, thus increasing toxicity in the host cell. Although antibiotic selection was applied to recover *Chlamydia* transformants [1–8], several alternative approaches have been employed to produce microbiological clones of *Chlamydia* species and to perform selection of transformants or mutants. These approaches include plaque cloning [9–11], focus cloning [12], and the isolation of single *Chlamydia*-infected cells using fluorescence-activated cell sorting [13]. We have recently reported application of laser microdissection for the isolation of single live cells infected with *C. trachomatis* [14].

Laser microdissection enables the contact- and contamination-free isolation and collection of homogeneous cell populations, single cells, and subcellular compartments from their native niches in tissues of multicellular organisms for the broad range of downstream molecular assays. Presented for the first time in 1996 [15], laser microdissection has being used for the identification of allelic loss or somatic mutations, analysis of cell type- or tissue-specific gene expression (real-time PCR, microarray, RNA-seq) or protein synthesis (western blot, 2D gel electrophoresis, SELDI-TOF and MALDI-TOF mass spectrometry), and metabolic profiling (reviewed in refs. 16 and 17).

There are several laser microdissection systems available from commercial sources. These systems basically use pulsed ultraviolet (UV) lasers, focused by the objective lens of a microscope, to separate samples from a tissue section attached to special slides equipped with the polymer foil (membrane), but they differ in their methods to capture the dissected samples. Laser microdissection is routinely used for the isolation of biomaterial from fixed tissue sections. However, we have recently reported the use of a laser microdissection system, in which the dissected samples are captured via gravity, for the isolation and collection of single live cells [18]. Using our findings, we have established a protocol for the isolation and propagation of single inclusion-derived *Chlamydia trachomatis* [14].

A workflow for the isolation of single inclusion-derived *C. trachomatis* by laser microdissection is as follows. Host cells are seeded onto a polyethylene naphthalate membrane of a special petri dish at low density to separate cells with an enough space. After the host cells have been attached to the membrane, the chlamydial elementary bodies (EBs) are inoculated into the petri dish at a low multiplicity of infection to infect host cells with no more than one infectious particle. To visualize the *Chlamydia*-infected cells, the cells in the petri dish undergo staining with a vital Golgi apparatus probe, BODIPY™ FL C5-Ceramide, approximately in the middle of the developmental cycle. This probe is a fluorescent sphingolipid that is routinely used for studying vesicular trafficking associated with the Golgi apparatus [19]. *Chlamydia* is known to redirect vesicular trafficking from the Golgi apparatus, leading to the accumulation of the fluorescent probe in a chlamydial inclusion, whereas noninfected cells loss the fluorescent signal because of the efflux of the probe through the exocytosis [13, 20, 21]. Next, samples contained single *Chlamydia*-infected cells undergo dissection from the rest of specimen (*see* Fig. 1a, b) followed by extraction into wells of the 8-well strips with the pregrown host cell monolayers (*see* Fig. 1c–f). Chlamydial inclusions maturate, and EBs are released initiating infection of the pregrown host cell monolayers (*see* Fig. 1g). Subsequent passaging of the individual wells enables propagation of single inclusion-derived *C. trachomatis* that can be used to obtain microbiological clones of *Chlamydia* and to perform selection of transformants and mutants. Single *Chlamydia*-infected cells can also undergo diverse downstream molecular assays to study features of the *Chlamydia* multiplication at a single inclusion level.

2 Materials

CAUTION! *C. trachomatis* is a human pathogen. When working with *C. trachomatis*, all procedures where aerosols can be produced may represent a significant risk of laboratory-acquired infection. Therefore, wear personal protective equipment (lab coat, gloves, safety glasses, and mask) and manipulate with *C. trachomatis* in a BSL-2 biosafety cabinet, unless otherwise indicated, as per local health and safety regulations.

CAUTION! Preparation of the paraformaldehyde solution and its subsequent use should be performed with the caution, because of its high toxicity. Wear personal protective equipment, use a hood and follow waste disposal regulations when working with paraformaldehyde.

General laboratory equipment, such as pipettes, centrifuges, BSL-2 biosafety cabinet, and CO_2 incubator, are not listed separately.

2.1 Preparation of Cells and Infection

1. Sucrose–phosphate–glutamate buffer: Mix 87 mL of 0.2 M Na_2HPO_4 (dibasic sodium phosphate) and 13 mL of 0.2 M NaH_2PO_4 (monobasic sodium phosphate) with 700 mL of distilled H_2O. Dissolve 75 g of sucrose and 0.72 g L-glutamic acid and adjust volume to 1 L. Adjust pH to 7.4 with 2 N NaOH if necessary. Autoclave the solution for 20 min or filter it through a membrane filter with a 0.22 μm pore size. The solution is stable up to 1 year at 4 °C [22].

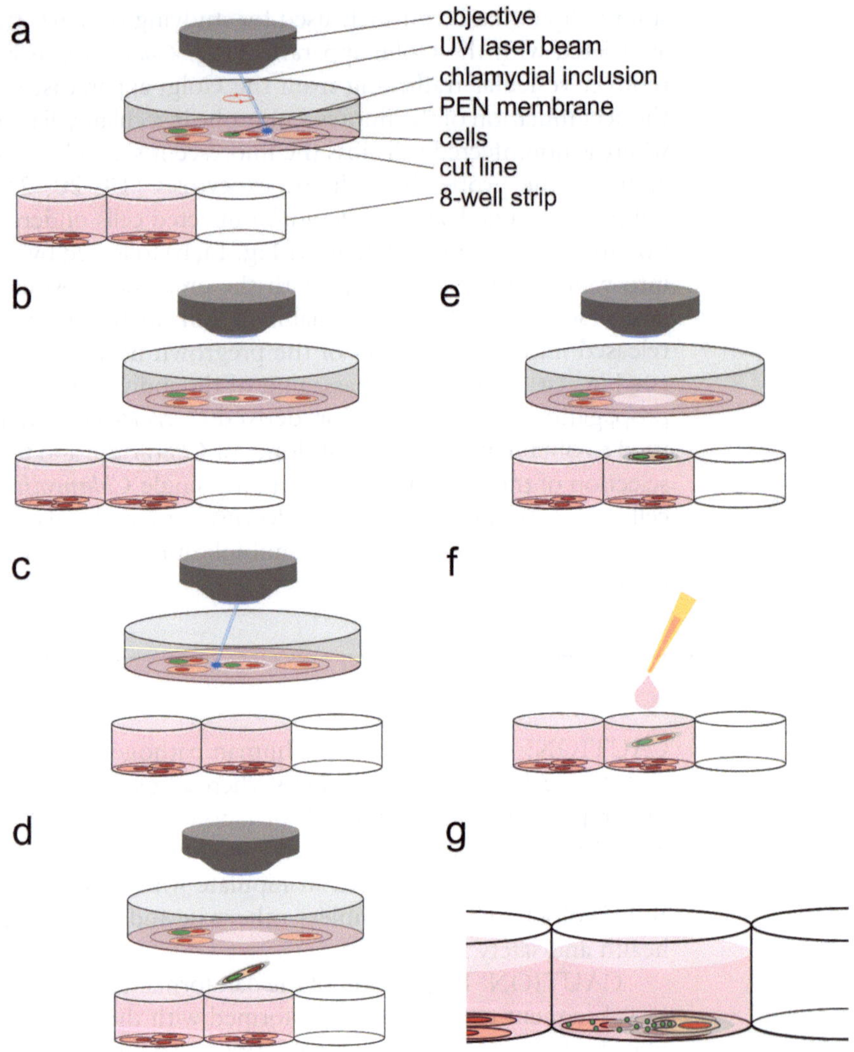

Fig. 1 A workflow of laser microdissection of single *Chlamydia*-infected cells. (**a**) A circular sample contained a single *Chlamydia*-infected cell is dissected from the rest of specimen by ablating the membrane using the pulsed UV laser (cutting step). There should an empty well of the 8-well strip under the specimen during this step (*see* **Note 10**). (**b**) After the sample has been dissected, it does not fall into a collector because the force of surface tension holds the sample in the plane of the membrane. (**c**) To extract the dissected sample, a short pulse of the UV laser is applied to edge of the sample (extraction step). Importantly, prior to the extraction step, the 8-well strip must be shifted to place a well with culture medium and a pregrown HeLa cell monolayer. (**d**, **e**) The extracted sample freely falls onto a surface of culture medium in the well placed under the specimen. (**f**) After laser microdissection has been completed, the isolated sample is submerged into culture medium by adding a droplet of culture medium above the floating sample. (**g**) A chlamydial inclusion maturates, and the EBs (shown in green) are released to infect the pregrown HeLa cell monolayer in the well within several consecutive days after a laser microdissection procedure

2. A stock of the EBs of *C. trachomatis* serovar E (strain Bour) with known titer in the sucrose–phosphate–glutamate buffer.
3. HeLa cell line ATCC® CCL-2™ (ATCC, USA).
4. 0.01% poly-L-lysine solution in water.
5. Complete culture medium for culturing HeLa cells and infecting them with *C. trachomatis*: High glucose (4.5 g/L) Dulbecco's modified Eagle medium (DMEM) containing 2 mM L-Glutamine, 10% fetal bovine serum, and 10 μg/mL gentamycin.
6. Cycloheximide stock solution (1 mg/mL): Dissolve 10 mg of cycloheximide into 10 mL of 95% ethanol. Store up to 6 months at −20 °C. Dilute 1:1000 into the complete culture medium (*see* **item 5**) for growth of *C. trachomatis*.
7. Hanks' balanced sault solution with calcium chloride, magnesium sulfate, and sodium carbohydrate (HBSS).
8. 50 mm plastic petri dishes (Leica Microsystems, Germany) equipped with polyethylene naphthalate (PEN) membrane (thickness: 2 μm) at the bottom (PEN-membrane petri dishes).
9. 8-well strips, flat bottom.

2.2 Vital Staining of the Chlamydial Inclusions

1. HBSS.
2. Complete culture medium (*see* Subheading 2.1, **item 5**).
3. BODIPY™ FL C5-Ceramide solution for vital staining of the *Chlamydia*-infected cells (*see* **Notes 1** and **2**): Reconstitute BODIPY™ FL C5-Ceramide complexed to bovine serum albumin (BSA) (ThermoFisher Scientific, #B22650) by dissolving 5 mg of the ready-made complex in 150 μL of sterile deionized water and prepare 5 μL aliquots for the long-term storage (*see* **Note 3**). To prepare a working solution of BODIPY™ FL C5-Ceramide, dissolve an aliquot (5 μL) of the stock solution in HBSS (995 μL) at a dilution of 1:200 immediately prior to the use (*see* **Note 4**).

2.3 Isolation of Single Chlamydia-Infected Cells by Laser Microdissection

1. HBSS.
2. Complete culture medium (*see* Subheading 2.1, **item 5**).
3. 8-well strips with the flat bottom.
4. Leica Laser Microdissection System LMD7 equipped with a climatic chamber (*see* **Note 5**).
5. Stereomicroscope with the upper light source.

2.4 Detection of Chlamydia trachomatis After Laser Microdissection

1. 10× phosphate buffered saline (10× PBS): Dissolve 80.04 g of NaCl, 2.00 g of KCl, 28.65 g of $Na_2HPO_4 \cdot 12H_2O$, 2.72 g of KH_2PO_4 in 1 L distilled water and store at the room temperature.

2. 4% paraformaldehyde (PFA) in 1× PBS (pH 7.4): Add 4 g of paraformaldehyde powder in 80 mL of distilled water. Add 15 μL of 10 N sodium hydroxide (NaOH) solution to enhance dissolving of the paraformaldehyde powder. Dissolve the paraformaldehyde powder by stirring on a magnetic hot-plate stirrer at 60–65 °C (*see* **Note 6**) until the solution clears. Cool the paraformaldehyde solution to room temperature. Add 10 mL of 10× PBS and adjust volume to 100 mL. Adjust pH, if necessary. Filter through a paper filter. Store at 4 °C for a week.
3. 1× PBS.
4. 1× PBS containing 0.2% Triton X100 and 3% BSA.
5. FITC-conjugated mouse monoclonal antibody against *C. trachomatis* lipooligosaccharide.
6. Epifluorescent microscope with the appropriate filter sets.

3 Methods

Laboratory maintenance of *C. trachomatis* including its isolation from clinical specimens, cultivation in cell monolayers, purification of the EBs, and tittering infectious forming units is described in detail in [22].

Methods of human cell culturing including HeLa cells and counting cell numbers using hemocytometer are described in detail in [23, 24].

Importantly, the protocol presented here has been adapted to the use of *Chlamydia trachomatis* serovar E (strain Bour) in a combination with the HeLa cell line. The use of distinct *C. trachomatis* serovars and host cell lines may require specific modifications of the protocol.

3.1 Preparation of Cells and Infection

All preparative procedures should be performed aseptically.

1. Insert a new PEN-membrane petri dish into a regular 90 mm plastic petri dish. This will protect the membrane external side from the contamination when the PEN-membrane petri dish is outside of the biosafety cabinet (*see* **Note 7**).
2. Add 1 mL of the ready to use 0.01% poly-L-lysine solution into a PEN-membrane petri dish using an automatic pipette, distribute liquid through the entire surface of the membrane (*see* **Note 8**), and leave the PEN-membrane petri dish at 37 °C for 30–60 min.
3. Aspirate the poly-L-lysine solution, thoroughly rinse the PEN-membrane petri dish with the sterile deionized water

and allow to dry at the aseptic environment at least 2 h before introducing cells and medium (*see* **Note 9**).

4. Seed 2 mL of HeLa cells at a cell density of 10,000 cells/mL in the complete culture medium into a PEN-membrane petri dish coated with poly-L-lysine and incubate cells at 37 °C in atmosphere containing 95% air and 5% CO_2 overnight.
5. Next day, inoculate a chlamydial stock containing pure EBs into the PEN-membrane petri dish at a multiplicity of infection (MOI) of less than 1. Mix gently the chlamydial stock with the complete culture medium to reach a uniform distribution of the EBs through the entire surface of the PEN-membrane petri dish and then centrifuge the EBs onto cells at 900 × *g* for 1 h with the subsequent incubation for another 2 h at 37 °C. Wash cells with 2 mL of HBSS once to remove the unbound EBs, add the complete culture medium supplemented with 1 μg/mL cycloheximide and place the PEN-membrane petri dish with the infected cells in the CO_2-incubator at 37 °C for additional 36 h.
6. A day before the laser microdissection procedure, prepare a HeLa cell suspension in the complete culture medium at a cell density of 100,000 cells/mL.
7. Seed 200 μL of the HeLa cell suspension in 7 wells of 8-well strips (*see* **Note 10**) and incubate cells at 37 °C in atmosphere containing 95% air and 5% CO_2 overnight. There should be almost a confluent cell monolayer next day.

3.2 Vital Staining of the Chlamydial Inclusions

1. Wash the infected cells with 2 mL of HBSS three times with the 5-min intervals and incubate them in 1 mL of the prewarmed BODIPY™ FL C5-Ceramide solution at 37 °C for 20 min.
2. Wash the cells with 2 mL of HBSS three times with the 5-min intervals and incubate them in 2 mL of the complete culture medium supplemented with 1 μg/mL cycloheximide at 37 °C and 5% CO_2 for 30 min. Replace the medium with 2 mL of fresh one supplemented with 1 μg/mL cycloheximide and incubate the cells at 37 °C and 5% CO_2 for additional 12 h to allow for transfer of the fluorescent dye from the Golgi apparatus to the chlamydial inclusion within infected cells or its efflux from noninfected cells through exocytosis (*see* **Note 11**).

3.3 Isolation of Single Chlamydia-Infected Cells by Laser Microdissection

1. Prewarm the climatic chamber of the laser microdissection instrument to 37 °C and clean all accessible surfaces of the climatic chamber and the stage and substage of the microscope with 70% ethanol.
2. Prior to the laser microdissection procedure, wash the *Chlamydia*-infected cells two times with 2 mL of HBSS with the

5-min intervals. Then, add 1 mL of fresh HBSS (*see* **Notes 12** and **13**).

3. Remove culture medium from those wells of the 8-well strips which contain the pregrown HeLa cell monolayer and add 350 μL of fresh complete culture medium supplemented with 1 μg/mL cycloheximide. Distribute liquid in each well using a pipette tip so that it should reach the edge of the well forming a meniscus (*see* **Note 14**). Wells should not be incompletely filled or overfilled with culture medium.
4. Transfer the PEN-membrane petri dish with *Chlamydia*-infected cells and the 8-well strips with the pregrown HeLa cell monolayer into the climatic chamber of the laser microdissection instrument.
5. Place the PEN-membrane petri dish without a cover on the stage of the laser microdissection instrument.
6. Insert two 8-well strips with the pregrown HeLa cell monolayer into the collector holder and place the holder on the substage of the laser microdissection instrument.
7. Select 20×/0.40 laser microdissection objective and adjust focus on cells.
8. Switch the laser microdissection instrument to the fluorescent mode and select a respective filter set for visualization of BODIPY™ FL C5-Ceramide (or BODIPY™ TR C_5-Ceramide).
9. Find a cell with the chlamydial inclusion by the inspection of cells on the PEN-membrane petri dish (*see* Fig. 2a). When a cell of interest is identified, switch back to the bright field mode of the laser microdissection instrument.
10. Circumscribe the cell by a circular shape with a diameter of 200 μm (*see* **Note 15**) so that the cell should be in the center of the shape. The shape should not contain unwanted cells.
11. Place an empty well of the 8-well strip under the specimen (*see* **Note 10**). Perform dissecting the 200 μm circular sample from the rest of specimen in the automatic mode "Draw and Cut" of the laser microdissection instrument (cutting step, *see* Figs. 1a and 2b). The laser settings for the cutting step are as follows: power—40 relative units (RU), an aperture—1 RU, a speed—6 RU, a head current—100%, a pulse frequency—120 Hz, an offset—120 RU (*see* **Notes 16** and **17**). After dissecting from the rest of the specimen, the sample does not fall into the collector because the surface tension holds it in plane of the PEN-membrane (*see* Figs. 1b and 2c).
12. To extract the dissected sample, shift the 8-well strip to place a well containing culture medium and a pregrown HeLa cell monolayer, and apply a short UV laser pulse to the edge of the sample in the manual mode "Move and Cut" of the laser

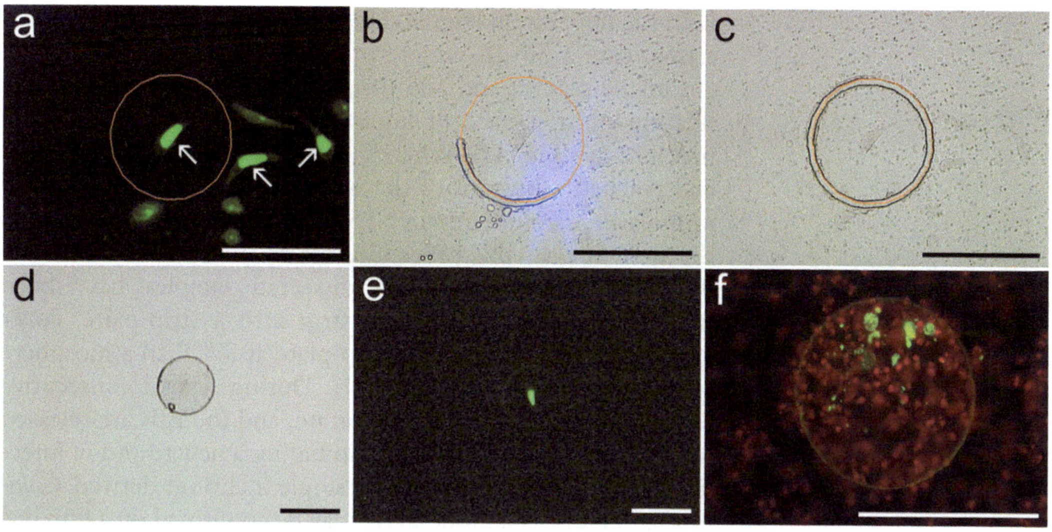

Fig. 2 Representative images of the isolation steps of a single *Chlamydia*-infected cell using laser microdissection. (**a**) A bright green fluorescent signal of the chlamydial inclusions (arrows) allows for discrimination of the *Chlamydia*-infected cells from noninfected cells with a weak fluorescent signal of the Golgi apparatus. (**b**) The same area shown in a bright field mode. Pulses of the UV laser dissect the sample with a single *Chlamydia*-infected cell from the rest of the specimen along the circular line (cutting step). (**c**) The dissected sample does not fall into the collector because the surface tension holds it in plane of the PEN-membrane. (**d**) The dissected sample floating on the liquid surface after extraction into the collector (inspection mode of the laser microdissection instrument, bright field). (**e**) A fluorescent signal of the inclusion of the single *Chlamydia*-infected cell isolated by laser microdissection (inspection mode of the laser microdissection instrument, epifluorescence mode). (**f**) Growth of the new inclusions after the isolation of a single *Chlamydia*-infected cell by laser microdissection has been confirmed using the FITC-conjugated mouse monoclonal anti-*C. trachomatis* lipooligosaccharide antibody (green fluorescence). The host cells were costained with Evans Blue dye (red fluorescence). The image has been reproduced from Podgorny et al., 2015 with permission from Elsevier. Scale bars: 200 μm

microdissection instrument (extracting step, *see* Fig. 1c). Extracted sample freely falls into the well (*see* Fig. 1d). The laser settings for the cutting step are as follows: power—40 relative units (RU), an aperture—20 RU, a speed—not applied for the extracting step, a head current—100%, a pulse frequency—10 Hz, an offset—120 RU (*see* **Notes 16** and **17**).

13. Inspect the well by switching the laser microdissection instrument to the inspection mode to check whether the sample has reached the well after the extraction step. It should be floating on the surface of culture medium (*see* Figs. 1e and 2d, e).
14. To collect the next single *Chlamydia*-infected cell switch the instrument back to the laser microdissection mode and shift the 8-well strip back to the empty well and repeat **steps 8–13** (*see* **Note 18**).
15. Remove the 8-well strips from the substage of the laser microdissection instrument and transfer them to a stereomicroscope

with the upper light source. Under control of the stereomicroscope, take a 50 μL sample of culture medium by a regular automatic pipette and add droplets above a dissected sample floating on a surface of liquid to submerge it into culture medium (*see* Fig. 1f). When the sample has been submerged and reached a bottom of the well with uninfected HeLa cell monolayer, carefully remove 150 μL of culture medium from the well. Repeat this procedure for each well that contains a dissected sample. When all dissected samples have been submerged, insert the 8-well strip into a strip-plate, cover with a lid, and incubate the strip-plate at 37 °C in atmosphere containing 95% air and 5% CO_2. During several consecutive days, chlamydial inclusions maturate, and the EBs are released to infect HeLa cell monolayers initiating a new round of infection (*see* Fig. 1g). Regrowth of single inclusion derived *Chlamydia* in HeLa cell monolayers may be confirmed on a fifth day after laser microdissection using an antibody specific to *Chlamydia trachomatis* antigen (*see* Fig. 2f).

3.4 Detection of Chlamydia trachomatis After Laser Microdissection

1. Prewarm 4% paraformaldehyde solution (*see* Subheading 2.4, **item 2**) to the room temperature.
2. Remove the complete culture medium from wells of the strip-plate and wash cells with 200 μL of HBSS two times with 5-min intervals.
3. After removal of HBSS, fix the cells with 200 μL of 4% paraformaldehyde solution for 20 min at room temperature.
4. Wash cells with 200 μL of 1× PBS three times with 5-min intervals.
5. Permeabilize the cells with 200 μL of 0.2% Triton X100 in 1× PBS for 30 min at room temperature.
6. Wash cells with 200 μL 1× PBS three times with 5-min intervals.
7. Incubate cells in 200 μL of FITC-conjugated mouse monoclonal antibody solution against *C. trachomatis* lipooligosaccharide for 1 h at room temperature with gentle shaking.
8. Wash cells with 200 μL 1× PBS three times with 5-min intervals.
9. Look on the wells using epifluorescent microscope with a respective filter set. There should be chlamydial inclusions in HeLa cell monolayers under or around the dissected samples.

4 Notes

1. Spectral characteristics of fluorescent sphingolipid BODIPY™ FL C_5-Ceramide: absorption—505 nm and emission—511 nm. The dye is exited by blue light and emits green light.
2. Optionally, red-fluorescent BODIPY™ TR C_5-Ceramide complexed to BSA may be used instead of green-fluorescent BODIPY™ FL C_5-Ceramide in those cases when green fluorescence interferes with another fluorescent dye. The stock solution of BODIPY™ TR C_5-Ceramide should be prepared according to the manufacturer's guidelines.
3. The reconstituted stock solution of BODIPY™ FL C_5-Ceramide complexed to BSA should be aliquoted and stored at −20 °C until needed.
4. The brightness of chlamydial inclusions and background staining depends on a serovar of *C. trachomatis*, host cells, and incubation time. Therefore, an optimal dilution of fluorescent BODIPY™ FL C_5-Ceramide (or BODIPY™ TR C_5-Ceramide) should be determined empirically for a definite *C. trachomatis* serovar combined with a definite host cell line.
5. Importantly, the protocol presented here has been adapted for the laser microdissection procedure using Leica LMD7 instrument only. Other laser microdissection systems use distinct principles for live cell isolation. Therefore, the protocol for the isolation of single *Chlamydia*-infected cells should be adapted for individual laser microdissection systems.
6. Do not heat the paraformaldehyde solution above 70 °C or boil it. This causes a break-down of paraformaldehyde.
7. Leaving the PEN-membrane petri dishes directly on any surfaces including the base of the biosafety cabinet can cause damage of the membrane as well as contamination of its external side. Therefore, always insert the PEN-membrane petri dishes into regular 90 mm plastic petri dishes.
8. All manipulations with the PEN-membrane petri dishes should be gentle to protect the membrane from damage. Do not allow contact of the membrane by a tip of an automatic pipette to avoid a rapture of the membrane.
9. The PEN-membrane petri dishes coated with poly-L-lysine stay good at the room temperature for at least a month.
10. There should be an empty well in each 8-well strip in which isolated single *Chlamydia*-infected cells are collected. It is necessary for the cutting step of the laser microdissection procedure (*see* Subheading 3.3, **step 11**). The use of an empty well during the cutting step preserves condenser from the

PEN-membrane and cellular fragments scattered during the ablation process.

11. Within host cells, *Chlamydia* acquire fluorescent ceramide within 30 min after administration and retain the probe for the entire duration of the intracellular growth [21]. Two hours after treatment with fluorescent ceramide, a bulk of the label has been exported from the Golgi apparatus, whereas the chlamydial inclusions remain intensely fluorescent.

12. A focus position of the UV laser beam along *Z*-axis is critical for its ablative ability. A liquid layer above cells in the membrane petri dish is an additional optical component, and its thickness determines a focus position of the UV laser beam along *Z*-axis. Once an optimal focus position has been empirically determined for a fixed volume (1 mL should be enough) of HBSS, use the same volume for all subsequent cutting sessions to reach reproducible results of laser microdissection. The use of a liquid volume distinct from the predetermined one reduces ablative ability of the UV laser that causes incomplete dissection of the area of interest from the rest of specimen.

13. To perform laser microdissection, a PEN-membrane petri dish contained live cells is placed on the stage of the laser microdissection instrument without a cover. Due to continuous evaporation of liquid in an open petri at 37 °C, it is necessary replace HBSS with 1 mL of fresh one every 40 min to avoid osmotic stress or drying of cells.

14. An extracting pulse of the UV laser (*see* Subheading 3.3, **step 12**) causes horizontal acceleration of the dissected sample. If a well is not entirely filled with culture medium, the dissected sample with a live *Chlamydia*-infected cell reaches the wall of the well where it attaches and undergoes drying causing cell death. Therefore, a meniscus reaching the edge of the well is an essential condition for collecting the dissected samples containing live cells.

15. We previously demonstrated that a circular shape with a diameter of 200 μm provides a safe distance from the UV laser spot to the cell in the center of the shape [14, 18].

16. Optimal laser settings for the cutting step as well as for the extracting step should be empirically determined for an individual laser microdissection instrument combined with the PEN-membrane petri dish filled with a fixed volume of culture medium. The values of the laser settings indicated in this chapter may be used as referential. Once determined, optimal laser settings for the cutting and the extracting steps should be applied for each subsequent session of laser microdissection. Specimen preparation, laser adjustment, overall workflow, and

limitations on single live cell isolation by laser microdissection are described in [25].

17. An offset determines a focus position of the laser beam along *Z*-axis and is used to compensate chromatic aberrations of lenses. The liquid layer above cells in the PEN-membrane petri dish serves as an additional optical component and enhances chromatic aberrations. Precise laser beam focusing is critical for the ablation process. An optimal value of the offset should be empirically determined for each objective combined with the PEN-membrane petri dish filled with a fixed volume of culture medium.

18. Optionally, *Chlamidia*-infected cells may be initially identified and undergo the cutting step only (dissection from the rest of the specimen). Then, all dissected samples undergo the extracting step during a single run. This reduces a time interval during which the dissected and collected samples are floating on the surface of culture medium in wells of the 8-well strips and undergo drying, or, in other words, a time interval between collecting the samples and submerging them into culture medium.

Acknowledgements

The development of a method for the isolation of single *Chlamydia*-infected cells using laser microdissection was supported by the Russian Science Foundation (projects no. 17-75-20099 and 14-14-00696). O.V.P was partially supported by the IDB RAS Government basic research program no. 0108-2019-0005. Laser microdissection was performed using equipment of the Core Facility of Koltzov Institute of Developmental Biology RAS.

References

1. Binet R, Maurelli AT (2009) Transformation and isolation of allelic exchange mutants of *Chlamydia psittaci* using recombinant DNA introduced by electroporation. Proc Natl Acad Sci U S A 106:292–297. https://doi.org/10.1073/pnas.0806768106
2. Demars R, Weinfurter J, Guex E et al (2007) Lateral gene transfer in vitro in the intracellular pathogen *Chlamydia trachomatis*. J Bacteriol 189:991–1003. https://doi.org/10.1128/JB.00845-06
3. Lenart J, Andersen AA, Rockey DD (2001) Growth and development of tetracycline-resistant *Chlamydia suis*. Antimicrob Agents Chemother 45:2198–2203. https://doi.org/10.1128/AAC.45.8.2198-2203.2001
4. Lowden NM, Yeruva L, Johnson CM et al (2015) Use of aminoglycoside 3′ adenyltransferase as a selection marker for *Chlamydia trachomatis* intron-mutagenesis and in vivo intron stability. BMC Res Notes 8:570. https://doi.org/10.1186/s13104-015-1542-9
5. Suchland RJ, Sandoz KM, Jeffrey BM et al (2009) Horizontal transfer of tetracycline resistance among *Chlamydia* spp. in vitro. Antimicrob Agents Chemother 53:4604–4611. https://doi.org/10.1128/AAC.00477-09
6. Tam JE, Davis CH, Wyrick PB (1994) Expression of recombinant DNA introduced into *Chlamydia trachomatis* by electroporation. Can J Microbiol 40:583–591. https://doi.org/10.1139/m94-093

7. Thompson CC, Griffiths C, Nicod SS et al (2015) The Rsb phosphoregulatory network controls availability of the primary sigma factor in *Chlamydia trachomatis* and influences the kinetics of growth and development. PLoS Pathog 11:e1005125. https://doi.org/10.1371/journal.ppat.1005125
8. Wang Y, Kahane S, Cutcliffe LT et al (2011) Development of a transformation system for *Chlamydia trachomatis*: restoration of glycogen biosynthesis by acquisition of a plasmid shuttle vector. PLoS Pathog 7(9):e1002258. https://doi.org/10.1371/journal.ppat.1002258
9. Banks J, Eddie B, Schachter J, Meyer KF (1970) Plaque formation by Chlamydia in L cells. Infect Immun 1(3):259–262
10. Matsumoto A, Izutsu H, Miyashita N, Ohuchi M (1998) Plaque formation by and plaque cloning of *Chlamydia trachomatis* biovar trachoma. J Clin Microbiol 36(10):3013–3019
11. Somboonna N, Mead S, Liu J, Dean D (2008) Discovering and differentiating new and emerging clonal populations of *Chlamydia trachomatis* with a novel shotgun cell culture harvest assay. Emerg Infect Dis 14(3):445–453. https://doi.org/10.3201/eid1403.071071
12. Gieffers J, Belland RJ, Whitmire W et al (2002) Isolation of *Chlamydia pneumoniae* clonal variants by a focus-forming assay. Infect Immun 70(10):5827–5834
13. Alzhanov DT, Suchland RJ, Bakke AC et al (2007) Clonal isolation of chlamydia-infected cells using flow cytometry. J Microbiol Methods 68(1):201–208. https://doi.org/10.1016/j.mimet.2006.07.012
14. Podgorny OV, Polina NF, Babenko VV et al (2015) Isolation of single Chlamydia-infected cells using laser microdissection. J Microbiol Methods 109:123–128. https://doi.org/10.1016/j.mimet.2014.12.018
15. Emmert-Buck MR, Bonner RF, Smith PD et al (1996) Laser capture microdissection. Science 274(5289):998–1001
16. Datta S, Malhotra L, Dickerson R et al (2015) Laser capture microdissection: big data from small samples. Histol Histopathol 30 (11):1255–1269. https://doi.org/10.14670/HH-11-622
17. Espina V, Wulfkuhle JD, Calvert VS et al (2006) Laser-capture microdissection. Nat Protoc 1(2):586–603
18. Podgorny OV (2013) Live cell isolation by laser microdissection with gravity transfer. J Biomed Opt 18(5):55002. https://doi.org/10.1117/1.JBO.18.5.055002
19. Pagano RE, Sleight RG (1985) Defining lipid transport pathways in animal cells. Science 229 (4718):1051–1057
20. Boleti H, Ojcius DM, Dautry-Varsat A (2000) Fluorescent labelling of intracellular bacteria in living host cells. J Microbiol Methods 40 (3):265–274
21. Hackstadt T, Scidmore MA, Rockey DD (1995) Lipid metabolism in *Chlamydia trachomatis*-infected cells: directed trafficking of Golgi-derived sphingolipids to the chlamydial inclusion. Proc Natl Acad Sci U S A 92 (11):4877–4881
22. Scidmore MA (2005) Cultivation and laboratory maintenance of *Chlamydia trachomatis*. Curr Protoc Microbiol Chapter 11:Unit 11A.1. https://doi.org/10.1002/9780471729259.mc11a01s00
23. Phelan MC (2006) Techniques for mammalian cell tissue culture. Curr Protoc Hum Genet Appendix 3:Appendix 3G. https://doi.org/10.1002/0471142727
24. Phelan K, May KM (2017) Mammalian cell tissue culture. Curr Protoc Hum Genet 94: A.3G.1–A.3G.22. https://doi.org/10.1002/cphg.41
25. Podgorny OV (2018) Live cell isolation by laser microdissection. Webinar available via BiteSizeBio. https://bitesizebio.com/webinar/live-cell-isolation-by-laser-microdissection/. Accessed 17 Oct 2018

Chapter 11

Genetic Manipulation of *Chlamydia trachomatis*: Chromosomal Deletions

Katerina Wolf, Mostafa Rahnama, and Kenneth A. Fields

Abstract

Progress in understanding molecular mechanisms contributing to chlamydial pathogenesis has been greatly facilitated by recent advances in genetic manipulation of *C. trachomatis*. Valuable approaches such as random, chemically induced mutagenesis or targeted, insertion-based gene disruption have led to significant discoveries. We describe herein a technique for generating definitive null strains via complete deletion of chromosomal genes in *C. trachomatis*. Fluorescence-reported allelic exchange mutagenesis (FRAEM), using the suicide vector pSUmC, enables targeted deletion of desired chromosomal DNA. The protocol provided here describes steps required to produce transformation competent chlamydiae, generate a specific allelic exchange plasmid construct, carry out mutagenesis, and isolate clonal populations of resulting mutant strains.

Key words Allelic exchange, FRAEM, Mutagenesis

1 Introduction

Allelic exchange by homologous recombination is a commonly used approach for generating targeted deletion mutants in genetically tractable bacteria [1]. One of the methods typically employed in producing these null mutations requires transformation of bacteria with a replicative-incompetent plasmid termed a suicide vector. This vector contains (a) chromosomal upstream and downstream sequences flanking a target gene to provide elements for homologous recombination within the bacterial chromosome and (b) an antibiotic cassette necessary for appropriate selection. The antibiotic cassette is generally placed between the upstream and downstream homologous sequences. Bacteria transformed with suicide vector constructs undergo two sequential rounds of recombination between the homologous plasmid-encoded and chromosomal sequences. The first recombination event (occurring within one of the homology arms) results in integration of plasmid DNA into the genome to yield a heterozygous merodiploid. To

Amanda Claire Brown (ed.), *Chlamydia trachomatis: Methods and Protocols*, Methods in Molecular Biology, vol. 2042,
https://doi.org/10.1007/978-1-4939-9694-0_11,

accomplish allelic exchange, the second recombination event must involve the opposite homology arm to result in resolution of the suicide vector backbone and replacement of the targeted gene with the antibiotic resistance cassette. Cultivation under conditions conducive for plasmid curing coupled with antibiotic selection provides a mechanism to recover successfully mutagenized strains. Although such genetic manipulation of numerous bacterial species has been feasible for decades, genetic manipulation of obligate intracellular bacteria has remained challenging. Here we demonstrate a novel approach for generation of chromosomal deletion mutants in *Chlamydia trachomatis*, serovar L2.

Chlamydiae are gram-negative, obligate intracellular bacteria that undergo a biphasic developmental cycle [2]. Eukaryotic host cells are infected with the elementary bodies (EBs), which are the chlamydial developmental forms specialized for transmission of the microorganism. Upon invasion, the EB differentiates into the intracellular, vegetative form termed a reticulate body (RB). Reticulate bodies grow and multiply in the host cytoplasm within a vacuole termed the chlamydial inclusion. After multiple rounds of bacterial cell division, the RBs asynchronously convert back to the infectious EBs, which are released upon exit from the infected eukaryotic cell. The entire developmental cycle during infection with *C. trachomatis*, L2 takes about 40–48 h. Overall, this biology poses significant barriers for successful genetic manipulation of chlamydiae—including generation of targeted deletion mutations in the chlamydial genome.

Many, but not all, chlamydial species contain a 7.5 kb cryptic plasmid that expresses at least eight genes [3]. *C. trachomatis* plasmid (pL2)-encoded *pgp6* is essential for maintenance of the plasmid by chlamydiae [4]. Deletion of *pgp6* leads to loss of the cryptic plasmid after a few rounds of chlamydial infection. We leveraged these findings to engineer a *Chlamydia*-specific conditionally replicative vector, pSUmC [5]. Initially, the pL2 plasmid isolated from *C. trachomatis*, serovar L2/434Bu was utilized to produce the chlamydial cloning plasmid pBOMB4-Tet-mCherry [6]. We used this expression plasmid as a backbone to generate pSUmC, where expression of *pgp6* was placed under control of a tetracycline (Tet) inducible promoter. The presence of anhydrotetracycline (ATc) inducer in the culture medium permits expression of *pgp6*, and confers the ability of *C. trachomatis* to maintain pSUmC. When ATc is removed, the Tet repressor binds to the operator positioned upstream of *pgp6*, and expression of this gene is inhibited, resulting in loss of pSUmC [5]. A constitutively expressed *bla* coding region is located downstream of *pgp6*, followed by elements encoding green-fluorescent protein (GFP), a pMB1 origin of replication, and red-fluorescent protein (mCherry). Although previous versions of pSUmC required iPCR-based techniques to insert chlamydial DNA representing homology arms [5, 7], the *bla-gfp* cassette in pSUmC described herein is flanked by unique SalI and SbfI sites. This feature enables convenient application of Gibson Assembly technology [8] to clone upstream and downstream homology arms into the 5′ and 3′ flanks of *bla-gfp*, respectively.

For mutagenesis, *C. trachomatis* L2 is transformed with a pSUmC-based allelic exchange construct using the $CaCl_2$ method [9], and cultures are maintained with penicillin G and ATc. Once transformants are detected, as determined by the appearance of red and green chlamydial inclusions, cultures are shifted to media lacking ATc. Loss of red fluorescence confirms plasmid curing whereas the presence of penicillin G selects for transformed chlamydiae that have undergone spontaneous homologous recombination, integrating the *bla-gfp* cassette into the chromosome in exchange for the chlamydial target gene. Transformants are typically bright red and green indicating the presence of multiple copies of pSUmC. Chlamydial mutants however will have lost the plasmid-encoded mCherry and only retain a single copy of the *bla-gfp* cassette in the genome, and therefore exhibit no red fluorescence and comparably dim green fluorescence. An isogenic population of mutant chlamydiae is obtained by the limiting dilutions method.

2 Materials

All solutions are prepared and handled in a class II biological safety cabinet. McCoy cells are routinely maintained according to the supplier's instructions (*see* **Note 1**). *Chlamydia trachomatis* serovar L2/434Bu is a Biosafety Level 2 pathogen. Follow all appropriate guidelines and regulations for the use and handling of pathogenic microorganisms. Use tissue culture grade water for all solutions prepared for cell culture experiments. We typically do not add antibiotics during routine maintenance of eukaryotic host cells.

2.1 Generation of Chlamydial Inocula for Transformation

1. RPMI-1640 tissue culture medium supplemented with 10% (vol/vol) heat-inactivated fetal bovine serum (FBS) (complete RPMI).
2. Hanks' Balanced Salt Solution (HBSS).
3. *Chlamydia trachomatis*, serovar L2/434Bu of known titer. Chlamydiae can be obtained from ATCC. The routine propagation and purification of *C. trachomatis* has been previously described in detail [10]. For this procedure high purity of the inocula is not critical.
4. 6-well plastic sterile tissue culture plates.
5. Cell scrapers (1.8 cm blade), sterile.
6. SPG (sucrose–phosphate–glutamic acid) buffer: Dissolve 75 g of sucrose, 2.47 g of Na_2HPO_4 and 0.36 g of NaH_2PO_4, and

0.72 g of L-glutamic acid in 900 mL of water, then adjust pH to 7.4, and adjust volume to 1 L. Filter sterilize and store SPG at 4 °C.

7. McCoy cells (mouse fibroblasts) are obtained from ATCC.
8. Temperature-controlled clinical centrifuge, sealed, with swinging bucket rotor and microplate buckets.
9. Temperature controlled benchtop centrifuge for microcentrifuge tubes.
10. Cycloheximide: Prepare 1000× stock solution by dissolving 10 mg of cycloheximide in 9 mL of water then adjust volume to 10 mL, filter sterilize, aliquot and store at −20 °C.
11. Methanol.
12. PBS (phosphate-buffered saline): Dissolve 7.89 g of NaCl, 0.201 g of KCl, 1.42 g of Na_2HPO_4, and 0.313 g of KH_2PO_4 in 900 mL of water, then adjust pH to 7.2 and adjust volume to 1 L. Autoclave to sterilize and store PBS at room temperature.
13. Primary monoclonal or polyclonal antibodies specific to *Chlamydia*.
14. Inverted fluorescence microscope.
15. 15-mL sterile conical tube.
16. $CaCl_2$ buffer: prepare 10 mM Tris pH 7.4 and 50 mM $CaCl_2$ in water, filter sterilize and store at 4 °C.
17. 2-mL Safe-Lock microcentrifuge tubes.
18. 1.5 mL sterile microcentrifuge tubes with silicone O-rings.
19. Tissue culture incubator.
20. Class II biological safety cabinet.

2.2 Generation of Chlamydial Suicide Vector

1. *C. trachomatis* genomic DNA.
2. Chlamydial suicide vector, pSUmC with SalI and SbfI restriction sites flanking the *bla-gfp* cassette. pSUmC-*sal-sbf* is available upon request from the corresponding author.
3. Q5® High-Fidelity DNA Polymerase.
4. Engineered primers to amplify *Chlamydia*-specific DNA and insert into vector via Gibson Assembly reaction.
5. SalI and SbfI restriction enzymes.
6. Gel DNA Recovery Kit for purification of PCR amplicons, such as QIAquick (Qiagen).
7. Agarose, molecular biology grade.
8. Gibson assembly cloning kit (New England Biolabs, Inc.).

9. Electrocompetent *E. coli* NEB10β (New England Biolabs, Inc.).
10. *dam*⁻/*dcm*⁻ Chemically competent *E. coli* (New England Biolabs, Inc.).
11. Plastic petri dishes for bacteria, sterile.
12. Carbenicillin: Prepare 1000× stock solution by dissolving 500 mg of carbenicillin in 9 mL of water then adjust volume to 10 mL, filter sterilize, aliquot and store at −20 °C.
13. LB broth, Miller (Luria–Bertani).
14. Agar, bacteriological, powder.
15. Plasmid miniprep kit, such as QIAprep Spin Mini Kit.
16. Plasmid maxiprep kit, such as QIAGEN Plasmid Plus Maxi Kit.
17. Sterile Erlenmeyer flasks for bacterial cultures.

2.3 Transformation of *Chlamydia trachomatis*

1. RPMI-1640 tissue culture medium supplemented with 10% (vol/vol) heat-inactivated FBS (complete RPMI).
2. Hanks' Balanced Salt Solution (HBSS).
3. *Chlamydia trachomatis*, crude inoculum of known titer.
4. 6-well plastic sterile tissue culture plates.
5. Sterile 1.5 mL microcentrifuge tubes.
6. McCoy cells.
7. Temperature controlled centrifuge, sealed, with swinging bucket rotor with microplate buckets.
8. Cycloheximide: 1000× stock solution (1 mg/mL).
9. Penicillin G: Prepare 10,000× stock solution (6 mg/mL) by dissolving 60 mg of the antibiotic in 9 mL of water then adjust volume to 10 mL, filter sterilize, aliquot and store at −20 °C.
10. Anhydrotetracycline (ATc): Prepare 10,000× stock solution by dissolving 500 μg of ATc in 1 mL of DMSO (cell culture grade). Aliquot (20 μL) and freeze the stock solution at −20 °C. Do not thaw and refreeze ATc for more than three times.
11. $CaCl_2$ buffer: prepare 10 mM Tris pH 7.4 and 50 mM $CaCl_2$ in water, filter sterilize and store at 4 °C.
12. Cell scrapers (1.8 cm blade).
13. 2-mL Safe-Lock microcentrifuge tubes.
14. Unmethylated plasmid DNA for deletion construct at 1.0–1.5 μg/μL.
15. Class II biological safety cabinet.

2.4 Isolation of Deletion Mutant by Limiting Dilution

1. RPMI-1640 tissue culture medium supplemented with 10% (vol/vol) heat-inactivated FBS (complete RPMI).
2. Hanks' Balanced Salt Solution (HBSS).
3. *Chlamydia trachomatis*, deletion mutant of known titer.
4. McCoy cells.
5. Temperature controlled clinical centrifuge, sealed, with swinging bucket rotor with microplate buckets.
6. Temperature controlled bench top centrifuge for microcentrifuge tubes.
7. Cycloheximide: 1000× stock solution (1 mg/mL).
8. Penicillin G: 10,000× stock solution (6 mg/mL).
9. Crude stock of *C. trachomatis* deletion mutant of known titer.
10. 384-well tissue culture plates.
11. Bio-Pure reagent reservoirs.
12. Multichannel pipette.
13. Sterile 50-mL conical tube.
14. Class II biological safety cabinet.

3 Methods

Except for specified centrifugation steps, all experiments are carried out at room temperature. *All experiments with McCoy cells and viable chlamydiae are performed in a Class II biological safety cabinet under aseptic conditions.*

3.1 Preparation of C. trachomatis Crude Stock for Transformation

1. Seed four 6-well plates with McCoy cells at 1 to 1.5 × 10^6 of cells per well, in complete RPMI medium. Incubate for 24 h at 37 °C in an atmosphere of 5% CO_2/95% humidified air (*see* **Note 2**).
2. Remove medium from confluent McCoy cultures and add *C. trachomatis* in 2 mL of HBSS at the MOI of 1 (2 to 3 × 10^6 of chlamydiae per well). Infect McCoy cells by centrifugation at 900 × *g* for 1 h at 20 °C in a centrifuge equipped with swinging bucket rotor and microplate buckets.
3. After centrifugation, remove inoculum, add 2 mL of complete RPMI medium supplemented with 1 μg/mL of cycloheximide per well, and incubate at 37 °C in an atmosphere of 5% CO_2/95% humidified air, for 40 h.
4. Harvest chlamydiae-infected McCoy cells by gentle scraping of cell cultures with a cell scraper and collect the contents of each well into 2-mL Safe-Lock sterile microcentrifuge tubes.

5. Pellet harvested cell material (24 tubes) at 20,000 × *g* for 30 min at 4 °C.
6. Remove supernatant and resuspend all 24 pellets in a total of 4 mL of SPG. Vortex well and centrifuge again at 200 × *g* for 5 min at 4 °C.
7. Finally, collect supernatant into a sterile 15-mL conical tube and disperse obtained inoculum into sterile microcentrifuge tubes with silicone O-rings in 50 μL aliquots. This crude chlamydial preparation is stored (in SPG) at −80 °C.
8. Prepare one 6-well plate with confluent McCoy cells as described above.
9. Perform a sham transformation reaction in order to determine the most accurate titer of the crude inoculum. To accomplish this set out 6 sterile microcentrifuge tubes with 50 μL of $CaCl_2$ buffer in each.
10. Thaw out an aliquot of the crude chlamydial inocula and add 0.5 μL into the first microcentrifuge tube, 1 μL into the second, 2 μL into the third tube and so on. Once 5 μL is added into the sixth tube, mix chlamydiae in the buffer by gently flicking each tube and incubate the bacteria in $CaCl_2$ buffer for 30 min.
11. Meanwhile, remove medium from one 6-well plate with McCoy cells and add 1 mL of HBSS per well.
12. The incubation of *C. trachomatis* in calcium buffer is stopped by adding 1 mL of HBSS into each reaction.
13. Pipet the contents of one microcentrifuge tube into one well in the 6-well plate and infect McCoy cells by centrifugation as described above. Incubate infected cells in complete RPMI supplemented with 1 μg/mL of cycloheximide as described above.
14. At ~24 h post infection, fix infected cell cultures with methanol for 5 min, wash wells three times with PBS and stain chlamydial inclusions with anti-*C. trachomatis* specific primary antibody followed by staining with a fluorophore conjugated secondary antibody according to the manufacturers' instructions.
15. The number of inclusion forming units (IFUs) per mL is obtained by enumeration of chlamydial inclusions using a fluorescent microscope as described by [10].

3.2 Generation of Chlamydial Suicide Vector

Our previously described pSUmC required multiple steps of iPCR-based cloning to generate vectors for allelic exchange mutagenesis [5, 7]. However, Gibson Assembly has gained traction as an effective molecular biology technique—particularly in the construction of allelic exchange constructs [11]. pSUmC-*sal-sbf* contains unique *Sal*I and *Sbf*I restriction sites immediately flanking the 5′ and 3′

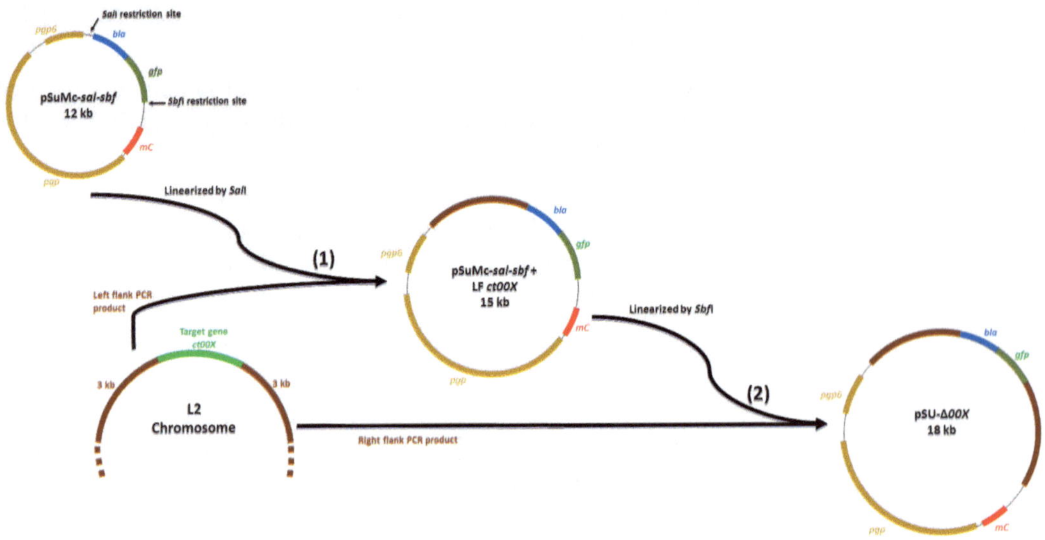

Fig. 1 pSUmC-*sal-sbf* contains unique *SalI* and *SbfI* restriction sites immediately flanking the 5′ and 3′ regions of the *bla-gfp* cassette, respectively. Two, sequential Gibson Assembly steps can be employed to insert (±) 3-kb left flank (upstream) and (±) 3-kb right flank (downstream) of the targeted locus (gene 00X) to the *SalI* and *SbfI* sites of the backbone plasmid, respectively

regions of the *bla-gfp* cassette, respectively. Two, sequential Gibson Assembly steps can be employed to insert (±) 3-kb left flank (upstream) and (±) 3-kb right flank (downstream) of the targeted locus (gene 00X) to the *SalI* and *SbfI* sites of the backbone plasmid, respectively (Fig. 1).

1. Design primers for amplifying (±) 3-kb left flank (upstream) of the targeted gene (gene 00X) that include specific overlap sequence (underlined) for the *SalI* digested site of the backbone plasmid (pSUmC-*sal-sbf*) (*see* **Note 4**).

   Forward and reverse *Chlamydia*-specific sequences should then be added to the 3′ end of X-LF-F and X-LF-R, respectively, to produce the complete primer sequences (*see* **Note 3**):

   00X-LF-F: *CGTCACTGCAGGTACCGGTCGA*- Specific sequence to the left flank of targeted gene.

   00X-LF-R: *GCGGATACATATTGAATGGTCGA*- Specific sequence to the left flank of targeted gene.

2. PCR-amplify left flank of 00X gene from chlamydial genomic DNA using X-LF-F and X-LF-R primers with Q5 High-Fidelity DNA Polymerase according to manufacturer's instructions. Separate the PCR reaction on a 0.8% agarose gel and extract the PCR product with a Gel DNA Recovery Kit.
3. Linearize backbone plasmid by digestion with SalI. Separate the linearized plasmid on a 0.8% agarose gel and extract the correct product with a Gel DNA Recovery Kit.

4. Assemble linearized backbone plasmid and left flank of 00X gene by a Gibson Assembly according to the manufacturer's instructions.
5. Transform NEB® 10β electrocompetent *E. coli* according to manufacturer's instructions. Spread the transformation on LB agar plates containing 50 μg/mL carbenicillin and incubate overnight at 37 °C.
6. Choose colonies that display both green and red florescence for screening and incubate positive colonies in 3 mL LB liquid medium containing 50 μg/mL carbenicillin overnight at 37 °C. Extract the plasmid DNA with a plasmid miniprep Kit and confirm the insertion by restriction digestion using appropriate restriction sites within the left flank DNA. Check the correct sizes by separating digested plasmids on a 0.8% agarose gel. PCR-based screening can also be employed depending on investigator preferences.
7. Design primers for amplifying (±) 3-kb right flank (downstream) of 00X that include specific overlap sequence (underlined) for the SbfI site of the backbone plasmid (pSUmC-*sal-sbf* + LF).

   Forward and reverse *Chlamydia*-specific sequences should then be added to the 3′ end of X-RF-F and X-RF-R, respectively, to produce the complete primer sequences:

   X-RF-F: *CTATACAAGTAACCTGCA*- Specific sequence to the downstream of targeted gene.

   X-RF-R: *GGTCTGACGCTCCCTGCA*- Specific sequence to the downstream of targeted gene.
8. PCR-amplify right flank of 00X gene from chlamydial genomic DNA using X-RF-F and X-RF-R primers with Q5 High-Fidelity DNA Polymerase according to manufacturer's instructions. Separate the PCR reaction on a 0.8% agarose gel and extract the PCR product with a Gel DNA Recovery Kit.
9. Linearize backbone plasmid by digestion with SbfI. Separate the linearized plasmid on a 0.8% agarose gel and extract the correct product with a Gel DNA Recovery Kit.
10. Assemble linearized backbone plasmid and right flank of 00X gene by a Gibson Assembly as described in **step 4**.
11. Transform *E. coli* and screen resulting colonies as described in **steps 5** and **6**. DNA sequence the engineered regions of final plasmids to confirm correct insertion of chlamydial DNA.

3.3 Transformation of *Chlamydia trachomatis*, L2

1. Prepare two 6-well plates (*see* **Note 2**) with confluent McCoy cells as described in Subheading 3.1. Based on chlamydial titer determined in the Subheading 3.1, calculate the number of chlamydiae required to infect 12 wells at the MOI of 2 (*see* **Note 5**).

2. Thaw out an aliquot(s) of crude *C. trachomatis* preparation, transfer required amount of chlamydial inocula into a 1.5 mL microcentrifuge tube and pellet bacteria by centrifugation at 20,000 × *g* for 30 min at 4 °C.
3. Meanwhile, thaw out the unmethylated deletion-construct plasmid DNA generated in Subheading 3.2, and set out 5 sterile 1.5 mL microcentrifuge tubes.
4. Next, resuspend the chlamydial pellet in 600 μL of $CaCl_2$ buffer and pipet 100 μL of the chlamydial inocula in each microcentrifuge tube (6 tubes total). Then quickly add 5 μg of the deletion construct DNA into each tube and mix contents by flicking each microcentrifuge tube every 10 min. Incubate a total of 30 min at room temperature.
5. Supplement each transformation reaction with 1 mL of HBSS.
6. Remove complete RPMI from the two 6-well plates and replace with 1.5 mL of fresh HBSS and 550 μL of the transformation reaction into each well (*see* **Note 6**).
7. Centrifuge plates at 900 × *g* for 1 h at 20 °C, remove chlamydial inoculum, and add 2 mL of complete RPMI per well supplemented with 1 μg/mL of cycloheximide. Incubate at 37 °C in an atmosphere of 5% CO_2 and 95% humidified air for 7 h (*see* **Note** 7).
8. Replace the medium with 2 mL of complete RPMI supplemented with 1 μg/mL cycloheximide, 0.6 μg/mL of penicillin G, and 50 ng/mL of anhydrotetracycline (ATc), and return plates back to the incubator (*see* **Note 8**).
9. Next day prepare two 6-well plates with fresh McCoy monolayers, as before.
10. At 48 h post infection repassage infected cells by harvesting all cultures, including media, using a cell scraper. Collect cultures into 2-mL Safe-Lock microcentrifuge tubes and centrifuge tubes at 20,000 × *g* for 30 min at 4 °C.
11. Pipet off supernatants and resuspend each pellet in 1 mL of HBSS, then vortex well and spin the harvested material down at 200 × *g* for 5 min at 4 °C to remove crude debris from eukaryotic host cells.
12. Meanwhile, remove complete RPMI from two plates containing new confluent McCoy cells and add 1 mL of fresh HBSS per well.
13. Next, add supernatants into this new cell culture, supernatant of one tube per well, and centrifuge 6-well plates at 900 × *g* for 1 h as described above.
14. Finally, remove HBSS and add complete RPMI supplemented with 1 μg/mL cycloheximide, 0.6 μg/mL of penicillin G, and

50 ng/mL of ATc. Incubate cells in the incubator for 48 h and then perform another repassage as described above.

15. Keep repassaging cells every 48 h until inclusions develop which exhibit both bright red- and green-fluorescence, as observed by microscopy. However, if after 5 passages, no fluorescent inclusions develop (*see* **Note 9**), restart the transformation protocol (beginning at **step 1** above).
16. Once fluorescent inclusions are detected, keep repassaging chlamydiae as described above but the infected cells are now cultured in complete RPMI lacking ATc. After several following passages, chlamydial inclusions will become apparent that are dim green and not red, as observed by microscopy, indicating the generation of the deletion mutant and loss of the suicide vector (*see* **Note 10**). Maintain an apparent MOI of 0.1 during these steps (*see* **Note 11**).
17. Sufficiently expand the number of dim green inclusions and prepare a small amount of a crude chlamydial stock and accurately determine the *C. trachomatis* titer as described in the Subheading 3.1.

3.4 Isolation of Deletion Mutant by Limiting Dilution

1. Prepare confluent monolayer of McCoy cells in a 384-well plate in complete RPMI.
2. Pipet 25 mL of HBSS into a 50-mL conical tube and add 50 chlamydiae. Vortex and transfer the diluted chlamydial mutant into a Bio-Pure reagent reservoir. Remove media from McCoy cells and apply 50 μL of chlamydial inoculum into each well employing multichannel pipet.
3. Centrifuge plate at 900 × *g* for 1 h at 20 °C.
4. Remove HBSS from the plate and add complete RPMI supplemented with 1 μg/mL cycloheximide, 0.6 μg/mL of penicillin G. Incubate the plate at 37 °C in an atmosphere of 5% CO_2/95% humidified air.
5. About 5–10 days post infection individual wells should contain McCoy cell monolayers containing multiple chlamydial inclusions. Identify wells containing *C. trachomatis* by fluorescence microscopy.
6. Depending on the number of positive wells, prepare confluent McCoy cells in an appropriate number of wells in a 6-well plate.
7. Then harvest these positive samples by scraping the wells with pipette tips and transferring the contents to 2-mL Safe-Lock microcentrifuge tubes containing 2 mL HBSS. Vortex obtained samples and apply each to one well in a 6-well plate.
8. Centrifuge plate(s) at 900 × *g* for 1 h at 20 °C. Incubate plate in complete RPMI supplemented with 1 μg/mL cycloheximide as described above.

9. A second round of isolation from a 384 well plate is typically employed to ensure clonal populations are obtained.
10. Once deletion mutants are isolated and purified, no selective pressure is required during chlamydial growth.

4 Notes

1. Although *C. trachomatis* is routinely propagated in human HeLa cells, murine McCoy cells are employed here because, in our hands, this cell line can handle the entire process of chlamydial transformation much better with regard to the cell viability and overall fitness of the host cell culture.
2. The number of 6-well plates used either in preparation of chlamydial crude stock or in chlamydial transformation can be adjusted.
3. Primers specific for left or right flank sequences (18–30 bp) should be designed having a melting temperature between 58 °C and 68 °C and lacking hairpin at temperatures not higher than 50 °C. For complete gene deletion, cloned 5′, left-flanking DNA should include up to the initiation codon for 00X, whereas 3′, right-flanking DNA should include up to the stop codon for 00X.
4. 3 kb flanking arms are routinely employed, but it is anticipated that shorter lengths are possible. Regardless, 3 kb arms are an efficient starting length unless cloning necessitates shorter DNA lengths.
5. When infecting McCoy cells with *C. trachomatis*, always check cell cultures for the desired level of infection and the condition of eukaryotic hosts by microscopy. If McCoy cells appear to be detaching and dying at 24 h post infection, there is no point in going on with the planned procedure. It is better to start the experiments over with a fresh monolayer of McCoy cells and readjusted number of chlamydial MOI.
6. During infection of McCoy cells with *C. trachomatis* by centrifugations in 6-well plates, 2 mL of HBSS per well is required, as lower volume often results in shifting of the cell monolayer onto one site and thus leads to an uneven chlamydial infection.
7. The 7 h recovery of chlamydiae after transformation has been determined to be the most optimal for transformation efficiency but could vary from lab to lab. In our hands, shorter recovery will result in low efficiency of bacterial transformation. Longer than 7 h recovery leads to the host cell lysis next day due to the inability of the selective antibiotic to sufficiently inhibit chlamydial propagation.

8. All antibiotics are aliquoted and stored at −20 °C however, once an aliquot is thawed out it is kept at 4 °C: Penicillin G for 1 week, whilst cycloheximide and carbenicillin for 1 month.
9. If the chlamydial transformation is unsuccessful, there are several steps that can be augmented. Instead of employing two 6-well plates, use ≥four 6-well plates with an MOI of 2 as described above. Next, check the pH of $CaCl_2$ buffer and make sure there are no precipitates in the buffer. The condition of the unmethylated suicide construct is also important. Run 0.8% agarose gel to confirm that the plasmid isn't nicked or otherwise damaged. Importantly, it is advisable to include a positive control, empty pSUmC or GFP::SW2 [9], in transformation experiments to ensure an adequate level of bacterial competence.
10. In some cases, merodiploids may not resolve to yield mutants, even after 10–20 passages. In this instance, DNA can be extracted from cultures and PCR used to confirm that crossover events are occurring in both homology arms. Design primer sets where one primer anneals within the *bla-gfp* cassette and the *Chlamydia*-specific primer anneals outside the engineered homology arm. Perform PCR and assess recombination as described [11]. If there is evidence for recombination within only one homology arm, verify that the allelic exchange construct is correct. If evidence indicates that recombination is occurring in both homology arms, it is likely that the targeted gene is essential and cannot be deleted via this protocol.
11. It is important to maintain a low MOI during passage of transformants, especially if the deletion results in significant attenuation of chlamydial fitness. In some cases, it may be necessary to isolate inclusions corresponding to mutant strains using micromanipulator or cloning cylinder approaches.

References

1. Nakashima N, Miyazaki K (2014) Bacterial cellular engineering by genome editing and gene silencing. Int J Mol Sci 15:2773
2. AbdelRahman Y, Belland R (2005) The chlamydial developmental cycle. FEMS Microbiol Rev 29:949–959
3. Zhong G (2017) Chlamydial plasmid-dependent pathogenicity. Trends Microbiol 25:141–152
4. Song L, Carlson JH, Whitmire WM, Kari L, Virtaneva K, Sturdevant DE et al (2013) *Chlamydia trachomatis* plasmid-encoded Pgp4 is a transcriptional regulator of virulence-associated genes. Infect Immun 81:636–644
5. Mueller KE, Wolf K, Fields KA (2016) Gene deletion by fluorescence-reported allelic exchange mutagenesis (FRAEM) in *Chlamydia trachomatis*. MBio 7:e01817–e01815
6. Bauler LD, Hackstadt T (2014) Expression and targeting of secreted proteins from *Chlamydia trachomatis*. J Bacteriol 196:1325–1334
7. Mueller KE, Wolf K, Fields KA (2017) *Chlamydia trachomatis* transformation and allelic exchange mutagenesis. Curr Protoc Microbiol 45:11A.3.1–11A.3.15
8. Gibson DG, Young L, Chuang R-Y, Venter JC, Hutchison Iii CA, Smith HO (2009)

Enzymatic assembly of DNA molecules up to several hundred kilobases. Nat Methods 6:343

9. Wang Y, Kahane S, Cutcliffe LT, Skilton RJ, Lambden PR, Clarke IN (2012) Development of a transformation system for *Chlamydia trachomatis*: restoration of glycogen biosynthesis by acquisition of a plasmid shuttle vector. PLoS Pathog 7:e1002258

10. Scidmore MA (2005) Cultivation and laboratory maintenance of *Chlamydia trachomatis*. In: Current protocols in microbiology. Wiley, New York

11. Silayeva O, Barnes AC (2018) Gibson assembly facilitates bacterial allelic exchange mutagenesis. J Microbiol Methods 144:157–163

Chapter 12

Mutagenesis of *Chlamydia trachomatis* Using TargeTron

Mary M. Weber and Robert Faris

Abstract

Chlamydia trachomatis is an important human pathogen that prior to 2011 was largely intractable to genetic manipulation. Here we describe the application of a group II intron, referred to as TargeTron, for site-specific insertional inactivation of target genetic loci in the obligate, intracellular bacteria *C. trachomatis*. In this chapter, we outline the methods for intron retargeting, chlamydia transformation, and mutant verification. We also outline a method for complementation of TargeTron mutants. Furthermore, we discuss potential pitfalls and alternative strategies for generating mutants with TargeTron technology.

Key words *Chlamydia trachomatis*, TargeTron, Group II intron, Insertional inactivation, Complementation

1 Introduction

Chlamydia spp. are obligate intracellular bacteria of human and veterinary importance [1]. *Chlamydia trachomatis* is the leading cause of bacterial sexually transmitted infections and blinding trachoma, whereas *Chlamydia pneumoniae* causes acute and chronic infections of the respiratory tract [2, 3]. The inability to genetically manipulate these important human pathogens until recently hindered identification and characterization of putative virulence factors. However, the development of a shuttle vector for stable transformation of *C. trachomatis* [4] and *C. pneumoniae* [5], as well as a defined *C. trachomatis* chemical mutant library [6, 7], has ushered in a new era of chlamydia research. Since the inception of these genetic tools, important modifications and new techniques have been developed, allowing for expression of epitope-tagged proteins in chlamydia [8–10], site-specific insertional inactivation using a group II intron (discussed in this chapter) [11–14] or allelic exchange [15, 16], and complementation of mutants [13–15] (discussed in this chapter).

Amanda Claire Brown (ed.), *Chlamydia trachomatis: Methods and Protocols*, Methods in Molecular Biology, vol. 2042, https://doi.org/10.1007/978-1-4939-9694-0_12,

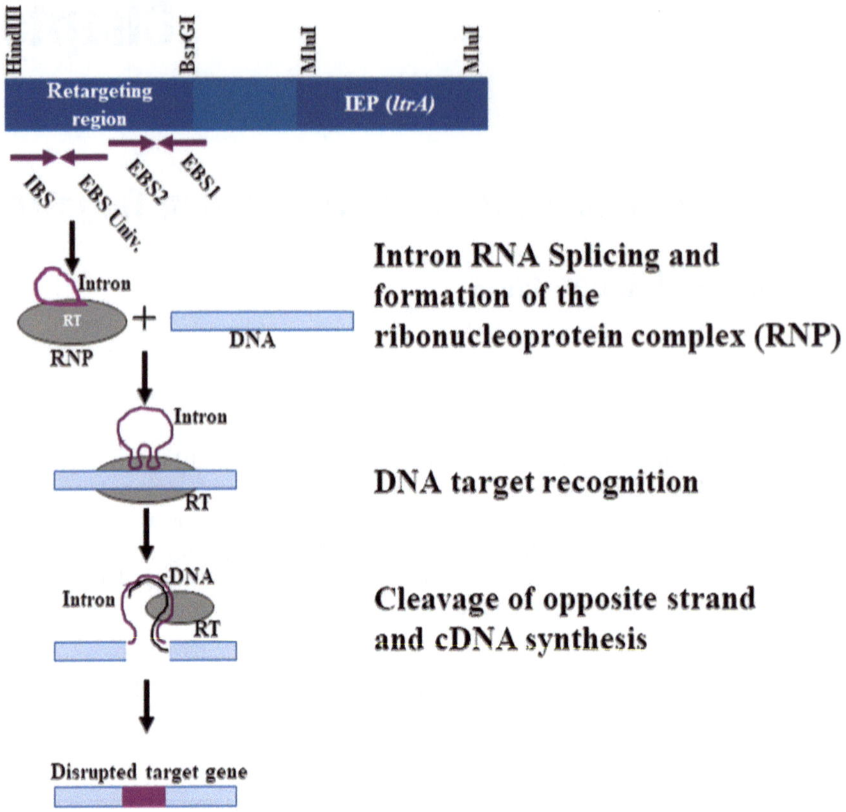

Fig. 1 GrpII intron retrohoming. The reverse transcriptase (RT) binds to the intron, promoting RNA splicing. The ribonucleoprotein (RNP) complex, consisting of the excised RNA lariat and the bound RT, is formed. DNA target sites are recognized by the RNP through binding of the RT and the intron. The intron integrates into the target site by reverse splicing and the opposite DNA strand is cleaved by the endonuclease activity of the RT, leaving a 3′ overhang that serves as a primer for cDNA synthesis by the RT. Intron cDNA is integrated into the genome using host repair enzymes. Image adapted from Enyeart et al. [17]

Mobile group II (GrpII) introns are bacterial retrotransposons, which, through a process termed retrohoming, insert into specific sites in DNA [17] (Fig. 1). Mobile GrpII introns consist of a catalytically active intron RNA and a region, which encodes a multifunctional protein capable of carrying out numerous DNA-manipulating functions, including acting as a reverse transcriptase (RT). GrpII introns can home to, and insert themselves in, target DNA sites by retrohoming. In the first stages of this process, the GrpII intron is transcribed and the RT is translated which, subsequently binds the larger precursor RNA molecule, allowing it to fold into a catalytically active RNA. The folded intron has an active site that uses bound Mg^{2+} to catalyze RNA splicing and through two transesterification reactions, yields an intron lariat RNA with two regions called EBS1 and EBS2 that play a major role in DNA binding and target recognition. It is primarily these regions that can be "reprogrammed" to recognize target genetic loci of interest. The ribonucleoprotein complex (RNP), consisting of the GrpII intron bound RT and the intron lariat, recognizes specific sequences in the target DNA. Recognition of DNA target sequences involves both the RT and base pairing with the intron RNA. Binding of the RT results in DNA melting, allowing the intron RNA to bind to the target sequences. The intron then integrates into the DNA by reverse splicing, creating a DNA-RNA hybrid, and the opposite DNA strand is cut by the endonuclease activity of the RT. Cutting of the opposite strand yields a 3′ overhang that is used as a primer for cDNA synthesis by the RT while host enzymes repair the cDNA.

Base pairing occurs between the target DNA and sequences within the intron RNA, thus GrpII introns can be engineered into gene targeting vectors or "TargeTrons" [18, 19]. The Ll.LtrB IIA, from *Lactococcus lactis* has been developed into a TargeTron that can inactivate genes of interest and has been successfully used to insertionally inactivate target genes in previously genetically intractable organisms including *Chlamydia trachomatis* [11, 12], *Francisella tularensis* subsp. *tularensis* [20], *Clostridium spp.* [21], *Rickettsia rickettsii* [22], and *Ehrlichia chaffeensis* [23]. The fact that L1.LtrB TargeTron works in most bacteria, functions in $RecA^+$ or $RecA^-$ bacteria, integrates into the genome with high efficiency, and can easily be retargeted to new DNA sites by PCR [17], makes the TargeTron an exceptional tool for gene targeting.

Site-specific insertional inactivation of target genes is facilitated by the development of three different vector systems for use in *C. trachomatis* [11–13] (Fig. 2). Insertion sites are predicted using the TargeTronics or Sigma algorithm and the intron is retargeted to the gene of interest using PCR (Figs. 3 and 4). The PCR products are digested and ligated into the linearized TargeTron vector. Successful retargeting is confirmed by DNA sequencing and the retargeted TargeTron vector is transformed into *C. trachomatis* L2 using chemical transformation with calcium chloride. Following four rounds of antibiotic selection, clonal transformants are isolated by plaque cloning or limiting dilution. Successful insertion is confirmed by various molecular approaches (PCR, sequence, western blotting, or Southern blotting). *C. trachomatis* mutants can then be complemented using the expression plasmids pBomb3 [13] or pBomb4 [8, 9] (Fig. 5) to verify that any noted phenotype of the mutant is due to loss of the target protein product.

2 Materials

2.1 Intron Template Synthesis (See Note 1)

1. Chlamydia TargeTron plasmid (Fig. 2): pACT [14], pDFTT3 [11], or pDFTT3-*aadA* [12].
2. Qiagen PCR core kit (Qiagen).

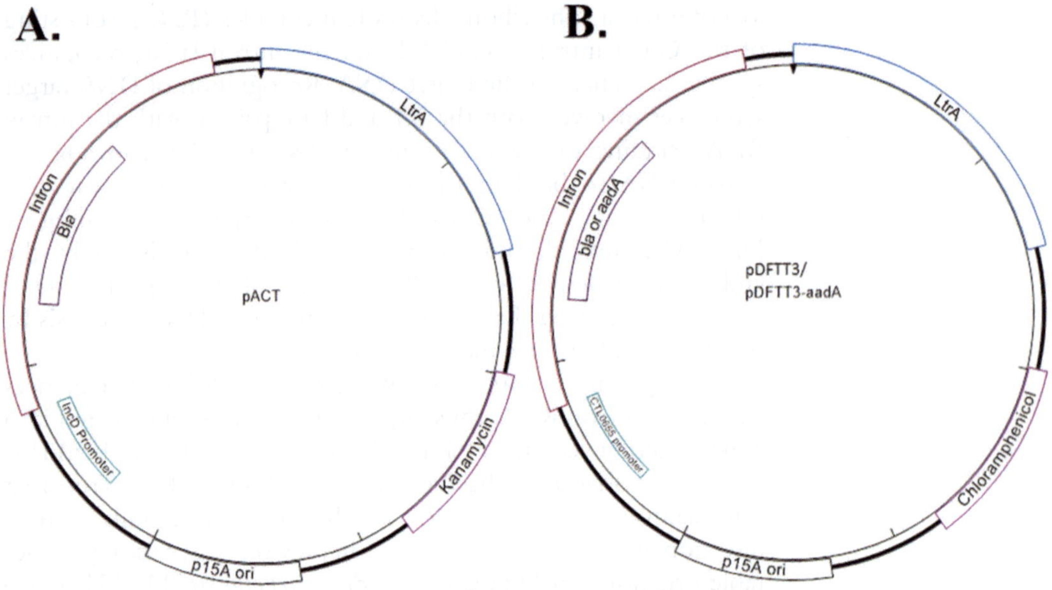

Fig. 2 Plasmid maps of *Chlamydia trachomatis* TargeTron suicide vectors. The commercially available vectors, pACDK4-C or pACD4K-K were modified to generate pACT or pDFTT3/pDFTT3-*aadA* respectively. (**a**) pACT contains a *bla* RAM cassette, driven by the *C. trachomatis* L2 IncD promoter, and a kanamycin resistance marker for selection in *E. coli.* (**b**) pDFTT3 contains a *bla* RAM cassette, whereas pDFTT3-*aadA* contains an *aadA* RAM cassette. The individual RAM cassettes are driven by the *C. trachomatis* CTL0655 promoter and a chloramphenicol resistance marker is used for selection in *E. coli*

3. TargeTron primers for intron synthesis: LtrB-IBS (5′-GTG CGC CCA GAT AGG GTG TTA AGT C-3′), EBS universal (5′-AAT TAG AAA CTT GCG TTC AGT AAA CAC AAC TTA TAC-3′), LtrB-EBS-T (5′-CGA TAG AGG AAA GTG TCT GAA ACC TCT AGT ACA AAG AAA GGT AAG TTA-3′), and LtrB-EBS-B (5′-TAA CTT ACC TTT CTT TGT ACT AGA GGT TTC AGA CAC TTT CCT CTA TCG-3′).
4. 50× TAE: 242 g Tris, 57.1 mL glacial acetic acid, 18.6 g EDTA, and ddH_2O up to 1 L.
5. 1% agarose gel (1 g agarose in 100 mL 1× TAE).
6. Ethidium bromide.
7. Gel extraction kit.
8. Sterile molecular biology grade water.
9. Sterile 0.2 mL PCR tubes.
10. Sterile 1.5 mL Eppendorf tubes.
11. Thermocycler.
12. DNA electrophoresis unit.
13. UV transilluminator.

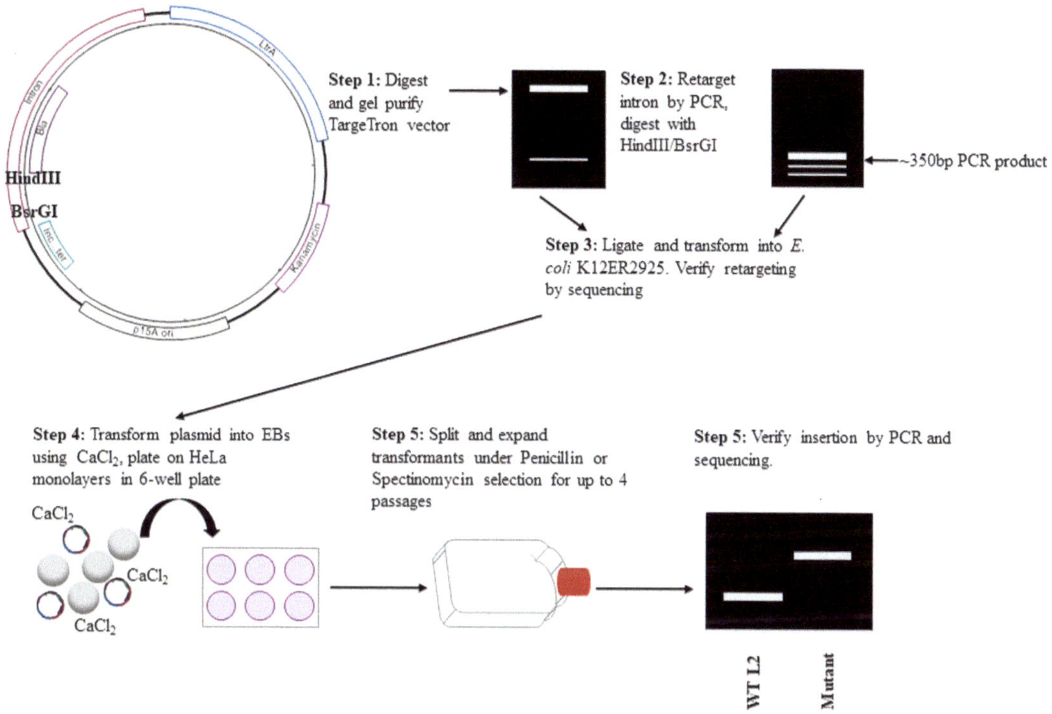

Fig. 3 Overview of intron retargeting and transformation of *Chlamydia trachomatis.* The GrpII intron is delivered into *C. trachomatis* L2 using the pACT, pDFTT3, or pDFTT3-*aadA* suicide vectors. The intron is retargeted to the gene of interest using PCR with primers designed using the TargeTronics or Sigma algorithm. The resulting PCR product is purified and digested with BsrGI and HindIII. The desired TargeTron vector is linearized using BsrGI and HindIII and the ~7.5 kb fragment is gel-extracted. The digested PCR product and linearized vector are ligated together and transformed into *E. coli.* Successful retargeting is confirmed by DNA sequencing. The resulting vector is transformed into purified *C. trachomatis* L2 elementary bodies (EB) using calcium chloride. Transformants are selected for using Penicillin G (pACT or pDFTT3) or Spectinomycin (pDFTT3-*aadA*). Following four rounds of selection, disruption of the target gene is confirmed by PCR and DNA sequencing

14. Microcentrifuge.
15. NanoDrop.
16. Razor blades.

2.2 Retargeting Intron to Target Gene

1. Chemically competent *Escherichia coli* K12 ER2925.
2. Chlamydia TargeTron plasmid (Fig. 2): pACT [14], pDFTT3 [11], or pDFTT3-*aadA* [12].
3. Qiagen PCR core kit (Qiagen).
4. IBS, EBS1, and EBS2 primer sequences based on your gene of interest.
5. EBS Universal Primer.
6. Primers for sequencing: pACT F sequencing primer (5′-CAGATAAAATATTTCTAGCTAGATTTCAGTGT-3′), T7

```
-1|+1          -30                              -1     LtrB   +1             +15    score
219|220s       ATCGCAACTGCCGGTGGATTAGGTCTTATA gtgc...tcac ACCATTGCCGCTGCT     5.9
226|227s       CTGCCGGTGGATTAGGTCTTATAACCATTG gtgc...tcac CCGCTGCTTTGATGG     6.8
307|308a       TTTCGATTTGATGAGAGCTTCCACAGCAGT gtgc...tcac GAGCTGGTAATTGGC     7.0

CTL300 length: 342 b.p.

Target sites are indicated by nucleotide residues bracketing the intron-insertion site, with "s" and "a"

PCR primers to modify the TargeTron for insertion into predicted sites:

219|220s       IBS1/2       AAAAAAGCTTATAATTATCCTTATTAGGCCTTATAGTGCGCCCAGATAGGGTG
219|220s       EBS1/delta   CAGATTGTACAAATGTGGTGATAACAGATAAGTCCTTATAACTAACTTACCTTTCTTTGT
219|220s       EBS2         TGAACGCAAGTTTCTAATTTCGATTCCTAATCGATAGAGGAAAGTGTCT


226|227s       IBS1/2       AAAAAAGCTTATAATTATCCTTATTATACCCATTGGTGCGCCCAGATAGGGTG
226|227s       EBS1/delta   CAGATTGTACAAATGTGGTGATAACAGATAAGTCCCATTGCCTAACTTACCTTTCTTTGT
226|227s       EBS2         TGAACGCAAGTTTCTAATTTCGATTTATAATCGATAGAGGAAAGTGTCT


307|308a       IBS1/2       AAAAAAGCTTATAATTATCCTTATTCCACAGCAGTGTGCGCCCAGATAGGGTG
307|308a       EBS1/delta   CAGATTGTACAAATGTGGTGATAACAGATAAGTCAGCAGTGATAACTTACCTTTCTTTGT
307|308a       EBS2         TGAACGCAAGTTTCTAATTTCGATTTGGAATCGATAGAGGAAAGTGTCT

EBS Universal primer        AATTAGAAACTTGCGTTCAGTAAACACAACTTATAC


gBlocks Gene Fragments to modify the TargeTron for insertion into predicted sites:

219|220s       gBlock       TTCCCCTCTAGAAAAAAAGCTTATAATTATCCTTATTAGGCCTTATAGTGCGCCCAGATAGGGTGTTAAGTCAAGT

226|227s       gBlock       TTCCCCTCTAGAAAAAAAGCTTATAATTATCCTTATTATACCCATTGGTGCGCCCAGATAGGGTGTTAAGTCAAGT

307|308a       gBlock       TTCCCCTCTAGAAAAAAAGCTTATAATTATCCTTATTCCACAGCAGTGTGCGCCCAGATAGGGTGTTAAGTCAAGT


Primers for gBlock amplification (optional):

LtrB-F                      TTCCCCTCTAGAAAAAAAGCTTATAATTATCCTTA

LtrB-R Universal primer     CAGATTGTACAAATGTGGTGATAACAGATAAGTC
```

Fig. 4 Example LtrB intron primer set for retargeting the GrpII intron using TargeTronics. The sequence of the target gene is entered into the TargeTronics or Sigma algorithm. The insert location is provided and the a (antisense) or s (sense) denotes the orientation of the intron relative to the gene. The primers for retargeting are listed along with the EBS Universal. Alternatively, the entire sequence can be synthesized and purchased as gBlocks through IDT

promoter primer for pDFTT3 or pDFTT3-*aadA* (5′-TAATACGACTCACTATAGGG-3′).

7. 50× TAE: 242 g Tris, 57.1 mL glacial acetic acid, 18.6 g EDTA, and ddH_2O up to 1 L.
8. Gel extraction kit.
9. PCR purification kit.
10. Plasmid miniprep kit.
11. 1% agarose gel (1 g agarose in 100 mL 1× TAE).
12. Ethidium bromide.

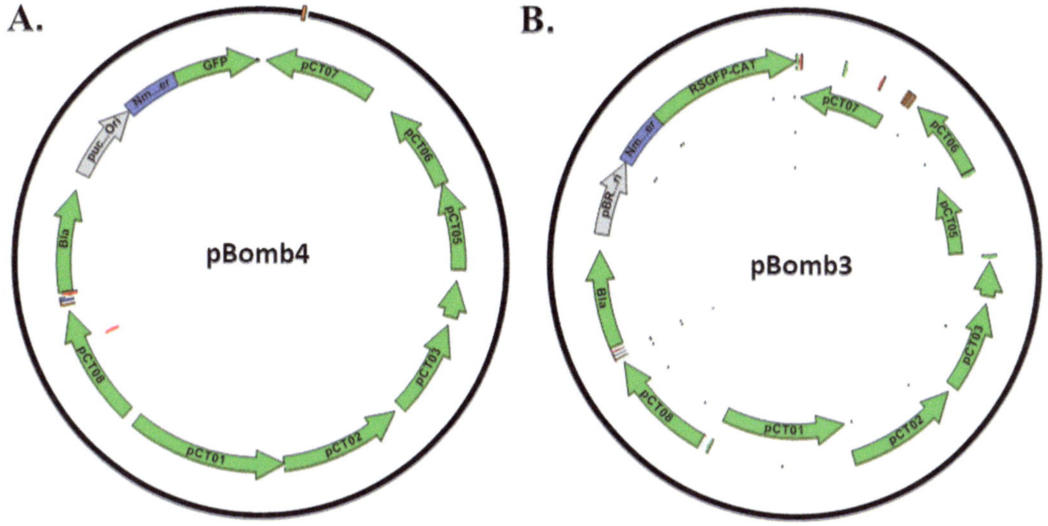

Fig. 5 Plasmid maps of *Chlamydia trachomatis* shuttle vectors used for mutant complementation. (**a**) pBomb4 contains a Tet inducible promoter and bacterial transformants are selected with using PenG. Transformed bacteria express GFP. (**b**) pBomb3 uses chloramphenicol resistance for selection of transformants. Transformed bacteria express rsGFP

13. BsrGI and HindIII restriction enzymes with 10× buffer for TargeTron.
14. T4 ligase and buffer.
15. LB broth.
16. LB agar plates with 100 μg/mL carbenicillin (pACT or pDFTT3) or 50 μg/mL spectinomycin (pDFTT3-*aadA*).
17. Sterile molecular biology grade water.
18. Sterile 0.2 mL PCR tubes.
19. Sterile 1.5 mL Eppendorf tubes.
20. Thermocycler.
21. Microcentrifuge.
22. DNA electrophoresis unit.
23. UV transilluminator.
24. NanoDrop.
25. 37 °C water bath.
26. 37 °C incubator.
27. Razor blades.

2.3 Transformation of Chlamydia trachomatis L2 with the TargeTron

1. Vero cells.
2. HeLa cells.
3. *Chlamydia trachomatis* L2.
4. TargeTron plasmid DNA.

5. Roswell Park Memorial Institute medium (RPMI) with L-glutamine.
6. Fetal bovine serum (FBS).
7. Gentamicin.
8. Penicillin G (pACT, pDFTT3, pBomb4) or Spectinomycin (pDFTT3-*aadA*).
9. Cycloheximide.
10. 0.5% agarose in sterile water.
11. Sucrose phosphate buffered glutamic acid (SPG): 75 g sucrose, 2.47 g Na_2HPO_4, 0.36 g NaH_2PO_4, up to 1 L water, pH to 7.4. Filter sterilize and store at 4 °C.
12. 1× Hank's balanced salt solution (HBSS).
13. 6-well plates.
14. T150 flasks.
15. T25 flasks.
16. Cell lifters.
17. Large and small cell scrapers.
18. 15 mL conical tube.
19. O-ring screw-capped 1.5 mL tubes.
20. Centrifuge with swinging bucket rotor for tissue culture plates and 15 mL conical tube.
21. Microcentrifuge.
22. Humidified CO_2 cell culture incubator set to 37 °C.

2.4 Clonal Expansion and Mutant Verification

1. Sterile water.
2. 15 mL conical tube.
3. Sucrose phosphate buffered glutamic acid (SPG): 75 g sucrose, 2.47 g Na_2HPO_4, 0.36 g NaH_2PO_4, up to 1 L water, pH to 7.4. Filter-sterilize and store at 4 °C.
4. 1× Hank's balanced salt solution (HBSS)
5. Vero cells.
6. HeLa cells.
7. Roswell Park Memorial Institute medium (RPMI) with L-glutamine.
8. Fetal bovine serum (FBS).
9. Gentamicin.
10. Penicillin G (pACT or pDFTT3) or Spectinomycin (pDFTT3-*aadA*).
11. Cycloheximide.
12. 0.5% agarose in sterile water.

13. 6-well plates.
14. T150 flasks.
15. Cell lifters.
16. Large cell scrapers.
17. O-ring screw-capped 1.5 mL tubes.
18. Centrifuge with swinging bucket rotor for tissue culture plates and 15 mL conical tube.
19. Microcentrifuge.
20. Humidified CO_2 cell culture incubator set to 37 °C.

2.5 Generation of Complementation Construct

1. Primers designed to express your gene of interest. Resuspend primers to 100 μM using molecular biology grade water.
2. Qiagen PCR core kit (Qiagen).
3. *C. trachomatis* L2 DNA.
4. Chemically competent *Escherichia coli* K12 ER2925.
5. Chlamydia complementation plasmid (Fig. 5): pBomb4 [9] or pBomb3 [13].
6. Primers for sequencing: pBomb R sequencing primer (5-′-GCAAAAACAGGAAGGCAAAATGCCGC-3′).
7. 50× TAE: 242 g Tris, 57.1 mL glacial acetic acid, 18.6 g EDTA, and ddH_2O up to 1 L.
8. Gel extraction kit.
9. PCR purification kit.
10. Plasmid miniprep kit.
11. 1% agarose gel (1 g agarose in 100 mL 1× TAE).
12. Ethidium bromide.
13. NotI and SalI or SacII and SalI restriction enzymes with 10× buffer.
14. T4 ligase and buffer.
15. LB broth.
16. LB agar plates with 100 μg/mL carbenicillin.
17. Sterile molecular biology grade water.
18. Sterile 0.2 mL PCR tubes.
19. Sterile 1.5 mL Eppendorf tubes.
20. Thermocycler.
21. DNA electrophoresis unit.
22. UV transilluminator.
23. NanoDrop.
24. 37 °C water bath.

25. 37 °C incubator.
26. Razor blades.

2.6 Complementation of TargeTron Mutants

1. HeLa cells.
2. Vero cells.
3. Purified TargeTron mutant EBs.
4. Roswell Park Memorial Institute medium (RPMI) with L-glutamine.
5. Fetal bovine serum (FBS).
6. Gentamicin.
7. Penicillin G or Chloramphenicol.
8. Cycloheximide.
9. 0.5% agarose in sterile water
10. Sterile water.
11. 6-well plates
12. T150 flasks.
13. Cell lifters.
14. Large and small cell scrapers.
15. 15 mL conical tube.
16. O-ring screw-capped 1.5 mL tubes.
17. Centrifuge with swinging bucket rotor for tissue culture plates and 15 mL conical tube.
18. Microcentrifuge.
19. Humidified CO_2 cell culture incubator set to 37 °C.

3 Methods

3.1 Intron Template Synthesis (See Note 1)

1. Setup a 50 μL PCR reaction using 2.5 μL LtrB-IBS (10 μM), 2.5 μL EBS universal (10 μM), 20 ng TargeTron vector, 5 μL Qiagen buffer, 1 μL dNTPs, up to 49.5 μL water, and 0.5 μL Taq polymerase.
2. Place samples in thermocycler: (1) initial denaturation at 98 °C for 30 s, (2) 25 cycles of denaturation at 98 °C for 10 s, annealing at 60 °C for 30 s, extension at 72 °C for 30 s and (3) a final extension at 72 °C for 10 min.
3. Pour a 1% agarose gel and allow the gel to solidify (~30 min). We add ethidium bromide to the liquid agarose prior to solidification. Mix 50 μL of PCR product with 10 μL of loading dye and load entire sample into gel. Run gel at 100 V for 45 min. Using a razor blade, cut out 216 bp PCR product and purify using a gel extraction kit following the manufacturer's

instructions. Elute the sample in 50 μL water. Read sample using NanoDrop to determine DNA concentration and dilute the PCR product to 20 ng/μL.

4. Prepare T+B primer mix by combining 22 μL water, 2 μL 10 μM LtrB-EBS-T primer (10 μM), and 2 μL 10 μM LtrB-EBS-B primer (10 μM).
5. To prepare intron template, mix PCR product (**step 3**) and T +B primer mix (**step 4**) 1:1.

3.2 Retargeting Intron to Target Gene

1. To retarget the intron to your gene of interest, design primers using TargeTronics, LLC (http://targetrons.com/targetron_pLtrB.php) or Sigma (http://www.sigmaaldrich.com/life-science/functional-genomics-and-rnai/targetron.html) (Fig. 3). Select two different insertion sites for your gene of interest (*see* **Note 2**) and order IBS, EBS1, EBS2, and EBS universal primers.
   (a) The IBS, EBS1, and EBS2 primer sequences will vary based on your gene of interest.
2. Resuspend primers to 100 μM using molecular biology grade water.
3. Prepare four primer mix using 2 μL IBS (100 μM), 2 μL EBS1 (100 μM), 2 μL EBS2 (20 μM), 2 μL EBS Universal (20 μM), and 12 μL molecular biology grade water.
4. Setup a 50 μL PCR reaction using 1 μL of the four-primer mix (**step 3**), 1 μL intron template (synthesized in Subheading 3.1 or purchased from Sigma), 5 μL Qiagen buffer, 1 μL dNTPs, 41.5 μL water, and 0.5 μL Taq polymerase (*see* **Note 3**).
5. Placc samples in Thermocycler: (1) initial denaturation at 98 °C for 30 s, (2) 25 cycles of denaturation at 98 °C for 10 s, annealing at 60 °C for 30 s, extension at 72 °C for 30 s, and (3) a final extension at 72 °C for 10 min.
6. Run 5 μL of PCR product on a 1% agarose gel and visualize using ethidium bromide staining and UV transillumination. The desired product should be ~350 bp (*see* **Note 4**).
7. Purify remaining DNA using a PCR purification kit following the manufacturer's instructions. Elute the DNA in 50 μL water.
8. Digest the entire PCR reaction by adding 6 μL 10× buffer, 1 μL BsrGI, and 1 μL HindIII to the purified DNA. Incubate the reaction at 37 °C for 1 h. Clean digested PCR product using a PCR purification kit following the manufacturer's instructions. Elute the DNA in 50 μL water.
9. Digest 5 μg TargeTron vector (pACT, pDFTT3, or pDFTT3-*aadA*) using 5 μL BsrGI, 5 μL HindIII, 6 μL 10× buffer, and up to 60 μL water at 37 °C for 2 h. Run entire reaction on a 1%

agarose gel at 100 V for 60 min. Gel purify the digested vector using a gel extraction kit. The linearized vector should be ~7.5 kb. Elute the sample in 50 μL water.

10. Ligate digested PCR product into linearized TargeTron vector using 2 μL vector (**step 9**), 15 μL digested PCR product (**step 8**), 2 μL ligase buffer, and 1 μL T4 ligase. Incubate at room temperature for 1 h or overnight at 16 °C.
11. Transform the entire ligation reaction into chemically competent *E. coli* K12 ER2925 (*see* **Note 5**). Plate entire transformation on LB agar plates with appropriate antibiotic. Incubate plates overnight at 37 °C.
12. Pick 5 colonies per transformation and isolate plasmid using a plasmid miniprep kit. Sequence plasmids using pACT seq F (pACT) or T7 promoter seq (pDFTT3 or pDFTT3-*aadA*). Analyze the sequences by aligning the IBS, EBS1, and EBS2 primers with the plasmid sequence. Sequences must align perfectly minus the nucleotide in the IBS that correspond to the restriction site.

3.3 Transformation of *Chlamydia trachomatis* L2 with the TargeTron

1. Seed ~8.0×10^5 HeLa cells into a single well of a 6-well plate (*see* **Note 6**). Use RPMI with 10% FBS and gentamicin. Incubate cells at 37 °C with 5% CO_2 for 24 h. Cells should be confluent.
2. Thaw 2.5×10^7 EBs. Add bacteria to a 15 mL conical tube containing 50 μL transformation mix (10 mM Tris pH 7.5, 50 mM $CaCl_2$ filter sterilized) and 3 μg TargeTron plasmid DNA. Mix gently by flicking the tube twice and incubate at room temperature for 30 min.
3. After 30 min, remove the media from the 6-well plate. Add 2 mL RPMI with 10% FBS and gentamicin to the 15 mL conical tube containing the transformation mix. Mix gently by pipetting 2–3 times and add the entire reaction to the HeLa monolayer in the 6-well plate.
4. Centrifuge for 30 min at $900 \times g$. Incubate at 37 °C with 5% CO_2 for 18 h.
5. At 18 h postinfection, remove the media from the 6-well plate and replace with 2 mL RPMI containing 0.1 U/mL Penicillin G (pACT or pDFTT3) or 200 μg/mL spectinomycin (pDFTT3-*aadA*) (*see* **Note 7**). Incubate transformants for an additional 24 h (this represents P0).
6. Harvesting and passing of transformants:
   (a) P1—At 40–48 h postinfection, remove media from the 6-well plate and wash the well with 1 mL sterile water, remove wash. Add 1 mL sterile water and scrape using a sterile cell lifter. Transfer to a 15 mL conical tube and

incubate at room temperature for 10 min. After 10 min incubation, vortex on highest setting for 10 s. Pellet debris at 1500 × *g* for 3 min. During spin remove media from HeLa monolayer in a T150 flask and replace with 30 mL RPMI with 10%FBS, 10 μg/mL cycloheximide, gentamicin, and 0.1 U/mL PenG (pACT or pDFTT3) or 200 μg/mL spectinomycin (pDFTT3-*aadA*). Add 400 μL of supernatant to the HeLa monolayer and incubate at 37 °C with 5% CO_2.

(b) P2—At 40–48 h postinfection, remove media from the flask and wash with 5 mL sterile water, remove wash. Add 1 mL sterile water and scrape using a sterile cell scraper. Transfer to a 15 mL conical tube and incubate at room temperature for 10 min. After 10 min incubation, vortex on highest setting for 10 s. Centrifuge at 1500 × *g* for 3 min. During spin remove media from HeLa monolayer in a T25 flask and replace with 10 mL RPMI with 10% FBS, 10 μg/mL cycloheximide, gentamicin, and 0.2 U/mL PenG (pACT or pDFTT3) or 400 μg/mL spectinomycin (pDFTT3-*aadA*). Add entire supernatant (being careful not to disturb the pellet) to the HeLa monolayer and incubate at 37 °C with 5% CO_2.

(c) P3 (*see* **Note 8**)—At 66–72 h postinfection, remove media from the flask and wash with 1 mL sterile water, remove wash. Add 1 mL sterile water and scrape using a sterile cell scraper. Transfer to a 15 mL conical tube and incubate at room temperature for 10 min. After 10 min incubation, vortex on highest setting for 10 s. Centrifuge at 1500 × *g* for 3 min. During spin remove media from HeLa monolayer in a T25 flask and replace with 10 mL RPMI with 10%FBS, 10 μg/mL cycloheximide, gentamicin, and 0.2 U/mL PenG (pACT or pDFTT3) or 400 μg/mL spectinomycin (pDFTT3-*aadA*). Add entire supernatant (being careful not to disturb the pellet) to the HeLa monolayer and incubate at 37 °C with 5% CO_2.

(d) P4—At 40–48 h postinfection, remove media from the flask and wash with 1 mL sterile water, remove wash. Add 1 mL sterile water and scrape using a sterile cell scraper. Transfer to a 15 mL conical tube and incubate at room temperature for 10 min. After 10 min incubation, vortex on highest setting for 10 s. Centrifuge at 1500 × *g* for 3 min. During spin remove media from HeLa monolayer in a T25 flask and replace with 10 mL RPMI with 10% FBS, 10 μg/mL cycloheximide, gentamicin, and 0.2 U/mL PenG (pACT or pDFTT3) or 200 μg/mL spectinomycin (pDFTT3-*aadA*). Add supernatant to the HeLa monolayer and incubate at 37 °C with 5% CO_2.

7. If after four passages no transformants are visible, the transformation was unsuccessful and should be terminated.

3.4 Clonal Expansion and Mutant Verification

1. From the P4 passage, remove media, and add 1 mL sterile water to wash. Remove wash, add 1 mL sterile water, and scrape using a sterile cell scraper. Transfer to a 15 mL conical tube and incubate at room temperature for 10 min. After 10 min incubation, vortex on highest setting for 10 s. Centrifuge at 1500 × *g* for 3 min. Transfer supernatant to an O-ring screw-top microcentrifuge tube and pellet bacteria at 12,000 × *g* for 20 min. Discard supernatant and resuspend pellet in 500 μL SPG and store at −80 °C in 100 μL aliquots.
2. Use one aliquot to verify mutant insertion by PCR and sequencing (*see* **Note 9**).
3. To plaque clone the mutant, add 1 μL of mutant frozen stock to 1 mL 1× HBSS. Vortex to mix and serially dilute 1:10 (6 dilutions total). Remove media from Vero cell monolayer in a 6-well plate (*see* **Note 10**) and add 1 mL of the individual dilutions to each well. Incubate at 37 °C with gentle rocking for 30 min--1 h. Remove inoculum and prepare agarose for overlay by adding 90 mL prewarmed (37 °C) RPMI with 10% FBS, 10 μg/mL cycloheximide, gentamicin, and 1 U/mL PenG (pACT or pDFTT3) or 500 μg/mL spectinomycin (pDFTT3-*aadA*) to a bottle containing 10 mL 0.5% agarose prewarmed to 56 °C (*see* **Note 11**). Mix by pipetting and add 5 mL per well. Let plate sit at room temperature for 30 min–1 h or until agarose has solidified. Incubate at 37 °C with 5% CO_2 for 7–10 days or until plaques appear.
4. Pick 3 plaques using a p200 tip. Inoculate onto a HeLa monolayer in a 6-well plate containing 2 mL RPMI with 10%FBS, 10 μg/mL cycloheximide, gentamicin, and 1 U/mL PenG (pACT or pDFTT3) or 500 μg/mL spectinomycin (pDFTT3-*aadA*). Incubate at 37 °C with 5% CO_2 for 48–72 h or until cells are heavily infected.
5. Expand one well of the 6-well plate by removing the media and washing the well with 1 mL sterile water, remove wash. Add 1 mL sterile water and scrape using a sterile cell lifter. Transfer to a 15 mL conical tube and incubate at room temperature for 10 min. After 10 min incubation, vortex on highest setting for 10 s. Pellet debris at 1500 × *g* for 3 min. During spin remove media from HeLa monolayer in a T150 flask and replace with 30 mL RPMI with 10%FBS, 10 μg/mL cycloheximide, gentamicin, and 1 U/mL PenG (pACT or pDFTT3) or 500 μg/mL spectinomycin (pDFTT3-*aadA*). Add entire supernatant to the HeLa monolayer and incubate at 37 °C with 5% CO_2.

6. At 40–48 h post-infection, remove media from the flask and wash with 5 mL sterile water, remove wash. Add 1 mL sterile water and scrape using a sterile cell scraper. Transfer to a 15 mL conical tube and incubate at room temperature for 10 min. After 10 min incubation, vortex on highest setting for 10 s. Centrifuge at 1500 × *g* for 3 min. Transfer supernatant to an O-ring screw-top microcentrifuge tube and pellet bacteria at 12,000 × *g* for 20 min. Discard supernatant and resuspend pellet in 1 mL SPG and store at −80 °C in 200 μL aliquots.
7. Use one aliquot to verify mutant insertion and purity by PCR (*see* **Note 12**).

3.5 Generation of Complementation Construct

1. Design primers to express your gene of interest (*see* **Note 13**). Resuspend primers to 100 μM using molecular biology grade water.
2. Setup a 50 μL PCR reaction using 2.5 μL (10 μM) F, 2.5 μL (10 μM) R, 1 μL (~20 ng) *C. trachomatis* L2 DNA, 5 μL Qiagen buffer, 1 μL dNTPs, up to 50 μL water, and 0.5 μL Taq polymerase.
3. Place samples in thermocycler: (1) initial denaturation at 98 °C for 30 s, (2) 25 cycles of denaturation at 98 °C for 10 s, annealing at 55 °C for 30 s, extension at 72 °C for 1 min/kb and (3) a final extension at 72 °C for 10 min.
4. Run 5 μL of PCR product on a 1% agarose gel and visualize using ethidium bromide staining and UV transillumination.
5. Purify remaining DNA using a PCR purification kit following the manufacturer's instructions. Elute the DNA in 50 μL water.
6. Digest the entire PCR reaction by adding 6 μL 10× buffer and 1 μL of each restriction enzyme. Incubate the reaction at 37 °C for 1 h. Clean digested PCR product using a PCR purification kit following the manufacturer's instructions. Elute the DNA in 50 μL water.
7. Digest 5 μg vector (pBomb4 or pBomb3) using 5 μL of each restriction enzyme, 6 μL 10× buffer, and up to 60 μL water. Incubate at 37 °C for 2 h. Run entire reaction on a 1% agarose gel at 100 V for 60 min. Gel purify the digested vector using a gel extraction kit (*see* **Note 14**).
8. Ligate digested PCR product into linearized vector using 2 μL vector (step 7), 6 μL digested PCR product (**step 6**), 2 μL ligase buffer, and 1 μL T4 ligase. Incubate at room temperature for 1 h or overnight at 16 °C.
9. Transform the entire ligation reaction into chemically competent *E. coli* K12 ER2925. Plate entire transformation on LB with appropriate antibiotic. Incubate plates overnight at 37 °C.

10. Pick 5 colonies per transformation and isolate plasmid using a plasmid miniprep kit. Sequence plasmids using pBomb R primers.

3.6 Complementation of TargeTron Mutants

1. Seed ~8.0×10^5 HeLa cells into a single well of a 6-well plate. Use RPMI with 10% FBS and gentamicin. Incubate cells at 37 °C with 5% CO_2 for 24 h. Cells should be confluent.
2. Thaw 2.5×10^7 purified TargeTron mutant EBs. Add bacteria to a 15 mL conical tube containing 50 μL transformation mix (10 mM Tris pH 7.5, 50 mM $CaCl_2$ filter sterilized) and 3 μg complementation plasmid DNA. Mix gently by flicking the tube twice and incubate at room temperature for 30 min.
3. After 30 min, remove the media from the 6-well plate. Add 2 mL RPMI with 10% FBS and gentamicin to the 15 mL conical tube containing the transformation mix. Mix gently by pipetting 2–3 times and add the entire reaction to the HeLa monolayer in the 6-well plate.
4. Centrifuge for 30 min at 900 × *g*. Incubate at 37 °C with 5% CO_2 for 18 h.
5. At 18 h postinfection, remove the media from the 6-well plate and replace with 2 mL RPMI containing 0.1 U/mL Penicillin G (pBomb4) or 0.2 μg/mL chloramphenicol (pBomb3). Incubate transformants for an additional 24 h (this represents P0).
6. Harvesting and passing of transformants (*see* **Note 15**):
   (a) P1—At 40–48 h post-infection, remove media from the 6-well plate and wash the well with 1 mL sterile water, remove wash. Add 1 mL sterile water and scrape using a sterile cell lifter. Transfer to a 15 mL conical tube and incubate at room temperature for 10 min. After 10 min incubation, vortex on highest setting for 10 s. Pellet debris at 1500 × *g* for 3 min. During the spin, remove media from HeLa monolayer in a T150 flask and replace with 30 mL RPMI with 10%FBS, 10 μg/mL cycloheximide, gentamicin, and 0.2 μg/mL chloramphenicol. Add 400 μL of supernatant to the HeLa monolayer and incubate at 37 °C with 5% CO_2.
   (b) P2—At 40–48 h post-infection, remove media from the flask and wash with 5 mL sterile water, remove wash. Add 1 mL sterile water and scrape using a sterile cell scraper. Transfer to a 15 mL conical tube and incubate at room temperature for 10 min. After 10 min incubation, vortex on highest setting for 10 s. Centrifuge at 1500 × *g* for 3 min. During spin remove media from HeLa monolayers grown in 2 T150 flasks and replace with 30 mL RPMI with 10%FBS, 10 μg/mL cycloheximide, gentamicin, and

0.3 μg/mL chloramphenicol. Add 500 μL supernatant to each flask and incubate at 37 °C with 5% CO_2.

(c) P3—At 66–72 h post-infection, remove media from the flask and wash each flask with 5 mL sterile water, remove wash. Add 1 mL sterile water per flask and scrape using a sterile cell scraper. Transfer to a 15 mL conical tube, combining samples from the two flasks, and incubate at room temperature for 10 min. After 10 min incubation, vortex on highest setting for 10 s. Centrifuge at 1500 × *g* for 3 min. During spin remove media from HeLa monolayer in 2 T150 flasks and replace with 30 mL RPMI with 10%FBS, 10 μg/mL cycloheximide, gentamicin, and 0.4 μg/mL chloramphenicol. Add 1 mL of supernatant to each flask and incubate at 37 °C with 5% CO_2.

(d) P4—At 40–48 h post-infection, remove media from flasks and wash each flask with 5 mL sterile water, remove wash. Add 1 mL sterile water per flask and scrape using a sterile cell scraper. Transfer to a 15 mL conical tube and incubate at room temperature for 10 min. After 10 min incubation, vortex on highest setting for 10 s. Centrifuge at 1500 × *g* for 3 min. During spin remove media from HeLa monolayer in a T150 flask and replace with 30 mL RPMI with 10%FBS, 10 μg/mL cycloheximide, gentamicin, and 0.4 μg/mL chloramphenicol. Add 1 mL supernatant to the HeLa monolayer and incubate at 37 °C with 5% CO_2 (*see* **Note 16**).

7. If after five passages no transformants are visible, the transformation was unsuccessful and should be terminated. If transformants are obtained, expression of the fusion protein should be verified by western blotting or immunofluorescence microscopy.

4 Notes

1. Subheading 3.1 details how to synthesize the Intron template. The intron template is included with the TargeTron Gene Knockout System kit (Sigma) and can be used in lieu of synthesizing your own. If the intron template is purchased, proceed directly to Subheading 3.2.

2. To design primers to retarget the intron to the gene of interest, the DNA sequence is submitted to the Sigma-Aldrich or TargeTronics algorithm, which will return primer sequences to retarget the intron to a desired gene (Fig. 4). We select primers that will result in an insertion as close as possible to the 5′ end since truncated protein products could still be produced. The

algorithm will return a list of predicted insertion sites with an insertion efficiency score, those with a high score should be used. Additionally, the intron can insert in either a sense (s) or antisense (a) configuration. While either can be selected, we have had the greatest success in obtaining mutants when the sense configuration is used. Thus, we select primer sets that will result in an insertion in the sense orientation near the beginning of the gene and have a high insertion score. We select two primer sets for each target gene because while it is predicted to insert in that site, we have had instances where we were unable to obtain a mutant in the first selected site but do obtain a mutant when it is targeted to another area of the gene. Therefore, the use of a second site increases the likelihood of obtaining a mutant. If no prediction sites are returned, an alternative mutagenesis strategy [15] must be used.

3. Other high-fidelity polymerases such as the Jumpstart RED Taq ReadyMix or Phusion high fidelity polymerase can be substituted for the Qiagen PCR core kit.
4. Three bands should be visible. The uppermost band (~350 bp) should be the most prominent and is the desired band. Two other bands (~250 bp and ~100 bp) may be visible, however they should not adversely affect the cloning reaction and therefore do not need to be removed using gel extraction.
5. Chemically competent *E. coli* can be substituted for electrocompetent *E. coli*. Other strains of *E. coli* such as DH5α or XL-1 Blue can be substituted for cloning; however, unmethylated plasmid DNA for transformation of *C. trachomatis* must be prepared from a *dam*$^-$ *E. coli* strain. Additionally, the *E. coli* K12ER2925 from NEB is resistant to spectinomycin and chloramphenicol and therefore cannot be used for cloning with pDFTT3-*aadA*.
6. Our laboratory uses HeLa cells for all transformations; however, other cell lines such as McCoy or Vero cells can be used.
7. We have found that starting with a lower antibiotic concentration (0.1 U PenG or 200 μg/mL spectinomycin) and steadily increasing the concentration over time increases the likelihood of obtaining a mutant. However, other research groups use a single, high concentration (1 U PenG or 500 μg/mL spectinomycin) throughout the transformation.
8. Depending on the transformation efficiency it may not be necessary to do P4. If many inclusions are apparent, we proceed to mutant verification and plaque cloning. If few inclusions are present, we do a fourth passage.
9. To isolate chlamydia genomic DNA, we use the Qiagen DNeasy blood and tissue kit following the cultured cell protocol. We use DNA isolated from wild-type *C. trachomatis* L2

and the TargeTron mutant as template for PCR using gene specific primers. Successful insertion of the TargeTron is evident by a ~1.8 kb shift in DNA isolated from the mutant relative to wild-type L2. We further confirm insertion by DNA sequencing.

10. We use Vero cells for plaque cloning as plaques are more readily visible. However, HeLa cells can still be used.
11. Agarose concentration can be increased to 1% if need be.
12. Following plaque cloning of the TargeTron mutant, we verify purity of the mutant by PCR. If wild-type contamination is present, indicated by two PCR bands (1.8 kb shifted mutant and normal sized wild-type) the mutant will need to be plaque cloned for a second time.
13. Both the pBomb4 and pBomb3 plasmids can be used to complement TargeTron mutants. The pBomb4 plasmid (use with mutants generated using pDFTT-*aadA*) possesses a Tet inducible promoter and transformants are selected for using PenG whereas the pBomb3 plasmid (use with mutants generated using pACT or pDFTT3) lacks a promoter and transformants are selected with chloramphenicol. For expression in pBomb4, we place our gene under the control of the Tet inducible promoter using the NotI/SalI site. For pBomb3, we place our gene under the control of the native promoter and clone our gene in using the SacII/SalI site. Alternatively, the gene of interest can be fused to the Tet promoter using overlap extension PCR and cloned into the pBomb3 plasmid. When designing primers, we add a Flag tag to the reverse primer for downstream visualization by immunofluorescence microscopy or western blotting.
14. When cloning inserts into the pBomb4 plasmid, we use the NotI/SalI site, which results in removal of mCherry. Therefore, during the gel purification step two bands will be visible, mCherry at ~750 bp and ~11 kb linearized vector.
15. TargeTron mutants can be transformed with pBomb4 plasmid as described in Subheading 3.3.
16. If numerous inclusions are present, this can be scaled down to 100 μL of lysate and the rest can be stored at −80 °C in SPG.

Acknowledgments

This work was supported by startup funds from the University of Iowa Carver College of Medicine Department of Microbiology and Immunology to M.M.W. We thank Shelby Andersen, Annie Holtz, and Stephanie Peterson for critical review of the manuscript.

References

1. Schachter J (1999) Infection and disease epidemiology, in chlamydia; intracellular biology, pathogenesis, and immunity. ASM Press, Washington, DC
2. Malhotra M, Sood S, Mukherjee A et al (2013) Genital *Chlamydia trachomatis*: an update. Indian J Med Res 138:303–316
3. Hu VH, Holland MJ, Burton MJ (2013) Trachoma: protective and pathogenic ocular immune responses to *Chlamydia trachomatis*. PLoS Negl Trop Dis 7(2):e2020
4. Wang Y, Kahane S, Cutcliffe LT et al (2011) Development of a transformation system for *Chlamydia trachomatis*: restoration of glycogen biosynthesis by acquisition of a plasmid shuttle vector. PLoS Pathog 7(9):e1002258
5. Shima K, Wanker M, Skilton RJ et al (2018) The genetic transformation of *Chlamydia pneumoniae*. mSphere 3:1–16
6. Kari L, Goheen MM, Randall LB et al (2011) Generation of targeted *Chlamydia trachomatis* null mutants. Proc Natl Acad Sci 108:7189–7193
7. Kokes M, Dunn JD, Granek JA et al (2015) Integrating chemical mutagenesis and whole-genome sequencing as a platform for forward and reverse genetic analysis of Chlamydia. Cell Host Microbe 17:716–725
8. Weber MM, Bauler LD, Lam J, Hackstadt T (2015) Expression and localization of predicted inclusion membrane proteins in *Chlamydia trachomatis*. Infect Immun 83:4710–4718
9. Bauler LD, Hackstadt T (2014) Expression and targeting of secreted proteins from *Chlamydia trachomatis*. J Bacteriol 196:1325–1334
10. Agaisse H, Derré I (2013) A *C. trachomatis* cloning vector and the generation of *C. trachomatis* strains expressing fluorescent proteins under the control of a *C. trachomatis* promoter. PLoS One 8(2):e57090
11. Johnson CM, Fisher DJ (2013) Site-specific, insertional inactivation of *incA* in *Chlamydia trachomatis* using a group II intron. PLoS One 8(12):e83989
12. Lowden NM, Yeruva L, Johnson CM et al (2015) Use of aminoglycoside 3′ adenyltransferase as a selection marker for *Chlamydia trachomatis* intron-mutagenesis and in vivo intron stability. BMC Res Notes 8:1–10
13. Weber MM, Noriea NF, Bauler LD et al (2016) A functional core of IncA is required for *Chlamydia trachomatis* inclusion fusion. J Bacteriol 198:1347–1355
14. Weber MM, Lam JL, Dooley CA et al (2017) Absence of specific *Chlamydia trachomatis* inclusion membrane proteins triggers premature inclusion membrane lysis and host cell death. Cell Rep 19:1406–1417
15. Mueller KE, Wolf K, Fields KA (2016) Gene deletion by fluorescence-reported allelic exchange mutagenesis in *Chlamydia trachomatis*. MBio 7:1–9
16. Keb G, Hayman R, Fields KA (2018) Floxed-cassette allelic exchange mutagenesis enables markerless gene deletion in *Chlamydia trachomatis* and can reverse cassette-induced polar effects. J Bacteriol 200(24):e00479–e00418
17. Enyeart PJ, Mohr G, Ellington AD, Lambowitz AM (2014) Biotechnological applications of mobile group II introns and their reverse transcriptases: gene targeting, RNA-seq, and non-coding RNA analysis. Mob DNA 5:1–19
18. Yao J, Lambowitz AM (2007) Gene targeting in gram-negative bacteria by use of a mobile group II intron ("Targetron") expressed from a broad-host-range vector. Appl Environ Microbiol 73:2735–2743
19. Zhong J, Karberg M, Lambowitz AM (2003) Targeted and random bacterial gene disruption using a group II intron (targetron) vector containing a retrotransposition-activated selectable marker. Nucleic Acids Res 31:1656–1664
20. Rodriguez SA, Yu J, Davis G et al (2008) Targeted inactivation of *Francisella tularensis* genes by group II introns. Appl Environ Microbiol 74:2619–2626
21. Heap JT, Pennington OJ, Cartman ST et al (2007) The ClosTron: a universal gene knock-out system for the genus Clostridium. J Microbiol Methods 70:452–464
22. Noriea NF, Clark TR, Hackstadt T (2015) Targeted knockout of the *Rickettsia rickettsii* OmpA surface antigen does not diminish virulence in a mammalian model system. MBio 6:1–9
23. Cheng C, Nair ADS, Indukuri VV et al (2013) Targeted and random mutagenesis of *Ehrlichia chaffeensis* for the identification of genes required for in vivo infection. PLoS Pathog 9 (2):e1003171

Chapter 13

Forward and Reverse Genetic Analysis of *Chlamydia*

Mateusz Kędzior and Robert J. Bastidas

Abstract

Chlamydia is a major etiological agent of human disease that affects millions of individuals worldwide. Historically, our understanding of the mechanisms that contribute to its pathogenesis has been limited. However, the recent development of powerful genetic tools for manipulating *Chlamydia* has resulted in significant gains in our ability to dissect its virulence mechanisms. These tools have overcome several barriers for manipulating intracellular pathogens and are amenable for the routine genetic engineering of *Chlamydia*. Here, we provide several detailed protocols for performing genetic analysis in *Chlamydia trachomatis* allowing investigators to elucidate how this obligate intracellular pathogen causes disease.

Key words *Chlamydia* genetics, *Chlamydia* chemical mutagenesis, *Chlamydia* DNA isolation, *Chlamydia* linkage analysis, *Chlamydia* plaque assay, *Chlamydia* plasmid transformation

1 Introduction

Chlamydia is an obligate intracellular pathogen that survives and thrives within human cells and is a major cause of sexually transmitted human disease [1–3]. Its life cycle alternates between an infectious elementary body (EB) and a vegetative, noninfectious reticulate body (RB), which replicates within the confines of a membrane bound pathogenic vacuole called the inclusion [4]. Understanding the mechanisms by which *Chlamydia* infections lead to pathogenesis is a topic of active investigation, and in the last few years, several factor important for its virulence and intracellular parasitism have been identified in part due to the development of genetic based approaches tailored for use in *Chlamydia*.

Several groups have harnessed chemical mutagenesis to generate diverse populations of *Chlamydia* mutant strains [5–9] that have been screened by TILLING [10] and Whole Genome Sequencing (WGS) based approaches. These screens have led to the identification of a *C. trachomatis* virulence factor (PmpD), a transcription regulator (ChxR) of virulence-associated genes, and a

Amanda Claire Brown (ed.), *Chlamydia trachomatis: Methods and Protocols*, Methods in Molecular Biology, vol. 2042, https://doi.org/10.1007/978-1-4939-9694-0_13,

metabolic factor (TrpB) required by *C. trachomatis* for avoiding IFN-γ-induced tryptophan starvation [5, 11, 12]. In addition, over 8000 single nucleotide variants, including nonsense inducing mutations in 80 *C. trachomatis* L2 ORFs, have been identified and mapped [7]. Forward genetic screens have also led to the discovery of *C. trachomatis* gene products required for *Chlamydia*'s intracellular growth (GlgB and GspE) [6], survival (CpoS) [13], and maintenance of its intracellular niche (CPAF, InaC) [7, 14]. Likewise, temperature sensitive conditional mutations [8] and *C. trachomatis* genes required for its survival during IFN-γ-driven persistence [15] have been discovered from forward genetic screens.

Tools for targeted genetic analysis have also been developed and are particularly useful for validating candidate loci identified from genetic screens. These include the use of a *C. trachomatis* adapted type II intron (TargeTron®, Sigma Inc) for targeted insertional mutagenesis [16, 17] and targeted gene replacement by Fluorescence Reported Allelic Exchange Mutagenesis (FRAEM) [18]. An upgrade to the FRAEM approach with a Cre-*loxP* system now allows for recycling of selectable markers during mutagenesis and disruption of individual genes within operons without incurring polar effects on transcription of downstream genes [19].

The development of tools for genetic analysis of *Chlamydia* now enables investigators to systematically map virulence factors in any *Chlamydia* species without major investments in time and resources. The genomes of *Chlamydia* species are relatively small (~1 MB) and mutations introduced by chemical mutagens into their genomes can be readily mapped with Next Generation Sequencing technologies. *Chlamydia* genetic exchange has been demonstrated in a coinfection setting with two different antibiotic-resistant strains giving rise to dual resistant recombinant progeny [20, 21]. Thus, lateral gene transfer events between wild type and mutant strains allows strong associations to be made between segregating mutations in a gene or genetic pathway and a phenotype. Genetic associations can be verified by either TargeTron or FRAEM based targeted mutagenesis. Here, we describe a methodology for performing genetic analysis in *Chlamydia* by both forward and reverse genetic approaches. In the following sections, we provide detailed protocols for generating populations of chemically mutagenized *Chlamydia*, mapping and linkage analysis of chemically induced mutations in strains of interest, and protocols for targeted mutagenesis that can be used to validate genetic linkages.

Note: *Chlamydia trachomatis* is a Biosafety Level 2 (BSL2) designated organism, and handling should be performed according to BSL2 safety guidelines.

2 Materials

2.1 Chemical Mutagenesis

1. Vero cells (ATCC® CCL-81).
2. Cell culture flasks for routine growth of Vero cells, 25 cm^2 cell culture flasks, and 6-well cell culture plates.
3. *C. trachomatis* L2 434 Bu rifampin-resistant (Rif^R) strain [6].
4. High-glucose Dulbecco's Modified Eagle Medium (DMEM) with L-glutamine and sodium pyruvate (ThermoFisher Scientific Inc).
5. Fetal bovine serum (FBS) (ThermoFisher Scientific Inc).
6. Ethyl methanesulfonate (EMS).
7. 1 M NaOH.
8. Phosphate-buffered saline (PBS) supplemented with 0.9 mM calcium chloride and 0.49 mM magnesium chloride.
9. Cycloheximide (100 mg/mL in DMSO) (Sigma-Aldrich Inc.)
10. UltraPure™ DNase/RNase-Free Distilled Water (ThermoFisher Scientific Inc.)
11. Sucrose–phosphate–glutamic acid (SPG) buffer (5×): 1.25 M sucrose, 50 mM sodium phosphate, 25 mM L-glutamic acid, pH 7.4. Filter-sterilize. Store at 4 °C.
12. Sucrose–phosphate–glutamic acid (SPG) buffer (1×): Dilute 5× stock with distilled water to 1× concentration and filter-sterilize. Store at 4 °C.
13. DMEM supplemented with sodium bicarbonate. Dissolve 13.5 g of DMEM high glucose, pyruvate, L-glutamine powder (ThermoFisher Scientific Inc.) and 1.85 g of sodium bicarbonate in 500 mL of distilled water to obtain 2× DMEM supplemented with 44 mM sodium bicarbonate, filter-sterilize, and store at 4 °C.
14. MEM nonessential amino acids (100×) (Sigma-Aldrich Inc.)
15. Gentamicin solution (50 mg/mL) (ThermoFisher Scientific Inc.)
16. Ofloxacin (1 mg/mL). Dissolve in 1 N NaOH, aliquot into single use aliquots, and store at −20 °C.
17. SeaKem® LE agarose (1.5% w/v) (Lonza.) Dissolve 7.5 g of agarose in 500 mL of distilled water, autoclave, and store at 4 °C. Melt agarose in a microwave prior to use.

2.2 Isolation of Chlamydia Enriched DNA

1. Vero cells (ATCC® CCL-81).
2. Cell culture flasks for routine growth of Vero cells and 6-well culture plates.
3. Chemically mutagenized *Chlamydia* strains.

4. High-glucose Dulbecco's Modified Eagle Medium (DMEM) with L-glutamine and sodium pyruvate (ThermoFisher Scientific Inc.)
5. Fetal Bovine Serum (FBS) (ThermoFisher Scientific Inc.)
6. UltraPure™ DNase/RNase-Free Distilled Water (ThermoFisher Scientific Inc.)
7. Sucrose–phosphate–glutamic acid (SPG) buffer (5×): 1.25 M sucrose, 50 mM sodium phosphate, 25 mM L-glutamic acid, pH 7.4. Filter-sterilize. Store at 4 °C.
8. Sucrose–phosphate–glutamic acid (SPG) buffer (1×): Dilute 5× stock with distilled water to 1× concentration and filter-sterilize. Store at 4 °C.
9. DNase I (New England Biolabs.)
10. DNeasy Blood and Tissue Kit (Qiagen Inc.)
11. Qubit dsDNA BR Assay Kit (ThermoFisher Scientific Inc.)
12. Qubit fluorometer (ThermoFisher Scientific Inc.)

2.3 Mapping Single Nucleotide Variants by Whole Genome Sequencing

1. Adaptive Focused Acoustics S220 instrument (Covaris) or NEBNext dsDNA Fragmentase (New England Biolabs).
2. Library prep kit (e.g., NEBNext Ultra II DNA Library Prep Kit for Illumina and NEBNext Multiplex Oligos for Illumina, New England Biolabs) and sequencing instrument (e.g., MiSeq, Illumina Inc.).

2.4 Linkage Analysis

1. Vero cells (ATCC® CCL-81).
2. Cell culture flasks for routine growth of Vero cells and 24-well cell culture plates.
3. High-glucose Dulbecco's Modified Eagle Medium (DMEM) with L-glutamine and sodium pyruvate (ThermoFisher Scientific Inc.)
4. Fetal Bovine Serum (FBS) (ThermoFisher Scientific Inc.)
5. Titered *C. trachomatis* Rif^R mutant strain and a titered *C. trachomatis* $Spec^R$ strain [6].
6. UltraPure™ DNase/RNase-Free Distilled Water (ThermoFisher Scientific Inc.)
7. Sucrose–phosphate–glutamic acid (SPG) buffer (5×): 1.25 M sucrose, 50 mM sodium phosphate, 25 mM L-glutamic acid, pH 7.4. Filter-sterilize. Store at 4 °C.
8. Sucrose–phosphate–glutamic acid (SPG) buffer (1×): Dilute 5× stock with distilled water to 1× concentration and filter-sterilize. Store at 4 °C.

2.5 Clonal Isolation of Strains by Plaque Assay

1. Vero cells (ATCC® CCL-81).
2. Cell culture flasks for routine growth of Vero cells, 6-well, and 96-well cell culture plates.
3. High-glucose Dulbecco's Modified Eagle Medium (DMEM) with L-glutamine and sodium pyruvate (ThermoFisher Scientific Inc.)
4. Fetal Bovine Serum (FBS) (ThermoFisher Scientific Inc.)
5. Crude lysates from cells coinfected with *C. trachomatis* Rif^R mutant strain and a *C. trachomatis* $Spec^R$ strain [6] (from Subheading 3.4).
6. UltraPure™ DNase/RNase-Free Distilled Water (ThermoFisher Scientific Inc.)
7. DMEM supplemented with sodium bicarbonate. Dissolve 13.5 g of DMEM high glucose, pyruvate, L-glutamine powder (ThermoFisher Scientific Inc.) and 1.85 g of sodium bicarbonate in 500 mL of distilled water to obtain 2×DMEM supplemented with 44 mM sodium bicarbonate, filter-sterilize, and store at 4 °C.
8. MEM nonessential amino acids (100×) (Sigma-Aldrich Inc.)
9. Gentamicin solution (50 mg/mL) (ThermoFisher Scientific Inc.)
10. Rifampin (25 mg/mL). Dissolve in DMSO, aliquot into single-use aliquots, and store at −20 °C protected from light.
11. Spectinomycin (50 mg/mL). Dissolve in distilled water, filter-sterilize, aliquot into single use aliquots and store at −20 °C.
12. SeaKem® LE agarose (1.5% w/v) (Lonza.) Dissolve 7.5 g of agarose in 500 mL of distilled water, autoclave, and store at 4 °C. Melt agarose in a microwave prior to use.
13. Sucrose–phosphate–glutamic acid (SPG) buffer (5×): 1.25 M sucrose, 50 mM sodium phosphate, 25 mM L-glutamic acid, pH 7.4. Filter-sterilize. Store at 4 °C.
14. Sucrose–phosphate–glutamic acid (SPG) buffer (1×): Dilute 5× stock with distilled water to 1× concentration and filter-sterilize. Store at 4 °C.

2.6 Genotyping Recombinant Progeny and Linkage Analysis

1. DNase-free distilled water.
2. Crude *C. trachomatis* lysates (from Subheading 3.5).
3. Primers designed to amplify each locus harboring a single nucleotide variant.
4. High-fidelity PCR amplification reagents and thermocycler.

2.7 Reverse Genetic Analysis

1. Primers required for retargeting the intron in pDFTT3-*aadA* for insertion at a specific genomic site (IBS, EBS1Δ, and EBS2

primers). Retargeting primers can be designed using software available at Targetronics, LLC (www.targetrons.com/targetron_pLtrB.php) (*see* **Note 17**).

2. *C. trachomatis* adapted TargeTron vector pDFTT3-*aadA* [17] (*see* **Note 18**).
3. EBS universal primer: 5′-AATTAGAAACTTGCGTTCAGTAAACACAACTTATAC-3′.
4. High-fidelity PCR amplification reagents and thermocycler.
5. QIAquick PCR Purification Kit (Qiagen Inc.)
6. *Hind*III and *BsrG*I, restriction enzymes.
7. DNA cloning reagents.
8. Qubit dsDNA BR Assay Kit (ThermoFisher Scientific Inc.)
9. Qubit fluorometer (ThermoFisher Scientific Inc.)
10. Chloramphenicol (25 mg/mL). Dissolve in ethanol and protect from light.
11. Titered *C. trachomatis* wild type strain.
12. Transformation buffer (5×): 250 mM $CaCl_2$, 50 mM Tris, pH 7.4. Filter-sterilize. Store at 4 °C.
13. UltraPure™ DNase/RNase-Free Distilled Water (ThermoFisher Scientific Inc.)
14. Vero cells (ATCC® CCL-81).
15. Cell culture flasks for routine growth of Vero cells and 6-well cell culture plates.
16. High-glucose Dulbecco's Modified Eagle Medium (DMEM) with L-glutamine and sodium pyruvate (ThermoFisher Scientific Inc.)
17. Fetal Bovine Serum (FBS) (ThermoFisher Scientific Inc.)
18. Spectinomycin (50 mg/mL). Dissolve in distilled water, filter-sterilize, aliquot into single use aliquots and store at −20 °C.
19. Sucrose–phosphate–glutamic acid (SPG) buffer (5×): 1.25 M sucrose, 50 mM sodium phosphate, 25 mM L-glutamic acid, pH 7.4. Filter-sterilize. Store at 4 °C.
20. Sucrose–phosphate–glutamic acid (SPG) buffer (1×): Dilute 5× stock with distilled water to 1× concentration and filter-sterilize. Store at 4 °C.

3 Methods

3.1 Chemical Mutagenesis

Currently, chemical mutagenesis remains as the only available tool for surveying *Chlamydia* gene function at a genome wide resolution due to the low transformation efficiency of *Chlamydia* with exogenous DNA. Chemical mutagens bypass this limitation and can yield large number of mutant strains containing a wide spectrum of mutations in their genomes. Mutagenized strains can be screened by defined assays for virulence traits either as pools or individually from arrays of individual clonal strains. The latter can be generated by plaque purification of clonal isolates our protocol described in Subheading 3.5. In this protocol we describe a procedure for generating pools of chemically mutagenized *Chlamydia* strains. This protocol is based on a previously established protocol [22].

Note: Ethyl methanesulfonate (EMS) is a powerful mutagen and carcinogen. Use appropriate personal protective equipment. EMS is a volatile liquid and should always be handled in a chemical safety hood. All EMS exposed waste (solid and liquid) should be treated with 1 M sodium hydroxide to neutralize EMS. Dispose of treated waste according to appropriate and approved procedures.

1. Seed Vero cells in DMEM supplemented with 10% FBS, at a density of 1×10^6 cells per well, in a 25 cm^2 flask, and culture overnight at 37 °C in a 5% CO_2 humidified incubator (*see* **Note 1**).
2. Use a titered stock of rifampin-resistant (Rif^R) *C. trachomatis* [6] and infect confluent Vero cell monolayers at a multiplicity of infection of 5 (*see* **Note 2**).
3. Spin down infected Vero monolayers at 2850 × *g* for 30 min at 10 °C to synchronize infection cycles (*see* **Note 3**).
4. Culture infected cell monolayers at 37 °C for 18 h in a 5% CO_2 humidified incubator.
5. Prepare fresh 20 mg/mL EMS in PBS supplemented with calcium chloride and magnesium chloride.
6. Wash infected Vero cell monolayers with 3 mL of PBS supplemented with calcium chloride and magnesium chloride.
7. Add 3 mL of EMS solution (**step 5**) to Vero cell monolayers and incubate cells in a chemical safety hood at room temperature for 1 h (*see* **Note 1**).
8. Remove EMS solution and wash cell monolayers 3 times with 3 mL of PBS supplemented with calcium chloride and magnesium chloride. All washes should be performed in a chemical safety hood. Treat all EMS exposed waste with 1 M sodium hydroxide.
9. Add 6 mL of DMEM supplemented with 10% FBS, 200 ng/mL cycloheximide, and 25 μg/mL gentamicin (*see* **Note 4**).
10. Culture infected cell monolayers at 37 °C for 72 h in a 5% CO_2 humidified incubator.
11. Harvest bacteria from infected cells by quickly washing infected cell monolayers with 10 mL of UltraPure water. Lyse cells by

adding 1 mL of UltraPure water and incubating cells for 20 min at room temperature with gentle agitation. Collect cell lysates in an Eppendorf tube and add 0.250 mL of 5× SPG buffer. Store lysates at −80 °C (*see* **Note 5**).

12. Determine the titer of viable *C. trachomatis* strains.
13. Repeat procedure to establish independent pools of mutagenized *C. trachomatis* strains.
14. Asses rates of mutagenesis by plaquing 7× tenfold serial dilutions of lysates containing 1×10^8 EMS mutagenized *C. trachomatis* inclusion forming units in the presence of 4 μg/mL ofloxacin (*see* **Note 6**). *See* Subheading 3.5 for plaquing protocol. Determine frequency of ofloxacin-resistant *C. trachomatis* plaque forming units and divide by total number of bacteria plated.
15. Screen mutants for phenotypes of interest.

3.2 Isolation of Chlamydia Enriched DNA

The relatively small size of *Chlamydia* genomes (~1 Mb) allows for physical mapping of single nucleotide variants by WGS in any mutant strain of interest. In this protocol, we describe a procedure for isolation and enrichment of *Chlamydia* DNA directly from crude lysates of Vero cells infected with a *Chlamydia* strain can yield several micrograms of DNA suitable for preparations of high-quality sequencing libraries for WGS.

1. Seed Vero cells in DMEM supplemented with 10% FBS, at a density of 6×10^5 cells per well, in all wells of two six well plates, and culture overnight at 37 °C in a 5% CO_2 humidified incubator (*see* **Note 7**).
2. Use a titered stock of a *C. trachomatis* mutagenized strain of interest and infect confluent Vero cell monolayers at a multiplicity of infection of 3.
3. Spin down infected Vero monolayers at 2850 × *g* for 30 min at 10 °C to synchronize infection cycles (*see* **Note 3**).
4. Culture infected cell monolayers at 37 °C for 40–48 h in a 5% CO_2 humidified incubator.
5. Assess efficiency of infection using an inverted light microscope. The majority of Vero cells should contain visible inclusions with mobile bacteria (*see* **Note 8**).
6. Harvest bacteria from infected cells by quickly washing infected cell monolayers with 3 mL of UltraPure water per well. Lyse cells by adding 400 μL of UltraPure water per well and incubating cells for 20 min at room temperature with gentle agitation. Collect cell lysates in Eppendorf tubes and centrifuge at 18,800 × *g* for 15 min at 4 °C. Discard supernatants and resuspend bacteria containing pellet in 1 mL of 1× SPG buffer.

7. Store crude lysates at −80 °C or proceed with DNA enrichment and isolation.
8. Sonicate crude lysates for 10 s to promote complete lysis of host cells.
9. Centrifuge crude lysates at 18,800 × *g* for 15 min at 4 °C and resuspend pellet in 100 μL of 1× NEB DNase I buffer.
10. Deplete host (Vero) DNA by adding 4 units of NEB DNase I and incubating at 37 °C for 30 min (*see* **Note 9**).
11. Add an additional 4 units of NEB DNase I and incubate at 37 °C for 30 min.
12. Remove DNase I by centrifuging crude lysates at 18,800 × *g* for 15 min at 4 °C and removing supernatant.
13. Wash pellet with 0.5 mL of PBS (removal of carry over DNase I).
14. Centrifuge crude lysates at 18,800 × *g* for 15 min at 4 °C and remove supernatant.
15. Resuspend pellet in 180 μL of buffer ATL (DNeasy Blood and Tissue Kit).

    Remainder of protocol follows instructions from Qiagen DNeasy Blood and Tissue Kit: Purification of Total DNA from Animal Tissues. All reagents are included in the kit.
16. Add 20 μL of proteinase K. Mix by vortexing and incubate at 55 °C overnight (*see* **Note 10**).
17. Add 4 μL of RNase A (100 mg/mL), mix by vortexing and incubate for 2 min at room temperature.
18. Vortex for 15 s and add 200 μL of buffer AL. Mix thoroughly by vortexing and incubate at 70 °C for 10 min.
19. Add 200 μL of 100% ethanol and mix by vortexing.
20. Pipet mixture into a DNeasy spin column and centrifuge at 6200 × *g* for 1 min.
21. Place column into new collection tube, add 500 μL of buffer AW1, and centrifuge at 6200 × *g* for 1 min.
22. Place column into new collection tube, add 500 μL of buffer AW2, and centrifuge at 18,800 × *g* for 3 min.
23. Place column into an Eppendorf tube, add 50 μL of buffer AW1, and centrifuge at 6200 × *g* for 1 min.
24. Discard column and measure total DNA concentration (*see* **Note 11**).
25. Run a 5 μL aliquot on a 1% agarose gel to assess integrity of isolated DNA.
26. Store isolated total DNA at −80 °C.

3.3 Mapping Single Nucleotide Variants by Whole Genome Sequencing

Several platforms can be used to sequence *Chlamydia* genomes. We routinely use the MiSeq (Illumina Inc.) platform which enables sequencing of up to 5 *Chlamydia* genomes at a time by multiplexing. Output reads provides over 25× coverage, which is more than adequate for genome assembly. Below we describe a basic protocol for DNA sequencing library preparation and single nucleotide variant mapping.

1. Fragment DNA to appropriate size (400–600 b.p. for Illumina-based sequencing) using an Adaptive Focused Acoustics S220 instrument (Covaris Inc.) according to manufacturer's settings. Fragmentation can also be achieved using Fragmentase (New England Biolabs) according to manufacturer's instructions.
2. Genomic sequencing library preparations can be constructed using commercially available kits (*see* **Note 12**). Libraries can be indexed using multiple barcoded primers, pooled, and sequenced simultaneously.
3. Sequencing is usually performed by core facilities or commercial services.
4. Illumina FASTQ reads can be used for reference-based assembly and single nucleotide variant mapping by using open source genome assemblers such as MAQ [23] and BWA [24]. Alternatively, user friendly commercial software programs are also available such as Geneious Prime from Biomatters Ltd. (https://www.geneious.com).

3.4 Linkage Analysis

Chemical mutagenesis can lead to the accumulation of multiple single nucleotide variants in the genomes of mutagenized strains. To identify causal variants in *Chlamydia* mutant strains, we exploit the high frequency at which *Chlamydia* strains exchange DNA [20, 21]. Coinfection of host cells with two antibiotic-resistant strains leads to production of recombinant progeny that are selected for by isolating strains exhibiting dual antibiotic resistance [20, 21]. Recombinant strains can then be screened for segregating variants that are linked to a phenotype. In this approach, chemical mutagenesis is performed in a rifampin-resistant (Rif^R) *C. trachomatis* L2 parental strain [6]. Mutants of interest are then coinfected with a spectinomycin-resistant ($Spec^R$) *C. trachomatis* L2 strain [6] according to the protocol detailed below.

1. Seed Vero cells in DMEM supplemented with 10% FBS, at a density of 5×10^4 cells per well in 1 well of a 24 well plate. Culture overnight at 37 °C in a 5% CO_2 humidified incubator (*see* **Note 7**).

2. Use a titered *C. trachomatis* Rif^R mutant strain and a titered *C. trachomatis* Spec^R strain [6] to coinfect the confluent Vero cell monolayer at a multiplicity of infection of 2 for each strain.
3. Spin down infected Vero cells at 2850 × *g* for 30 min at 10 °C to synchronize infection cycles (*see* **Note 3**).
4. Culture the infected cell monolayer at 37 °C for 36–40 h in a 5% CO_2 humidified incubator.
5. Assess infection efficiency using an inverted light microscope. The majority of Vero cells should contain visible inclusions with mobile bacteria (*see* **Note 8**).
6. Harvest bacteria from infected cells by quickly washing infected cell monolayer with 1 mL of UltraPure water. Following wash, lyse cells by adding 400 μL of UltraPure water and incubating cells for 20 min at room temperature with gentle agitation. Add 100 μL of 5× SPG buffer and store crude lysates at −80 °C or proceed with clonal isolation of doubly antibiotic-resistant strains.

3.5 Clonal Isolation of Strains by Plaque Assay

A plaque assay enables the isolation of a clonal bacterial population derived from a single EB. In this protocol, we utilize a plaque assay to isolate individual doubly antibiotic-resistant recombinant progeny strains. Clonal isolation by limiting dilution is an alternative assay that can also be employed. A detailed protocol has been previously published [25] and will not be discussed here.

1. Seed Vero cells in DMEM supplemented with 10% FBS, at a density of 6 × 10^5 cells per well, in all wells of two six well plates, and culture overnight at 37 °C in a 5% CO_2 humidified incubator (*see* **Note 7**).
2. Infect confluent Vero cell monolayers in one plate with 10 μL of 6× 1:10 serial dilutions of crude cell lysates from coinfected cells. In the second plate, infect cell monolayers in one well with the Rif^R mutant strain, and cell monolayers in a second well with the Spec^R strain. Infect each strain at a multiplicity of infection of 2 (*see* **Note 13**).
3. Spin down the infected cell monolayers at 2850 × *g* for 30 min at 10 °C to synchronize the infection cycles (*see* **Note 3**).
4. Culture infected cell monolayers at 37 °C for 2 h in a 5% CO_2 humidified incubator to allow for bacterial internalization.
5. Remove growth medium and add 6 mL of DMEM/agarose overlay media (Table 1) per well (*see* **Note 14**).
6. Cool plates outside of laminar flow hood at room temperature for 10 min (*see* **Note 15**).
7. Place plates inside laminar flow hood, remove lids, and dry overlays for 15 min.

Table 1
Recipe for DMEM/agarose overlay media (*see* Subheading 2.1 for preparation of reagents)

Reagent	Final concentration
2× DMEM with sodium bicarbonate	1×
FBS	10%
MEM nonessential amino acids (100×)	1×
Gentamycin solution (50 mg/mL)	0.025 mg/mL
Rifampin (25 mg/mL)	200 ng/mL
Spectinomycin (50 mg/mL)	150 μg/mL
UltraPure water	Up to final volume
SeaKem® agarose (1.5% w/v)	0.54%

8. Culture the cells at 37 °C for 10–14 days in a 5% CO_2 humidified incubator until plaques become visible by light microscopy.
9. Select a well with clearly separated plaques and using an inverted light microscope and a permanent ink pen, circle 24 *Chlamydia* bearing plaques for harvesting and expansion in Vero cells.
10. Seed a 96-well plate with Vero cells at a density of 2×10^4 cells per well and culture overnight at 37 °C in DMEM supplemented with 10% FBS.
11. When cells are fully confluent, replace growth medium with DMEM supplemented with 10% FBS and 200 ng/mL rifampin plus 150 μg/mL spectinomycin.
12. Core individual plaques using a p1000 pipette tip and transfer to one well of confluent Vero cell monolayers seeded in a 96-well plate (*see* **Note 16**).
13. Spin down infected Vero cells at 2850 × *g* for 30 min at 10 °C to synchronize the infection cycles.
14. Culture the infected cells at 37 °C for 48 h in a 5% CO_2 humidified incubator.
15. Harvest monoclonal *Chlamydia* populations from infected cells in each well of the 96-well plate by washing cell monolayers with 200 μL of UltraPure water per well. Lyse cells by adding 80 μL of UltraPure water per well and incubating cells for 20 min at room temperature with gentle agitation. Add 20 μL of 5× SPG buffer and store crude lysates at −80 °C.
16. Save a small aliquot of crude lysate for segregation analysis of single nucleotide variants by PCR.

3.6 Genotyping Recombinant Progeny and Linkage Analysis

1. Prepare a 1:10 dilution of crude lysates from infected cells with DNase-free distilled water.
2. Use 1 μL of diluted crude lysate as template in a standard PCR reaction containing primers designed to amplify each locus harboring a single nucleotide variant (PCR amplicon should be between 600 and 1000 b.p. in length for optimal amplification).
3. During PCR cycling, include an initial step at 94 °C for 10 min for lysing *Chlamydia* and exposing DNA.
4. Verify if each amplified locus harbors a single nucleotide variant by Sanger sequencing.
5. Screen each recombinant strain for desired phenotype and use genotyping results to establish genetic linkage.

3.7 Reverse Genetics Analysis

Genetic links can be validated by generating an independent *Chlamydia* strain harboring a single null allele of the locus exhibiting linkage in an otherwise wild type background. For single nucleotide variants residing in essential genes, linkage can be validated by complementing recombinant strains with *C. trachomatis*-ColE1 shuttle vectors [26] expressing a wild type ORF under native or inducible promoters. Null alleles can be generated by gene replacement using fluorescence-reported allelic exchange mutagenesis (FRAEM) [18]. A detailed protocol describing gene exchange by FRAEM in *Chlamydia* has previously been published [25] and will not be mentioned here. Null alleles within operons can be generated without incurring polar effects on downstream genes by utilizing a modified Cre-*loxP* FRAEMs approach [19]. Here, we describe a protocol for generating loss of function alleles in nonessential ORFs using a *C. trachomatis* adapted Group II intron (TargeTron system) [17].

1. Design IBS, EBS1Δ, and EBS2 retargeting primers (*see* **Note 17**).
2. Prepare 100 μM working stocks of IBS and EBS1Δ retargeting primers and 20 μM working stocks of EBS2 and universal EBS primers (*see* Subheading 2.6 for EBS primer sequence).
3. Combine 2 μL of IBS primer stock, 2 μL of EBS1Δ primer stock, 2 μL of EBS2 primer stock, 2 μL of EBS universal primer stock, and 12 μL of UltraPure water.
4. Generate retargeting DNA fragments by mixing 1 μL of the four-primer master mix prepared in **step 3** and 1 μL of pDFTT3-*aadA* plasmid DNA [17] (~50 ng/μL) (*see* **Note 18**) in a standard high-fidelity PCR reaction mix. Amplify fragments by PCR based amplification.

5. Verify amplification by gel electrophoresis. Retargeting amplicons should be 350 b.p. in size.
6. Isolate retargeting fragments by gel extraction.
7. Digest retargeting amplicons and pDFTT3-*aadA* plasmid with *Hind*III and *Bsr*GI.
8. Purify digested products using a QIAquick PCR Purification Kit.
9. Measure DNA concentrations of each purified sample.
10. Ligate digested retargeting amplicons and digested pDFTT3-*aadA* and transform into competent *E. coli* cells. Select transformants by plating cells on LB-agar supplemented with 25 μg/mL chloramphenicol. Protect plates from light (chloramphenicol is light sensitive) and incubate at 37 °C until colonies are observed.
11. Screen for plasmids containing the retargeted 350 b.p. insert in pDFTT3-*aadA*.

    The next step consists of transforming *Chlamydia* with suicide TargeTron vectors to generate null alleles. The transformation protocol described below can also be used to transform FRAEM based suicide plasmids and *Chlamydia* shuttle plasmids for complementation analysis.
12. Seed Vero cells in DMEM supplemented with 10% FBS, at a density of 6×10^5 cells per well, in all wells of two six well plates, and culture overnight at 37 °C in a 5% CO_2 humidified incubator.
13. Mix 1×10^6 *C. trachomatis* IFUs, 10 μg of retargeted pDFTT3-*aadA* plasmid DNA, 50 μL of transformation buffer, and UltraPure water up to 250 μL (*see* **Note 19**).
14. Prepare a mock transformation control by repeating **step 2** and replacing pDFTT3-*aadA* plasmid DNA with UltraPure water.
15. Incubate transformation mixtures for 30 min at room temperature.
16. Use entire transformation mixture to infect confluent Vero cell monolayers seeded in all wells of a six well plate. Distribute transformation mixture equally among all wells.
17. Spin down infected Vero cell monolayers at 2850 × *g* for 30 min at 10 °C to synchronize the infection cycles.
18. Culture infected cell monolayers at 37 °C for 12–14 h in a 5% CO_2 humidified incubator.
19. Replace growth medium with DMEM supplemented with 10% FBS and 150 μg/mL spectinomycin (*see* **Note 20**).
20. Culture cell monolayers at 37 °C for an additional 28–34 h in a 5% CO_2 humidified incubator.

21. Harvest bacteria by quickly washing infected cell monolayers with 3 mL of UltraPure water. Lyse cells by adding 400 μL of UltraPure water per well and incubating cells for 20 min at room temperature with gentle agitation. Collect cell lysates in Eppendorf tubes and centrifuge at 18,800 × *g* for 15 min at 4 °C. Discard supernatants and resuspend pellet in a total of 1 mL of 1× SPG buffer. Use lysates immediately in **step 11**. Do not freeze or store.
22. Replace growth medium on Vero cell monolayers previously seeded as described in **step 1** with DMEM supplemented with 10% FBS and 150 μg/mL spectinomycin. Infect cell monolayers with crude cell lysates from **step 10**. Distribute entire cell lysates equally among all wells.
23. Spin down the infected Vero cells at 2850 × *g* for 30 min at 10 °C to synchronize the infection cycles.
24. Culture infected cell monolayers at 37 °C for 40–48 h in a 5% CO_2 humidified incubator (*see* **Note 21**).
25. Check for presence of inclusions using an inverted light microscope. If inclusions are not observed, repeat **steps 10–13** until inclusions are observed (*see* **Notes 8** and **22**).
26. Once inclusions are observed, harvest bacteria as described in **step 10** and store at −80 °C (*see* **Note 23**).
27. Verify integration of Group II intron at targeted locus (*see* **Note 24**).
28. Isolate clonal populations by plaque purification and complement.

4 Notes

1. Vero cells can tolerate high levels of EMS mutagenesis and can survive short periods of incubation outside of a humidified CO_2 incubator [22]. Follow proper sterile technique and approved safety procedures for handling cell cultures. All work should be done in a sterile laminar flow hood.
2. We recommend performing EMS mutagenesis on a *Chlamydia* antibiotic-resistant strain for facilitating downstream linkage analysis.
3. Centrifugation of *Chlamydia* is required for optimal and synchronized development of bacterial inclusions. Some mutant strains are unable to infect host cells in the absence of centrifugation.
4. Inhibition of eukaryotic protein translation by cycloheximide can increase recovery of slow growing mutant *Chlamydia* strains.

5. EMS mutagenesis will decrease recovery of viable bacteria. At 72 h post EMS treatment, roughly 10% of *C. trachomatis* inclusions will have a wild type appearance while remaining inclusions will be devoid of bacteria.
6. If ofloxacin-resistant *C. trachomatis* plaque-forming units are not observed, determine ofloxacin minimum inhibitory concentration and repeat plaque assay.
7. Vero cells can be cultured for extended periods of time without passaging and are ideal for routine growth and transformation of *Chlamydia* mutant strains.
8. Vero cells occasionally form large vacuoles upon stressful conditions that resemble *C. trachomatis* inclusions. Typically, these vacuoles are devoid of moving bacteria. In some instances, *Chlamydia* mutant strains can also form inclusions with non-mobile bacteria.
9. DNase I should only target host genomic DNA. *Chlamydia* EBs are osmotically stable and intact at this junction of the protocol.
10. Overnight incubation is optimal for proteinase K induced lysis of EBs and degradation of residual DNase I. Alternatively, incubate for a minimum of 3 h.
11. Accurate measurement of total DNA is critical for preparing high quality sequencing libraries. Quantify DNA by Qubit fluorometric quantification or with a Bioanalyzer.
12. Sequencing library preparation kits from New England Biolabs (NEB) and from Kapa Biosystems yield high-quality sequencing libraries that are compatible with Illumina sequencing platforms.
13. Infection with the antibiotic-resistant strains serves as a control for antibiotic selection efficacy. Plaques should not be observed if antibiotics are active during plaque assay.
14. Agarose cools and solidifies quickly. To avoid solidification during preparation of overlay, remove growth medium from infected Vero cell monolayers prior to melting agarose. Boil agarose in a microwave oven and quickly add to DMEM/agarose overlay already containing all other components (Table 1). Do not let agarose cool. Mix and pour immediately.
15. Airflow inside laminar flow hood causes surface vibrations in the DMEM/agarose overlay which disturbs solidification of the overlay.
16. To core a desired marked plaque, use a p1000 pipette with an attached 1 mL sterile tip, align the tip perpendicularly to the plate surface, and insert into the agarose/DMEM overlay directly over the plaque until reaching the bottom of the

plate. Gently pipet up and down and transfer the agarose plug to a well of a 96-well plate containing a confluent Vero cell monolayer. Resuspend the plug briefly in the growth medium by pipetting up and down.

17. Targetronics provides two different algorithms for designing targeting primers for LtrB and EcI5 introns. The LtrB intron in pDFTT3-*bla* and pDFTT3-*aadA* recognizes different target sequences than the EcI5 intron, and the appropriate algorithm (www.targetrons.com/targetron_pLtrB.php) should be used when designing retargeting primers. Primer design requires a paid subscription (http://www.targetrons.com/targetron_buy.php). Design several primer sets. Choose primer sequences based on the highest probability of insertion (reported *p*-value) and predicted location of integration. Output primer sequences will contain *HindIII* and *BsrGI* restriction site sequences.
18. pDFTT3-*aadA* [17] is a derivate of pDFTT3-*bla* [16] which contains a Group II intron expressing a *bla* cassette under the control of a *C. trachomatis* promoter. In pDFTT3-*aadA*, the *bla* cassette has been replaced with the spectinomycin resistance cassette *aadA*. Both plasmids are available upon request.
19. *Chlamydia* preparations from crude cell lysates of infected host cells exhibit higher transformation efficiencies relative to gradient purified EB preparations. Plasmid DNA isolated from common *E. coli* laboratory strains such as DH5alpha is perfectly suitable for transformation.
20. Determine optimal antibiotic concentration for selection by performing a growth curve assay.
21. This step is considered passage number 1.
22. Lack of observable inclusions is typical at passage number 1. Repeat passaging infected cells at most until passage number 6. If inclusions are not observed by then, repeat transformation.
23. Presence of inclusions in the mock control can result from growth of spontaneously spectinomycin-resistant *C. trachomatis* strains. In this case, repeat transformation with new source of *Chlamydia* parental strain.
24. Determine integration of Group II intron by PCR genotyping (*see* Fig. 1) and single genomic integration by Southern blot analysis.

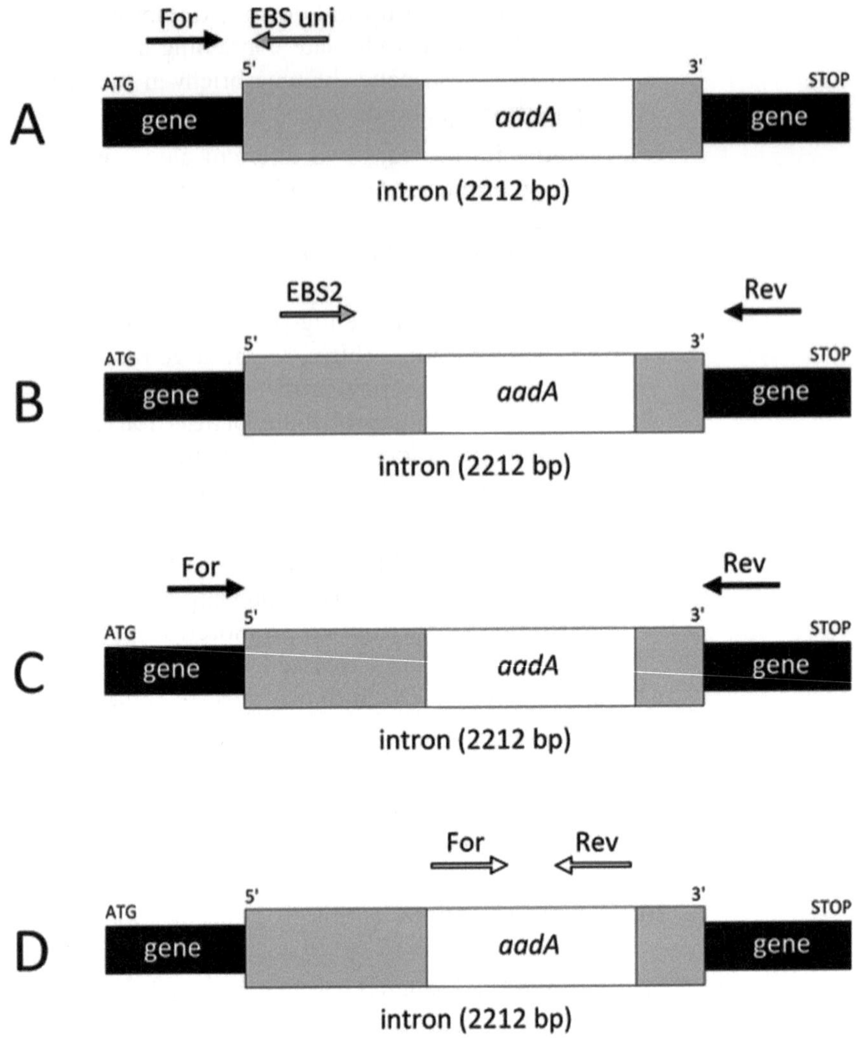

Fig. 1 PCR based strategy for verifying insertional inactivation of an ORF by TargeTron based mutagenesis. (**a**) Use a gene specific forward primer and an intron specific reverse primer (e.g., EBS universal primer) to amplify across the 5′ gene–intron junction. (**b**) Use an intron specific forward primer (e.g., EBS2 primer) and a gene specific reverse primer to amplify across the 3′ gene–intron junction. (**c**) Use gene specific primers flanking the insertion site to verify intron insertion and ORF disruption. (**d**) Use *aadA* cassette specific primers to detect the spectinomycin resistance marker

Acknowledgments

This work was supported by NIH grant 1R21AI140019-01 to R.J.B. We thank Raphael H. Valdivia and members of the Valdivia laboratory for their contributions to the development of this work. We also thank Lee Dolat, Victor J. Ocasio, and Samantha E. Bowen for critical feedback.

References

1. Peipert JF (2003) Clinical practice. Genital chlamydial infections. N Engl J Med 349:2424–2430. https://doi.org/10.1056/NEJMcp030542
2. Burton MJ (2007) Trachoma: an overview. Br Med Bull 84:99–116. https://doi.org/10.1093/bmb/ldm034
3. Mabey D, Peeling RW (2002) Lymphogranuloma venereum. Sex Transm Infect 78:90–92. https://doi.org/10.1136/sti.78.2.90
4. Moulder JW (1991) Interaction of *Chlamydiae* and host cells in vitro. Microbiol Rev 55:143–190
5. Kari L, Goheen MM, Randall LB et al (2011) Generation of targeted *Chlamydia trachomatis* null mutants. Proc Natl Acad Sci U S A 108:7189–7193. https://doi.org/10.1073/pnas.1102229108
6. Nguyen BD, Valdivia RH (2012) Virulence determinants in the obligate intracellular pathogen *Chlamydia trachomatis* revealed by forward genetic approaches. Proc Natl Acad Sci U S A 109:1263–1268. https://doi.org/10.1073/pnas.1117884109
7. Kokes M, Dunn JD, Granek JA et al (2015) Integrating chemical mutagenesis and whole-genome sequencing as a platform for forward and reverse genetic analysis of *Chlamydia*. Cell Host Microbe 17:716–725. https://doi.org/10.1016/j.chom.2015.03.014
8. Brothwell JA, Muramatsu MK, Toh E et al (2016) Interrogating genes that mediate *Chlamydia trachomatis* survival in cell culture using conditional mutants and recombination. J Bacteriol 198:2131–2139. https://doi.org/10.1128/JB.00161-16
9. Rajaram K, Giebel AM, Toh E et al (2015) Mutational analysis of the *Chlamydia muridarum* plasticity zone. Infect Immun 83:2870–2881. https://doi.org/10.1128/IAI.00106-15
10. Colbert T, Till BJ, Tompa R et al (2001) High-throughput screening for induced point mutations. Plant Physiol 126:480–484. https://doi.org/10.1104/pp.126.2.480
11. Kari L, Southern TR, Downey CJ et al (2014) *Chlamydia trachomatis* polymorphic membrane protein D is a virulence factor involved in early host-cell interactions. Infect Immun 82:2756–2762. https://doi.org/10.1128/IAI.01686-14
12. Yang C, Kari L, Sturdevant GL et al (2017) *Chlamydia trachomatis* ChxR is a transcriptional regulator of virulence factors that function in in vivo host-pathogen interactions. Pathog Dis 75. https://doi.org/10.1093/femspd/ftx035
13. Sixt BS, Bastidas RJ, Finethy R et al (2017) The *Chlamydia trachomatis* inclusion membrane protein CpoS counteracts STING-mediated cellular surveillance and suicide programs. Cell Host Microbe 21:113–121. https://doi.org/10.1016/j.chom.2016.12.002
14. Snavely EA, Kokes M, Dunn JD et al (2014) Reassessing the role of the secreted protease CPAF in *Chlamydia trachomatis* infection through genetic approaches. Pathog Dis 71:336–351. https://doi.org/10.1111/2049-632X.12179
15. Muramatsu MK, Brothwell JA, Stein BD et al (2016) Beyond tryptophan synthase: identification of genes that contribute to *Chlamydia trachomatis* survival during gamma interferon-induced persistence and reactivation. Infect Immun 84:2791–2801. https://doi.org/10.1128/IAI.00356-16
16. Johnson CM, Fisher DJ (2013) Site-specific, insertional inactivation of *incA* in *Chlamydia trachomatis* using a group II intron. PLoS One 8:e83989. https://doi.org/10.1371/journal.pone.0083989
17. Lowden NM, Yeruva L, Johnson CM et al (2015) Use of aminoglycoside 3′ adenyltransferase as a selection marker for *Chlamydia trachomatis* intron-mutagenesis and in vivo intron stability. BMC Res Notes 8:570. https://doi.org/10.1186/s13104-015-1542-9
18. Mueller KE, Wolf K, Fields KA (2016) Gene deletion by fluorescence-reported allelic exchange mutagenesis in *Chlamydia trachomatis*. MBio 7:e01817–e01815. https://doi.org/10.1128/mBio.01817-15
19. Keb G, Hayman R, Fields KA (2018) Floxed-cassette allelic exchange mutagenesis enables markerless gene deletion in *Chlamydia trachomatis* and can reverse cassette-induced polar effects. J Bacteriol 200. https://doi.org/10.1128/JB.00479-18
20. Demars R, Weinfurter J, Guex E et al (2007) Lateral gene transfer in vitro in the intracellular pathogen *Chlamydia trachomatis*. J Bacteriol 189:991–1003. https://doi.org/10.1128/JB.00845-06
21. DeMars R, Weinfurter J (2008) Interstrain gene transfer in *Chlamydia trachomatis* in vitro: mechanism and significance. J Bacteriol 190:1605–1614. https://doi.org/10.1128/JB.01592-07
22. Nguyen BD, Valdivia RH (2013) Forward genetic approaches in *Chlamydia trachomatis*.

J Vis Exp (80):e50636. https://doi.org/10.3791/50636

23. Li H, Ruan J, Durbin R (2008) Mapping short DNA sequencing reads and calling variants using mapping quality scores. Genome Res 18:1851–1858. https://doi.org/10.1101/gr.078212.108
24. Li H, Durbin R (2009) Fast and accurate short read alignment with burrows-wheeler transform. Bioinformatics 25:1754–1760. https://doi.org/10.1093/bioinformatics/btp324
25. Mueller KE, Wolf K, Fields KA (2017) *Chlamydia trachomatis* transformation and allelic exchange mutagenesis. Curr Protoc Microbiol 45:11A.3.1–11A.3.15. https://doi.org/10.1002/cpmc.31
26. Bastidas RJ, Valdivia RH (2016) Emancipating *Chlamydia*: advances in the genetic manipulation of a recalcitrant intracellular pathogen. Microbiol Mol Biol Rev 80:411–427. https://doi.org/10.1128/MMBR.00071-15

Chapter 14

A Coinfection Model to Evaluate Chlamydia Inc Protein Interactions

Rachel Ende and Isabelle Derré

Abstract

Chlamydia trachomatis resides and replicates within a membranous vacuole, termed the inclusion. A group of Type III secreted effector proteins, the inclusion membrane proteins (Inc), are embedded within the inclusion membrane and facilitate the interaction of the inclusion with host cell organelles. These interactions are vital for bacterial replication and allow for the acquisition of essential nutrients from the host cell. However, it is not known if Inc proteins function independently or require interactions with other Inc proteins to function. This chapter describes a system to test the homotypic/heterotypic interactions of Inc proteins through the coinfection of *Chlamydia* strains expressing differently tagged inclusion membrane proteins. Our approach takes advantage of the natural homotypic fusion of inclusions and allows for the study of Inc protein interactions when they are embedded within the inclusion membrane.

Key words *Chlamydia trachomatis*, Inclusion, Inclusion membrane protein, Homo/heterotypic interaction, *Chlamydia* transformation, Coimmunoprecipitation, Coinfection

1 Introduction

Chlamydia trachomatis is an obligate intracellular pathogen and the causative agent of the most commonly reported sexually transmitted infection of bacterial origin [1]. If left untreated, genital tract infections can lead to pelvic inflammatory disease and infertility. Upon infection of genital tract epithelial cells, the bacteria reside and replicate within a membranous vacuole, termed the inclusion [2]. As an intracellular pathogen, *C. trachomatis* has evolved mechanisms to manipulate host cell organelles and cellular pathways to acquire essential nutrients and maintain proper bacterial replication [3]. *C. trachomatis* utilizes a bacterial Type III secretion system (T3SS) to translocate bacterial effector proteins into the host cell [4]. One major group of T3SS effector proteins is the inclusion membrane proteins (Inc), which are translocated and embedded into the inclusion membrane [5–7]. Inc proteins are characterized by one or more bi-lobed transmembrane domains

Amanda Claire Brown (ed.), *Chlamydia trachomatis: Methods and Protocols*, Methods in Molecular Biology, vol. 2042, https://doi.org/10.1007/978-1-4939-9694-0_14,

[8] and cytosolic tails that allow for interactions with host cell molecules. The hydrophobic transmembrane domains make Inc proteins difficult to purify and few Inc proteins have been functionally characterized [9–25]. It is not yet clear whether Inc proteins function independently, or if they act in concert with one another, as Inc proteins can be evenly distributed within the inclusion membrane or found in distinct patches, as shown for IncB, CT101, Inc222, and Inc850 [26]. Moreover, a study, that used a heterologous bacterial two-hybrid system, indicated that Inc proteins may directly interact with one another and predicted that IncD interacts with itself as well as six other Inc proteins [27]. To investigate Inc protein interactions in the context of *C. trachomatis* infection, we have developed a system to test the homotypic/heterotypic interactions of Inc proteins where cells are coinfected with strains of *C. trachomatis* that express differently tagged Inc proteins. *C. trachomatis* inclusions naturally undergo homotypic fusion [28], thus upon coinfection inclusions harbor a mixed population of the strains of interest and display the differently tagged Inc proteins on their surface. Unlike other methods, this system allows for the study of Inc protein interactions when they are present in their native environment. With this system we have validated the IncD self-interaction purposed by Gauliard et al. [27]. Here we describe a protocol that can be used to study the interaction between any Inc proteins of interest.

2 Materials

2.1 *Chlamydia* Transformation

1. 75 cm^2 tissue culture flask (Corning).
2. HeLa cells (ATCC CCL-2).
3. BSL2 biological safety cabinet.
4. Humidified CO_2 cell culture incubator.
5. 1× sterile phosphate-buffered saline (PBS).
6. 0.05% sterile trypsin–EDTA (Thermo Scientific).
7. Disposable, sterile polypropylene 15 mL conical tubes.
8. Hemocytometer.
9. Dulbecco's Modified Eagle Medium (DMEM), high glucose (Thermo Scientific).
10. Heat-inactivated fetal bovine serum (FBS) (Invitrogen).
11. Tissue culture treated 100 mm × 20 mm petri dishes.
12. Sterile 1.5 mL Eppendorf tubes.
13. *Chlamydia trachomatis* (L2/434/Bu ATCC VR-902B).
14. Plasmid vectors for transformation into *Chlamydia trachomatis* (*see* **Note 1**).

15. Calcium Transformation Buffer: 10 mM Tris pH 7.4, 50 mM $CaCl_2$ in ddH_2O. Filter-sterilized and stored at room temperature.
16. Penicillin G (10,000 U/mL in ddH_2O). Stored at −20 °C.
17. Cyclohexamide (1 mg/mL in dimethyl sulfoxide (DMSO)). Stored at 4 °C.
18. Microcentrifuge.
19. Sucrose, phosphate buffered glutamic acid (SPG): 7.5% sucrose, 17 mM Na_2HPO_4, 3 mM NaH_2PO_4, 5 mM L-glutamic acid, pH 7.4 in ddH_2O. Filter-sterilize. Store at 4 °C.
20. Micropipettes, sterile pipette tips, and disposable sterile serological pipettes.

2.2 Coinfection

1. Transformant *Chlamydia* strains expressing differently tagged Inc proteins.
2. Tissue culture treated, sterile 24-well plates.
3. Tissue culture treated, sterile 6-well plates.
4. Glass coverslips, round, sterile.
5. DMEM, 10% FBS, with freshly made aTc (4 ng/mL, Sigma).
6. Fixing solution: 1× PBS containing 4% paraformaldehyde.
7. Forceps.
8. Mounting media (DABCO): 200 mM [1,4-diazabicyclo [2.2.2]octane] (Sigma), 86% Glycerol, 20 mM Tris–HCl pH 8 in ddH_2O.
9. Glass microscope slide.
10. Nail polish.

2.3 Coimmunoprecipitation and Western Blot (See Note 2)

1. Lysis buffer: 20 mM Tris pH 7.5, 150 mM NaCl, 2 mM EDTA, 1%Triton X100, 1 mM PMSF and protease inhibitor cocktail (Roche).
2. 0.5 mL sterile Eppendorf tubes.
3. Anti-FLAG M2 agarose beads (Sigma).
4. Mini tabletop centrifuge.
5. Wash buffer: 20 mM Tris pH 7.5, 150 mM NaCl, 2 mM EDTA, and 1% Triton X-100.
6. Elution buffer: 20 mM Tris pH 7.5, 150 mM NaCl, 2 mM EDTA, 100 μg/mL 3XFLAG peptide (Sigma).
7. 6× sample buffer: 300 mM Tris pH 6.8, 12% SDS, 0.6% Bromophenol blue, 60% glycerol, 300 mM DTT.
8. 10× Tris–glycine electrophoresis buffer: 25 mM Tris, 250 mM glycine, 0.1% SDS in ddH_2O.
9. 10× Transfer buffer: 25 nM Tris, 192 mM glycine in ddH_2O.

10. 0.2 μm nitrocellulose blotting membrane (GE Healthcare Life Sciences).
11. 5% Nonfat milk in 1× PBS-0.05% Tween (PBS-T).
12. *Antibodies:*
    (a) Mouse monoclonal anti-Flag M2 antibody (Sigma).
    (b) Mouse monoclonal anti-Myc antibody (Cell Signaling).
    (c) Goat anti-mouse Alexa Fluor 488 (Molecular Probes).
    (d) Goat anti-mouse Alexa Fluor 594 (Molecular Probes).
    (e) Peroxidase-conjugated goat anti-mouse IgG (Jackson ImmunoResearch).

3 Methods

3.1 *Chlamydia* Transformation

1. Plate cells for transformation:
    (a) Wash HeLa cells grown to confluency in a 75 cm^2 tissue culture flask three times with 5 mL of sterile 1× PBS.
    (b) After the last wash, aspirate the PBS and pipette 1 mL of 0.05% Trypsin onto the cell monolayer with a 5 mL serological pipette. Incubate for 3–5 min at 37 °C with 5% CO_2 until the cells have detached from the plastic.
    (c) Pipette 9 mL of prewarmed DMEM, 10% FBS against the cell monolayer to resuspend the cells and inactivate the trypsin. Transfer the resuspended cells into a 15 mL conical tube.
    (d) Use a hemocytometer to count the cells and determine the number of cells/mL.
    (e) Seed a 10 cm dish with 3×10^6 HeLa cells per dish in 10 mL DMEM, 10% FBS and incubate at 37 °C with 5% CO_2 until confluent (~1 day) (*see* **Note 3**).
2. Once the HeLa cells are confluent, set up the *Chlamydia* transformation reaction. For each 10 cm dish (*see* **Note 4**):
    (a) Pipette (in the order given here) 6 μg of *Chlamydia* transformation vector plasmid DNA diluted in 10 μL ddH_2O, *Chlamydia* elementary bodies (such that the multiplicity of infection equals 1) diluted in 10 μL of SPG (*see* **Note 5**) and 200 μL of calcium transformation buffer into a 1.5 mL Eppendorf tube and mix by pipetting up and down.
    (b) Incubate the transformation reaction at room temperature for 30 min.
3. Following the 30 min incubation, pipette the transformation reaction into a 15 mL conical tube containing 10 mL of

prewarmed (37 °C) DMEM, 10% FBS and mix by pipetting up and down with a 10 mL serological pipette.

4. Aspirate the media from the 10 cm dish and immediately pipette the DMEM/transformation reaction mixture from **step 3** into the 10 cm dish. Incubate the cells at 37 °C with 5% CO_2 overnight (*see* **Note 6**).
5. At 18 h postinfection, aspirate the media from the 10 cm dish and add 10 mL of prewarmed DMEM, 10% FBS containing 1 U/mL of Penicillin G (*see* **Note** 7) and return the cells to the incubator.
6. Passaging of transformants: (*see* **Note 8**).
   (a) At 48 h postinfection, aspirate the media from the 10 cm dish and pipette 2 mL of sterile ddH_2O onto the infected monolayer of cells.
   (b) Use a cell scraper to scrape the infected cells of the bottom of the 10 cm dish and pipette the water–cell mixture into a sterile 15 mL conical tube.
   (c) Centrifuge the samples at 290 × *g* for 5 min at room temperature to pellet nuclei and unlysed cells.
   (d) Collect the supernatant with a 5 mL serological pipette and transfer the supernatant to a 15 mL conical tube containing 10 mL of DMEM, 10% FBS containing 1 μg/mL cycloheximide and 1 U Penicillin G and mix by pipetting up and down.
   (e) Pipette the medium from **step 6d** onto a confluent monolayer of HeLa cells in a 10 cm dish and incubate at 37 °C with 5% CO_2 for 48 h. This completes passage 1 (P1).
7. Repeat **steps 6a–e** through passage 3 (P3) (*see* **Note 9**).
8. Harvest transformants using **steps 6a–c**. After **step 6c**, collect the supernatant with a 5 mL serological pipette and split into two 1.5 mL sterile Eppendorf tubes (1 mL per tube).
9. Centrifuge the samples for 20 min at ~13,500 × *g* at 4 °C.
10. Aspirate the supernatant and resuspend the pellets in 100 μL of SPG each. Combine the resuspended pellets, aliquot into 10 μL aliquots, and store at −80 °C (*see* **Note 10**).
11. Titrate the preparation of the transformant bacteria from **step 10** to determine the concentration of bacteria that results in an MOI of 1.
    (a) Seed six wells of a 24-well tissue culture dish with 2×10^5 HeLa cells per well and incubate at 37 °C with 5% CO_2 until confluent (~1 day).
    (b) Once the cells are confluent, set up six 1.5 mL Eppendorf tubes. One that contains 1 mL of DMEM, 10% FBS and 5 that contain 500 μL of DMEM, 10% FBS.

(c) Pipette 1 μL of the bacterial preparation from **step 10** into the Eppendorf tube containing 1 mL of DMEM, 10% FBS and perform 1:2 serial dilutions using the remaining Eppendorf tubes.

(d) Aspirate the media from the 24-well dish and pipette each of the dilutions from **step 11c** into a separate well to infect the established monolayers. Incubate at 37 °C with 5% CO_2 for 24 h.

(e) Visualize inclusions by epifluorescence confocal microscopy to determine the number of infectious particle per mL.

12. Once the transformant strain has been obtained and titrated, clones should be isolated and amplified by plaque purification assay (*see* **Note 11**).

3.2 Coinfection

Once the transformant strains have been generated and titrated, coinfection conditions can be established where cells are infected, at the same time, with a *C. trachomatis* strain expressing a Flag-tagged Inc protein and a strain expressing a Myc-tagged Inc protein. It is also necessary to include single infection controls for both the Flag- and Myc-tagged strains.

1. Plate cells for infection:

(a) Lift HeLa cells from 75 cm^2 tissue culture flask and determine the number of cells/mL as indicated in Subheading 3.1, **steps 1a–d**).

(b) Place sterile coverslips in each of the wells of a 24-well plate and pipette 300 μL of DMEM, 10% FBS per well. The coinfection conditions are done in duplicate and one condition for each of the single infections should be included.

(c) Pipette an additional 500 μL of DMEM, 10% FBS containing 2×10^5 HeLa cells per well and incubate at 37 °C with 5% CO_2 until confluent (~1 day).

2. Prepare the infection media:

(a) For coinfection conditions, determine the amount of EBs needed for a *combined* MOI of 5 for the Flag- and Myc-tagged Inc-expressing *Chlamydia* strains generated through *Chlamydia* transformation. For single infection control conditions, determine the amount of EBs needed for an MOI of 5 for *either* the Flag- *or* Myc-tagged Inc-expressing *Chlamydia* strain alone.

(b) Pipette the determined amount of EBs into a 15 mL conical tube containing 500 μL per well of DMEM, 10% FBS. Mix by pipetting up and down with a 5 mL serological pipette.

3. Aspirate the media from the cells plated in **step 1**. Using a 5 mL serological pipette, pipette 500 μL of infection media per well and incubate for 24 h at 37 °C with 5% CO_2.
4. At 7 h postinfection (*see* **Note 12**), pipette an additional 500 μL of DMEM, 10% FBS containing 4 ng/mL of aTc.
5. At 24 h postinfection, aspirate the media and pipette 500 μL of the 4% PFA fixing solution per well. Incubate for 30 min at room temperature.
6. Aspirate the fixing solution and wash the coverslips three times with 500 μL of 1× PBS.
7. After the last wash, use forceps to remove the coverslips from the wells and place them on the staining platform (*see* **Note 13**). Carefully cover the coverslips with 50 μL of primary antibody diluted in 0.1% Triton-X100 and incubate at room temperature for 1 h. For each coinfection condition, one coverslip is stained with the mouse anti-Flag primary antibody and the other coverslip with the mouse anti-Myc primary antibody (*see* **Note 14**).
8. Remove the primary antibody, return the coverslips to the wells and wash three times with 500 μL of 1× PBS.
9. After the last wash, remove the coverslips from the wells and place them on the staining platform and carefully cover the coverslips with 50 μL of secondary antibody diluted in 0.1% Triton-X100 and incubate at room temperature for 1 h.
10. Remove the secondary antibody, return the coverslips to the wells and wash three times with 500 μL of 1× PBS.
11. Mounting coverslips: (*see* **Note 15**).
    (a) Pipette 4 μL of mounting media per coverslip onto a glass microscope slide.
    (b) Remove the coverslip from the well and place the coverslip onto the mounting media with the cells facing the glass microscope slide.
    (c) Aspirate any excess mounting media around the edges of the coverslips and seal the coverslip by applying a thin layer of nail polish around the coverslip.
12. Visualize the expression of the tagged Inc protein via epifluorescence or confocal microscopy. If the coinfection is successful, 100% of the inclusions should show positive staining with both the anti-Flag and the anti-Myc antibody (Fig. 1).

3.3 Coimmunoprecipitation

1. Once the coinfection and induction conditions have been optimized the co-infection can be scaled up to a 6-well dish by using five times the number of cells, media, and bacteria (*see* **Note 16**).

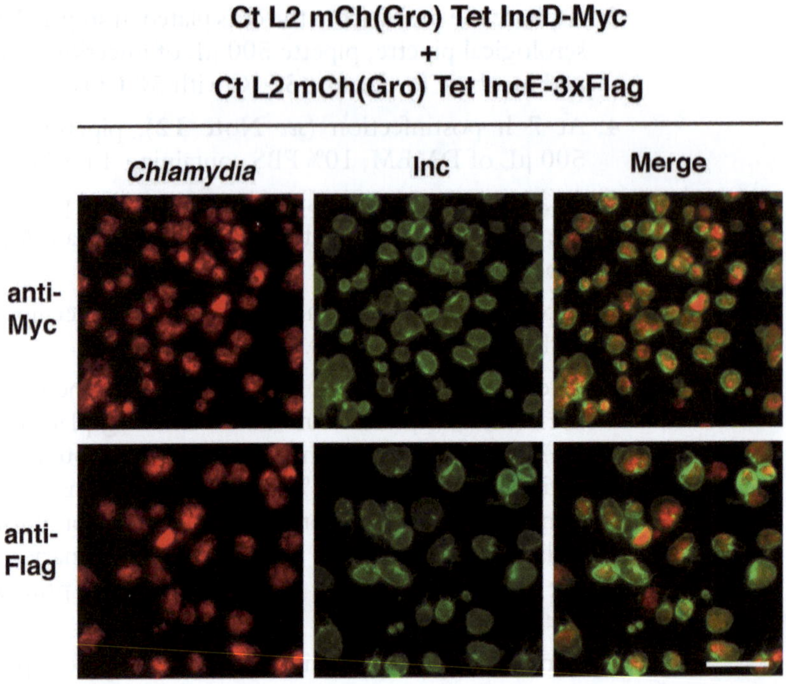

Fig. 1 Inclusion localization of Inc proteins expressed by *C. trachomatis* strains in a coinfection model system. Epifluorescence micrographs of HeLa cells coinfected with strains of *C. trachomatis* expressing mCherry constitutively (Left Panels) and IncE-3xFLAG or IncD-Myc under the control of the aTc inducible promoter (Middle Panels). The cells were infected and Inc expression was induced with 2 ng/mL aTc, fixed 24 h postinfection, immunostained with mouse anti-Myc antibodies (Top Panels) or mouse anti-FLAG antibodies (Bottom Panels) and imaged using an epifluorescence microscope. The merge is shown on the right. Scale bar: 50 μm. Reproduced from Han and Derré (2017) with permission from Frontiers [29]

2. Pipette 1 mL of 1× PBS per well to wash the infected cell monolayer.
3. Aspirate the PBS and add 300 μL of cold lysis buffer per well. Collect the samples with a pipette, transfer to 1.5 mL Eppendorf tubes and incubate on a rotator for 20 min at 4 °C.
4. During the 20 min incubation at 4 °C, equilibrate the anti-FLAG M2 agarose beads in lysis buffer.
   (a) Pipette enough slurry to equal 10 μL of beads/sample.
   (b) Centrifuge the slurry in a mini-tabletop centrifuge to pellet the beads and aspirate off the supernatant.
   (c) Wash the beads three times by pelleting the beads and resuspending in 500 μL of lysis buffer.
   (d) Resuspend the pelleted beads in enough lysis buffer to pipette 50 μL of beads/lysis buffer per sample, and place on ice.
5. After the 20 min incubation at 4 °C, centrifuge the samples at 15,682 × *g* for 10 min at 4 °C and pipette the supernatant into a 0.5 mL Eppendorf tube (*see* **Note 17**).

6. Transfer 20 μL of the supernatant into a fresh 1.5 mL Eppendorf tube containing 4 μL of 6× Sample Buffer and boil for 5 min (this will be your total lysate).
7. Pipette 50 μL of the equilibrated anti-Flag M2 beads (*see* **step 4**) (*see* **Note 18**) into the 0.5 mL Eppendorf tubes containing the supernatant from **step 5** and incubate on a rotator at 4 °C for 2 h.
8. Centrifuge the samples in a mini tabletop centrifuge for 30 s to pellet the beads. Aspirate the supernatant then resuspend the beads in 500 μL of wash buffer by inverting the tube several times.
9. Repeat **step 8** two more times, then aspirate all of the wash buffer and add 25 μL of elution buffer.
10. Gently flick the tube 30 times. If preparing multiple samples, set on ice before continuing with the other samples.
11. Repeat **step 10** two more times.
12. Centrifuge the sample in a microcentrifuge for 1 min at 15,682 × *g* (room temperature).
13. Collect 20 μL of the elution fraction (*see* **Note 19**) with a pipette and transfer to an Eppendorf tube containing 4 μL of 6× sample buffer (these will be your IP samples). Boil the samples for 5 min.
14. Load 10 μL of the total lysate samples and 10 μL of IP sample into an SDS page electrophoresis gel and electrophorese until the dye front has reached the bottom of the gel (*see* **Note 20**).
15. Transfer to a nitrocellulose membrane at 4 °C (*see* **Note 2**).
16. Block in PBS-T, 5% nonfat milk for 1 h at room temperature.
17. Pipette 10 mL of primary antibody diluted in PBS-T, 5% nonfat milk (mouse anti-Flag or mouse anti-Myc) onto the membrane and incubate on shaker overnight at 4 °C.
18. Perform three 10 min washes with 10 mL of PBS-T.
19. Pipette 10 mL of secondary antibody (peroxidase-conjugated goat anti-mouse IgG) onto the membrane and incubate on shaker for 1 h at room temperature.
20. Perform three 10 min washes with 10 mL of PBS-T.
21. If the Inc proteins interact, the Myc-tagged Inc proteins should immunoprecipitate with the Flag-tagged Inc proteins and the lanes corresponding to the IP sample of the coinfection condition should have a band present when probed with the anti-Flag *and* anti-Myc antibodies. The lanes corresponding to the single infections should only have a band present when probed with the antibody against the respective tag (Fig. 2).

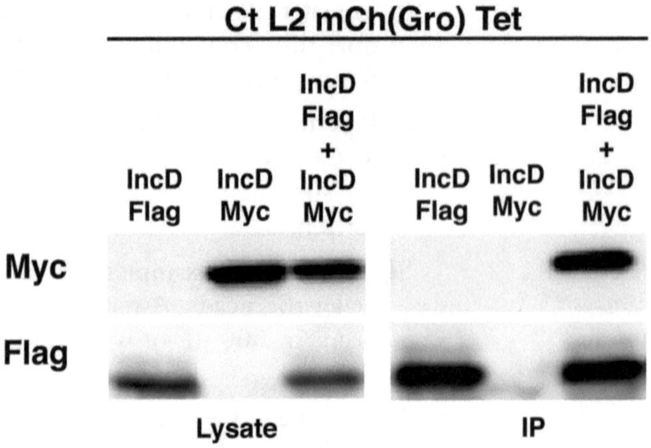

Fig. 2 Example of coimmunoprecipitation results showing that IncD self-interacts at the inclusion membrane. HeLa cells were infected with a strain of *C. trachomatis* expressing IncD-Flag or IncD-Myc under the control of the aTc inducible promoter or coinfected with these two strains (IncD-Flag+IncD-Myc). IncD-Flag was immunoprecipitated with anti-Flag M2 beads. A portion of the cell lysate (Left Panel, Lysate) and the immunoprecipitated proteins (Right Panel, IP) were separated by SDS-PAGE and analyzed by western blot with antibodies against Myc (Top Panels) and Flag (Bottom Panels). Reproduced from Han and Derré (2017) with permission from Frontiers [29]

4 Notes

1. There have been several plasmid vectors published that allow for transformation into *C. trachomatis.* Examples include p2TK2-SW2 mCh(Gro) [13] and its derivatives, the pBOMB4-tet vector [19] and its derivatives, as well as the pASK_GFP_mKate2_L2 vector [30] and its derivatives. The plasmids mentioned above encode the beta-lactamase enzyme and can be selected for with ampicillin or penicillin as described in this chapter. However, plasmids that contain an *aadA* gene encoding an aminoglycoside 3′ adenyltransferase enzyme are also available and selection should be performed with 200–500 μg/mL of spectinomycin [31].
2. For a more detailed description of our western blot protocol please *see* [29].
3. Cells can be seeded the day before passaging the transformants, or multiple dishes can be seeded at the same time but at lower densities to allow for the immediate passage of transformants once they are harvested. Cells should be seeded so that confluent 10 cm dishes are available every 2 days.
4. Transformations in our lab are done under selection with both 0.5 U and 1 U of Penicillin G. Thus, two 10 cm dishes per transformation (and for the subsequent passages) are needed.

We observe that although we often obtain transformants under both conditions, there are times when transformants are present under only one condition, but the condition they are found in varies.

5. Our lab uses crude extract rather than purified EBs. It is important to have titrated this crude extract and determined the amount of bacteria required to achieve an MOI of 1 prior to beginning this step. Having a lower infection will reduce the probability of generating a transformant strain.
6. We suggest planning to perform this step in the afternoon so that the selection with antibiotic at 18 h postinfection (*see* **Note 5**) will take place the following morning.
7. The penicillin G is very sensitive to freeze/thaw and will lose its activity if freeze-thawed multiple times. We recommend that penicillin G aliquots not be freeze-thawed more than twice and that they are stored at −20 °C. Also, when a new stock of penicillin G is made, a titration curve to determine its efficacy should be performed.
8. Before continuing with the passaging of transformants it is important to inspect the morphology of the inclusions with light microscopy to verify that the selection with the penicillin G is efficient. The inclusions should be aberrant and the bacteria should be swollen and somewhat un-uniform in shape.
9. Throughout the passaging of the transformants it is important to observe them by light microscopy. With each passage the number of aberrant inclusions should decrease and the number of wildtype-appearing inclusions should increase. In early passages some inclusions may contain both wildtype and transformant bacteria, and the transformant bacteria can allow for the survival of the wildtype bacteria in the presence of penicillin G. If this is the case, it is important to perform additional passages (at an MOI < 1) to obtain inclusions that contain only transformant bacteria. Utilizing a transformation vector that contains a fluorescent marker allows for the visualization of transformant bacteria and a quick assessment of the amount of wildtype contamination. An inclusion that appears only faintly fluorescent, but is full of wildtype bacteria, is indicative of an inclusion containing both wildtype and transformant bacteria. Also, if there are a large number of wildtype inclusions, it is important to decrease the amount of supernatant (i.e., the MOI) that is passaged so that the cell monolayer is not destroyed by an extremely high infection. On average, we perform between 3 and 4 passages of our transformant bacteria.
10. This step can serve as stopping point, as the 10 μL aliquots can be kept at −80 °C until ready to be used for infection.

11. For description of clone isolation by plaque purification *see* Chapter 12, Subheading 3.4, or [32].
12. The time postinfection at which the expression of the tagged Inc is induced by the addition of aTc can be changed and may need to be adjusted for various Incs. When longer induction periods are used, lower amounts of aTc are needed. We recommend that you determine the optimal induction conditions via immunofluorescence prior to performing this step. We perform overnight inductions with 0.5–2 ng/mL aTc at 7 h postinfection and 4–6 h inductions with 20 ng/mL aTc at 18–20 h postinfection.
13. Our lab uses an overturned 24-well plate covered in Parafilm as a staining platform. We remove the coverslips from the wells and place them, with the cells facing up, on the staining platform prior to adding the 50 μL of antibody directly on top of the coverslip. We then return the coverslips to their respective wells for the washes with 1× PBS.
14. It is possible to costain the same coinfected coverslip with a rabbit anti-Flag antibody and a mouse anti-Myc antibody. However, the rabbit anti-Flag antibody can lead to uneven staining of inclusions even at high concentrations of the antibody. Thus, we use duplicate coverslips of the same coinfection condition and stain them individually with mouse anti-Flag and mouse anti-Myc antibodies.
15. Four coverslips can be mounted onto 1 glass microscope slide.
16. We recommend that coverslips be added to each 6-well and stained with anti-Flag and anti-Myc antibodies to verify that 100% of the inclusions are positive for both of the tagged Inc constructs each time the experiment is performed.
17. Be sure to promptly remove the Eppendorf tubes from the centrifuge so that the pellet does not have time to loosen, and be careful not to disturb the pellet while collected the supernatant.
18. The beads will have settled at the bottom of the tube and should be resuspended by pipetting up and down.
19. Be extremely careful to avoid picking up any beads when you are collecting your elution fraction. Any beads that are collected will result in bands corresponding to the heavy chain and light chain of the anti-Flag antibody in your western blot, which could interfere with the interpretation of the data.
20. The samples should be loaded to create two duplicate gels. Following transfer of the gels to nitrocellulose membranes, one membrane should be probed with the anti-Flag M2 antibody and the other membrane with the anti-Myc antibody.

Acknowledgments

This work was supported by NIAID grant R01AI101441 to ID. We thank Rebecca Stanhope, Maria Cortina, and Ralph Bishop for reviewing the manuscript.

References

1. Malhotra M, Sood S, Mukherjee A, Muralidhar S, Bala M (2013) Genital *Chlamydia trachomatis*: an update. Indian J Med Res 138:303–316
2. Moulder JW (1991) Interaction of chlamydiae and host cells in vitro. Microbiol Rev 55:143–190
3. Elwell C, Mirrashidi K, Engel J, Francisco S (2016) Chlamydia cell biology and pathogenesis. Nat Rev Microbiol 14:385–400. https://doi.org/10.1038/nrmicro.2016.30
4. Mueller KE, Plano GV, Fields KA (2014) New frontiers in type III secretion biology: the chlamydia perspective. Infect Immun 82:2–9. https://doi.org/10.1128/IAI.00917-13
5. Dehoux P, Flores R, Dauga C, Zhong G, Subtil A (2011) Multi-genome identification and characterization of chlamydiae-specific type III secretion substrates: the Inc proteins. BMC Genomics 12:109. https://doi.org/10.1186/1471-2164-12-109
6. Lutter EI, Martens C, Hackstadt T (2012) Evolution and conservation of predicted inclusion membrane proteins in chlamydiae. Comp Funct Genomics 2012:362104. https://doi.org/10.1155/2012/362104
7. Moore ER, Ouellette SP (2014) Reconceptualizing the chlamydial inclusion as a pathogen-specified parasitic organelle: an expanded role for Inc proteins. Front Cell Infect Microbiol 4:1–10. https://doi.org/10.3389/fcimb.2014.00157
8. Bannantine JP, Griffiths RS, Viratyosin W, Brown WJ, Rockey DD (2000) A secondary structure motif predictive of protein localization to the chlamydial inclusion membrane. Cell Microbiol 2:35–47. https://doi.org/10.1046/j.1462-5822.2000.00029.x
9. Hackstadt T, Scidmore-Carlson MA, Shaw EI, Fischer ER (1999) The *Chlamydia trachomatis* IncA protein is required for homotypic vesicle fusion. Cell Microbiol 1:119–130. https://doi.org/10.1046/j.1462-5822.1999.00012.x
10. Scidmore MA, Hackstadt T (2001) Mammalian 14-3-3β associates with the *Chlamydia trachomatis* inclusion membrane via its interaction with IncG. Mol Microbiol 39:1638–1650. https://doi.org/10.1046/j.1365-2958.2001.02355.x
11. Rzomp KA, Moorhead AR, Scidmore MA (2006) The GTPase Rab4 interacts with *Chlamydia trachomatis* inclusion membrane protein CT229. Infect Immun 74:5362–5373. https://doi.org/10.1128/IAI.00539-06
12. Derré I, Swiss R, Agaisse H (2011) The lipid transfer protein CERT interacts with the Chlamydia inclusion protein IncD and participates to ER-Chlamydia inclusion membrane contact sites. PLoS Pathog 7:e1002092. https://doi.org/10.1371/journal.ppat.1002092
13. Agaisse H, Derré I (2014) Expression of the effector protein IncD in *Chlamydia trachomatis* mediates recruitment of the lipid transfer protein CERT and the endoplasmic reticulum-resident protein VAPB to the inclusion membrane. Infect Immun 82:2037–2047. https://doi.org/10.1128/IAI.01530-14
14. Lutter EI, Barger AC, Nair V, Hackstadt T (2013) *Chlamydia trachomatis* inclusion membrane protein CT228 recruits elements of the myosin phosphatase pathway to regulate release mechanisms. Cell Rep 3:1921–1931. https://doi.org/10.1016/j.celrep.2013.04.027
15. Aeberhard L, Banhart S, Fischer M, Jehmlich N, Rose L, Koch S, Laue M, Renard BY, Schmidt F, Heuer D (2015) The proteome of the isolated *Chlamydia trachomatis* containing vacuole reveals a complex trafficking platform enriched for retromer components. PLoS Pathog 11:1–25. https://doi.org/10.1371/journal.ppat.1004883
16. Kokes M, Dunn JD, Granek JA, Nguyen BD, Barker JR, Valdivia RH, Bastidas RJ (2015) Integrating chemical mutagenesis and whole-genome sequencing as a platform for forward and reverse genetic analysis of Chlamydia. Cell Host Microbe 17:716–725. https://doi.org/10.1016/j.chom.2015.03.014
17. Mirrashidi KM, Elwell CA, Verschueren E, Johnson JR, Frando A, Von Dollen J, Rosenberg O, Gulbahce N, Jang G, Johnson T, Jager S, Gopalakrishnan AM, Sherry J, Dan Dunn J, Olive A, Penn B, Shales M, Cox JS, Starnbach MN, Derre I,

Valdivia R, Krogan NJ, Engel J (2015) Global mapping of the inc-human interactome reveals that retromer restricts chlamydia infection. Cell Host Microbe 18:109–121. https://doi.org/10.1016/j.chom.2015.06.004

18. Mital J, Lutter EI, Barger AC, Dooley CA, Hackstadt T (2015) *Chlamydia trachomatis* inclusion membrane protein CT850 interacts with the dynein light chain DYNLT1 (Tctex1). Biochem Biophys Res Commun 462:165–170. https://doi.org/10.1016/j.bbrc.2015.04.116
19. Weber MM, Noriea NF, Bauler LD, Lam JL, Sager J, Wesolowski J, Paumet F, Hackstadt T (2016) A functional core of IncA is required for *Chlamydia trachomatis* inclusion fusion. J Bacteriol 198:1347–1355. https://doi.org/10.1128/JB.00933-15
20. Elwell CA, Czudnochowski N, von Dollen J, Johnson JR, Nakagawa R, Mirrashidi K, Krogan NJ, Engel JN, Rosenberg OS (2017) Chlamydia interfere with an interaction between the mannose-6-phosphate receptor and sorting nexins to counteract host restriction. elife 6:1–17. https://doi.org/10.7554/eLife.22709
21. Paul B, Kim HS, Kerr MC, Huston WM, Teasdale RD, Collins BM (2017) Structural basis for the hijacking of endosomal sorting nexin proteins by *Chlamydia trachomatis*. elife 6:1–23. https://doi.org/10.7554/eLife.22311
22. Sixt BS, Bastidas RJ, Finethy R, Baxter RM, Carpenter VK, Kroemer G, Coers J, Valdivia RH (2017) The *Chlamydia trachomatis* inclusion membrane protein CpoS counteracts STING-mediated cellular surveillance and suicide programs. Cell Host Microbe 21:113–121. https://doi.org/10.1016/j.chom.2016.12.002
23. Weber MM, Lam JL, Dooley CA, Noriea NF, Hansen BT, Hoyt FH, Carmody AB, Sturdevant GL, Hackstadt T (2017) Absence of specific *Chlamydia trachomatis* inclusion membrane proteins triggers premature inclusion membrane Lysis and host cell death. Cell Rep 19:1406–1417. https://doi.org/10.1016/j.celrep.2017.04.058
24. Stanhope R, Flora E, Bayne C, Derré I (2017) IncV, a FFAT motif-containing *Chlamydia* protein, tethers the endoplasmic reticulum to the pathogen-containing vacuole. Proc Natl Acad Sci USA 114:12039–12044. https://doi.org/10.1073/pnas.1709060114
25. Nguyen PH, Lutter EI, Hackstadt T (2018) *Chlamydia trachomatis* inclusion membrane protein MrcA interacts with the inositol 1,4,5-trisphosphate receptor type 3 (ITPR3) to regulate extrusion formation. PLoS Pathog 14:e1006911. https://doi.org/10.1371/journal.ppat.1006911
26. Mital J, Miller NJ, Fischer ER, Hackstadt T (2010) Specific chlamydial inclusion membrane proteins associate with active Src family kinases in microdomains that interact with the host microtubule network. Cell Microbiol 12:1235–1249. https://doi.org/10.1111/j.1462-5822.2010.01465.x
27. Gauliard E, Ouellette SP, Rueden KJ, Ladant D (2015) Characterization of interactions between inclusion membrane proteins from *Chlamydia trachomatis*. Front Cell Infect Microbiol 5:1–11. https://doi.org/10.3389/fcimb.2015.00013
28. Ridderhof JC, Barnes RC (1989) Fusion of inclusions following superinfection of HeLa cells by two serovars of *Chlamydia trachomatis*. Infect Immun 57:3189–3193
29. Han Y, Derré I (2017) A co-infection model system and the use of chimeric proteins to study *Chlamydia* inclusion proteins interaction. Front Cell Infect Microbiol 7:1–9. https://doi.org/10.3389/fcimb.2017.00079
30. Wickstrum J, Sammons LR, Restivo KN, Hefty PS (2013) Conditional gene expression in *Chlamydia trachomatis* using the tet system. PLoS One 8:1–11. https://doi.org/10.1371/journal.pone.0076743
31. Lowden NM, Yeruva L, Johnson CM, Bowlin AK, Fisher DJ (2015) Use of aminoglycoside 3′ adenyltransferase as a selection marker for *Chlamydia trachomatis* intron-mutagenesis and in vivo intron stability. BMC Res Notes 8:1–10. https://doi.org/10.1186/s13104-015-1542-9
32. Nguyen BD, Valdivia RH (2012) Virulence determinants in the obligate intracellular pathogen *Chlamydia trachomatis* revealed by forward genetic approaches. Proc Natl Acad Sci USA 109:1263–1268. https://doi.org/10.1073/pnas.1117884109

Chapter 15

Rodent Infections for *Chlamydia* spp.

Charles W. Armitage, Alison J. Carey, and Kenneth W. Beagley

Abstract

Chlamydia spp. infections cause immunopathology of the male and female urogenital tracts and incidence continues to rise across the globe. Animal models offer the opportunity to study the host: pathogen relationship, with rodent models being an attractive first step in studying immune interactions, genetic knockout, as well as bacterial inhibitor and vaccine trials. Here we describe the methodology to infect both male and female rodents at various mucosal sites, with a particular focus on the reproductive tracts.

Key words *Chlamydia trachomatis*, *Chlamydia muridarum*, *Chlamydia caviae*, Reproductive tract, Urogenital tract

1 Introduction

Chlamydia trachomatis is the most common bacterial sexually transmitted infection worldwide [1], it is the most common cause of infectious blindness [2] and has also been associated with the development of reactive arthritis [3]. *Chlamydia pneumoniae* is a common cause of community-acquired pneumonia and has been associated with a variety of conditions including asthma, Alzheimer's disease and atherosclerosis/cardiovascular disease [4]. As with most infectious diseases, human studies are restricted by ethical constraints and as a result research has relied on animal models to supplement in vitro studies. Animal models of chlamydial infections have been developed in nonhuman primates, pigs, mini-pigs, guinea pigs, and mice. For practical and cost reasons, nonhuman primate and pig models are not available to most researchers and will not be considered in this chapter. Animal models are only of value if they recapitulate the conditions seen in human disease and very rarely does a single model reproduce all of the clinical manifestations. With regard to genital chlamydial infections, infection of mice with the natural murine chlamydial species *C. muridarum* [5]

Charles W. Armitage and Alison J. Carey contributed equally to this work.

Amanda Claire Brown (ed.), *Chlamydia trachomatis: Methods and Protocols*, Methods in Molecular Biology, vol. 2042, https://doi.org/10.1007/978-1-4939-9694-0_15, © Springer Science+Business Media, LLC, part of Springer Nature 2019

is the most commonly used model. Vaginal *C. muridarum* infections are self-limiting, ascend to the upper reproductive tract resulting in salpingitis and infertility, and infection induces long-lasting immunity [6]. Mice must however, be hormonally manipulated in order that 100% of animals can be infected, which prevents the use of mice for sexual transmission studies. Genital *C. muridarum* infections also disseminate to other tissues (ovary, blood, spleen, brain, joints) facilitating studies of nongenital chlamydial disease. *C. muridarum* is also able to establish an infection in male mice allowing studies of effects on male fertility. Female mice can also be infected with human strains of *C. trachomatis*; however vaginal infections do not ascend to the uterine horns or oviducts, and upper tract disease only occurs following direct infection into the uterine horns or oviducts with high numbers of bacteria. The major strengths of the mouse model are small size and ease of handling, low cost, availability of many genetically modified knockout and transgenic mouse strains and the abundance of immunological reagents. For these reasons the mouse is the model of choice for most studies, particularly immunopathogenesis and vaccine studies. Guinea pigs can be infected with *C. caviae*, a chlamydial species originally isolated from guinea pigs with conjunctivitis [7]. Both female and male guinea pigs can be infected [8], and importantly, infection of females does not require hormonal manipulation, facilitating studies of natural sexual transmission. Guinea pigs also have a longer estrus cycle (14–19 days) than mice (4–5 days) that more closely approximates the human cycle. Guinea pigs are also easy to handle and relatively low cost; however, studies may be limited by availability of immunological reagents, although this situation is improving. In this chapter we provide methods for infection of mice with both *C. muridarum* and *C. trachomatis* and infection of guinea pigs with *C. caviae*.

2 Materials

General laboratory equipment and consumables, such as micropipettes, sterile pipette tips, culture flasks, media etc., as well as required PPE is assumed and therefore not listed here.

2.1 Female Genital Infections of Rodents

1. Adolescent (6–8 weeks) female mice. Generally, inbred BALB/c, C57BL/6, or C3H mice are used due to availability and widespread use (*see* **Note 1**).
2. Adolescent female Lewis, Wistar, and/or Sprague-Dawley rats (*see* **Note 1**).
3. Sexually mature female guinea pigs (random outbred) (weighing 600–800 g). Guinea pigs are fed standard guinea pig

pellets, and drinking water requires supplementation with 4 mg/20 L ascorbic acid (vitamin c) (*see* **Note 1**).

4. *Chlamydia trachomatis* (urogenital strains = Serovar D, E, F, G, H, I, J, K, L2), *Chlamydia muridarum* (Nigg), and *Chlamydia caviae*. All can be purchased from American Type Culture Collection (ATCC). *Chlamydia* spp. are generally propagated in mouse McCoy fibroblast or human HeLa cells lines, and purified using Renograffin, by centrifugation on continuous isopycnic Renografin density gradients, or Ultravist as described elsewhere [9].
5. Sucrose–phosphate–glutamine (SPG) buffer: 75 g sucrose, 0.52 g Na_2HPO_4, 1.22 g NaH_2PO_4 in 1 L deionized H_2O, pH adjusted to 7. Sterilized by autoclaving at 121 °C for 15 min. Add 5 mM L-glutamine prior to use. Keep at 4 °C for up to 1 month. SPG aliquots for vaginal swabs are prepared by placing three glass beads in a 1.5 mL microfuge tube and adding 500 μL of SPG.
6. Nonsurgical embryo transfer device (NSET; ParaTechs Corp., Lexington, KY).
7. Ketamine (ketamine injection as ketamine hydrochloride) (*see* **Note 2**).
8. Xylazine (Ilium Xylazil as hydrochloride) (*see* **Note 2**).
9. Depo Provera® (medroxyprogesterone acetate) (*see* **Note 3**).
10. Appropriately sized speculum (ParaTechs Corp., Lexington, KY).
11. PCR assays for detection of *Chlamydia* species (*see* **Note 4**).
12. McCoy B or HeLa cells for determination of recoverable *Chlamydia*.
13. Heating pad.
14. Clippers (Wahl, Sterling, IL).
15. 10% povidone-iodine *or* 2% chlorhexidine gluconate
16. Sterile gauze.
17. Sterile scalpels.
18. 30G needles
19. Dissolvable suture materials and surgical wound closure clips.
20. Sterile saline.
21. Sterile swabs.
22. 70% ethanol
23. Cotton balls.
24. Sterile microfuge tubes.

2.2 Male Urogenital Infections

1. Adolescent (6–8 weeks) male mice. Generally, inbred BALB/c, C57BL/6, or C3H mice are used due to availability and widespread use (*see* **Note 1**).
2. Adolescent male Lewis, Wistar, and/or Sprague-Dawley rats (*see* **Note 1**).
3. Sexually mature male guinea pigs (random outbred) (weighing 600–800 g). Guinea pigs are fed standard guinea pig pellets, and drinking water requires supplementation with 4 mg/20 L ascorbic acid (vitamin c) (*see* **Note 1**).
4. *Chlamydia trachomatis* (urogenital strains = Serovar D, E, F, G, H, I, J, K, L2), *Chlamydia muridarum* (Nigg), and *Chlamydia caviae.* All can be purchased from American Type Culture Collection (ATCC). *Chlamydia* spp. are generally propagated in mouse McCoy fibroblast or human HeLa cells lines, and purified using Renograffin or Ultravist as described elsewhere [9].
5. Sucrose–phosphate–glutamine (SPG) buffer: 75 g sucrose, 0.52 g Na_2HPO_4, 1.22 g NaH_2PO_4 in 1 L deionized H_2O, pH adjusted to 7. Sterilized by autoclaving at 121 °C for 15 min. Add 5 mM L-glutamine prior to use. Keep at 4 °C for up to 1 month.
6. Ketamine (ketamine injection as ketamine hydrochloride) (*see* **Note 2**).
7. Xylazine (Ilium Xylazil as hydrochloride) (*see* **Note 2**).
8. Isoflurane (*see* **Note 2**).
9. PCR assays for detection of *Chlamydia* species (*see* **Notes 4** and **5**).
10. McCoy B cells for determination of recoverable *Chlamydia.*
11. Dacron swabs.

2.3 C. muridarum and C. pneumoniae Respiratory Infection

1. Adolescent (6–8 weeks) mice. Generally, inbred BALB/c, C57BL/6, or C3H mice are used due to availability and widespread use (*see* **Note 1**).
2. *Chlamydia pneumoniae* or *Chlamydia muridarum* (Nigg). All can be purchased from American Type Culture Collection (ATCC). *Chlamydia* spp. are generally propagated in mouse McCoy fibroblast or human HeLa cells lines, and purified using Renograffin or Ultravist as described elsewhere [9], (*see* **Note 6**).
3. Sucrose–phosphate–glutamine (SPG) buffer: 75 g sucrose, 0.52 g Na_2HPO_4, 1.22 g NaH_2PO_4 in 1 L deionized H_2O, pH adjusted to 7. Sterilized by autoclaving at 121 °C for 15 min. Add 5 mM L-glutamine prior to use. Keep at 4 °C for up to 1 month.

4. 4% Isoflurane (*see* **Note 2**).
5. Ear tags.
6. Digital balance.

2.4 C. muridarum and C. trachomatis Gastrointestinal Infections

1. Adolescent (6–8 weeks) mice. Generally, inbred BALB/c, C57BL/6, or C3H mice are used due to availability and widespread use (*see* **Note 1**).
2. *Chlamydia trachomatis* (urogenital strains = Serovar D, E, F, G, H, I, J, K, L2), and *Chlamydia muridarum* (Nigg). All can be purchased from American Type Culture Collection (ATCC). *Chlamydia* spp. are generally propagated in mouse McCoy fibroblast or human HeLa cells lines, and purified using Renograffin or Ultravist as described elsewhere [9] (*see* **Note 7**).
3. Sucrose–phosphate–glutamine (SPG) buffer: 75 g sucrose, 0.52 g Na_2HPO_4, 1.22 g NaH_2PO_4 in 1 L deionized H_2O, pH adjusted to 7. Sterilized by autoclaving at 121 °C for 15 min. Add 5 mM L-glutamine prior to use. Keep at 4 °C for up to 1 month.
4. 4% Isoflurane (*see* **Note 2**).
5. PCR assays for detection of *Chlamydia* species (*see* **Note 4**).
6. 20–22G Jorgensen feeding needle attached to a syringe.
7. PCR assays for detection of *Chlamydia* species (*see* **Notes 4** and **5**).
8. McCoy B cells for determination of recoverable *Chlamydia*.

2.5 Ocular Infections

1. Adolescent (6–8 weeks) mice. Generally, inbred BALB/c, C57BL/6, or C3H mice are used due to availability and widespread use (*see* **Note 1**).
2. *Chlamydia trachomatis* (urogenital strains = Serovar D, E, F, G, H, I, J, K, L2), and *Chlamydia muridarum* (Nigg). All can be purchased from American Type Culture Collection (ATCC). *Chlamydia* spp. are generally propagated in mouse McCoy fibroblast or human HeLa cells lines, and purified using Renograffin or Ultravist as described elsewhere [9] (*see* **Note 8**).
3. Ketamine (ketamine injection as ketamine hydrochloride) (*see* **Note 2**).
4. Xylazine (Ilium Xylazil as hydrochloride) (*see* **Note 2**).

2.6 Systemic Infections

1. Adolescent (6–8 weeks) mice. Generally, inbred BALB/c, C57BL/6, or C3H mice are used due to availability and widespread use (*see* **Note 1**).

2. *Chlamydia trachomatis* (urogenital strains = Serovar D, E, F, G, H, I, J, K, L2), and *Chlamydia muridarum* (Nigg). All can be purchased from American Type Culture Collection (ATCC). *Chlamydia* spp. are generally propagated in mouse McCoy fibroblast or human HeLa cells lines, and purified using Renograffin or Ultravist as described elsewhere [9] (*see* **Notes 9** and **10**).
3. Heating lamp.
4. Tailvein Restrainer (Braintree Scientific).
5. 29–30G needle and syringe.
6. Sterile gauze or absorbent material.
7. 4% isoflurane.

3 Methods

3.1 Female Genital Infections of Rodents

Mice: *C. muridarum* and *C. trachomatis* Vaginal Infection (*See* **Notes 3**, **11**, and **12**)

1. Following 1 week of Depo Provera® pretreatment (*see* **Note 3**), anesthetize female mice are using ketamine (100 mg/kg) and xylazine (10 mg/kg) diluted to 200 μL in PBS and delivered via intraperitoneal injection.
2. Gently place anesthetized mice on their backs on a slight incline with the reproductive tract facing upward. Using a sterile 20 or 200 μL pipette tip and a P20 pipette, deliver 20 μL of *C. muridarum* or *C. trachomatis* in SPG into the vaginal vault (Fig. 1). Keep mice slightly elevated for 5 min to prevent the inoculum from draining out the vaginal vault. Swab external genitalia with 70% ethanol to minimize respiratory infection of other mice due to inhalation during normal mouse activity.
3. Return mice to cages and observe for recovery from anesthesia.
4. Take vaginal swabs throughout the duration of infection to track the kinetics of the infection by culture or PCR (*see* **Note 4**); generally, every 3 days to detect *Chlamydia* spp. that have gone through a round of infection and release. Infection usually peaks around day 6–9.

Mice: *C. trachomatis* Transcervical Infection (*See* **Note 13**)

1. Anesthetize mice using ketamine (100 mg/kg) and xylazine (10 mg/kg) diluted to 200 μL in PBS and delivered via intraperitoneal injection.
2. Attach a nonsurgical embryo transfer device (NSET; ParaTechs Corp., Lexington, KY) to a pipette and load with the required infectious dose of *C. trachomatis*.

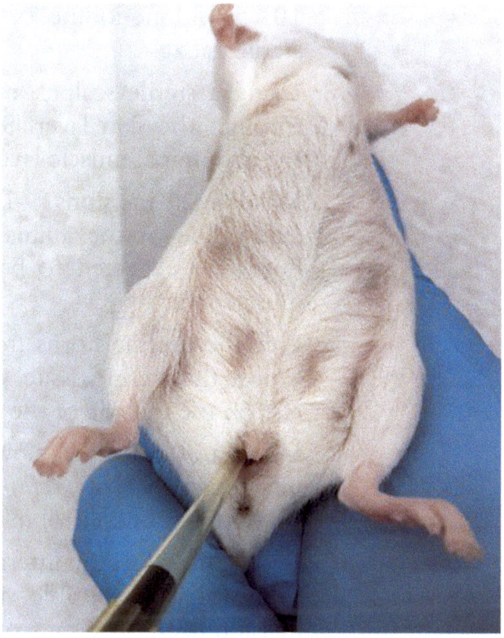

Fig. 1 Intravaginal infection of mice

3. Insert an appropriately sized speculum into the mouse vaginal vault.
4. Insert the NSET through the speculum into the uterine horn, bypassing the cervical os, directing the device to either the left or ride side.
5. Pipette the infectious dose into the uterine horn and remove the NSET tip and the speculum. If infection of both horns is required repeat **steps 2–5**.
6. Return the mouse to the cage and allow it to recover from the anesthetic.
7. To monitor infection whole tissue is required to be collected from euthanized animals, homogenized, and sub-cultured onto McCoy B or HeLa cells to determine the level of recoverable *Chlamydia*.

Rats: *C. muridarum* and *C. trachomatis* Intrauterine Infection (*See* **Note 14**)

1. Anesthetize female rats using ketamine (100 mg/kg) and xylazine (10 mg/kg) diluted to 200 μL in PBS and delivered via intraperitoneal injection.
2. Using clippers, remove approximately a 3 cm^2 area of fur from the dorsal surface (back), halfway between the tail and the center of the back. Disinfect the shaved surface with either

10% povidone-iodine or 2% chlorhexidine gluconate using sterile gauze.

3. Using a sterile scalpel, make a 1.5–2.5 cm midline incision through the skin layer. Slightly separate the skin layer from the abdominal muscle layer.
4. Make small incisions (~1 cm) through the muscle layer on either side of the animal directly above where the uterine horns are expected to be (they may be visible through the muscle layer).
5. Free the uterine horns from the periovarian fatty tissues and expose the uterine tissue. Gently pull the uterine horns out and rest on the outside of the muscle layer.
6. Using a 30G needle, slowly inject 25–50 μL of the required infectious dose of *Chlamydia* into each horn. Inject sterile PBS/saline for control mice.
7. Return horns to the internal cavity and close the internal incisions using dissolvable suture materials. Close the external incision using surgical wound closure clips.
8. Pain relief may be required as per your Institutions' ethical guidelines and should be administered as required.

Guinea Pigs: *C. caviae* Vaginal Infection (*See* **Note 15**)

1. Anesthetize Guinea pigs by intraperitoneally injecting anesthetic into the posterior quadrant of the abdomen, with xylazine (5 mg/kg) and ketamine (40 mg/kg) diluted in PBS a total volume of 325 μL (*see* **Note 16**).
2. Rupture vaginal membrane using a swab moistened with sterile saline.
3. Infect Guinea pig by inserting a micropipette tip containing 5×10^3 IFUs of *C. caviae* in 25 μL of SPG against the cervix (approximately 3 cm).
4. Remove the pipette tip and clean the external genitals with a cotton ball soaked in 70% ethanol.
5. Return Guinea pigs to cage and monitor until they recover from the anesthesia.
6. Vaginal swabs can be taken every 3 days by restraining the animal as previously described, and using a swab presoaked in 500 μL of SPG, swab the vaginal vault (three times clockwise, three times counterclockwise). Place vaginal swab in labeled microfuge tube and cut off the tip.

3.2 Male Urogenital Infections

Mice: *C. muridarum* Penile Infection (*See* **Note 17**)

1. Restrain and anesthetize male mice with ketamine (100 mg/kg) and xylazine (10 mg/kg) via the intraperitoneal route.

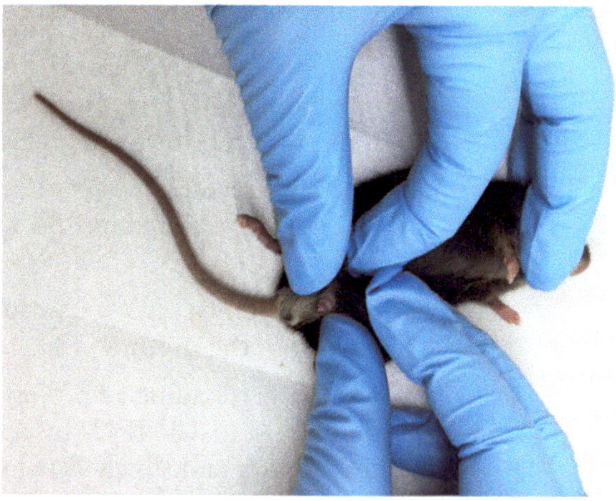

Fig. 2 Retracting the penile foreskin to expose the meatus urethra (glans), for penile infection

2. Using an index finger and thumb, retract the penile foreskin, exposing the glans (Fig. 2).
3. Pipette 5 μL of concentrated *C. muridarum* (10^6 IFUs to guarantee 100% infection [10]) onto the glans. Capillary action will facilitate entry into the penile urethra. If the chlamydial seed is not concentrated enough, two doses of 5 μL can be applied with a 5-min rest.
4. Return animals to cages and monitor until they recover from the anesthetic.

Rats: *C. muridarum* Penile Infection

1. Anesthetize male rats with ketamine (100 mg/kg) and xylazine (10 mg/kg) via the intraperitoneal route.
2. Lay rats on their back, and pull back the prepuce to expose the meatus urethra.
3. Place 5–10 μL of SPG containing 10^7 IFU of *Chlamydia* on the end of the meatus urethra, keeping animals on their back. The inoculum will reflux into the urethra.
4. Repeat this infection protocol for three consecutive days to ensure an infection is established [11, 12].
5. To monitor infection levels, semen and tissues (urethra, bladder, seminal vesicles, prostate, testis, epididymis) can be collected and either sub-cultured on McCoy B cells or examined via PCR using the chlamydial 16S rRNA gene (*see* **Note 4**). Tissues should be homogenized in 500–1000 μL of diluent (SPG (for culture), TRIzol, RNAlater). Prostatic fluid can be collected under isoflurane anesthesia via electrostimulation of the prostate gland with a probe and square wave generator [13].

Guinea Pigs: *C. caviae* Penile Infection

1. To reduce the risk of urination, and therefore expulsion of the infectious inoculum, remove water bottles from cage a minimum of 3 h prior to the infection procedure.
2. Anesthetize Guinea pigs with xylazine (5 mg/kg) and ketamine (40 mg/kg) diluted in PBS a total volume of 325 μL (*see* **Note 16**).
3. Lay the guinea pig on its back and retract the external meatus to expose the urethra [14].
4. Load a gel-loading micropipette tip with 10 μL of SPG containing up to 10^5 IFU. Insert the tip approximately 2 cm into the urethra and slowly expel the inoculum.
5. Keep animals on their backs until they begin to recover from the anesthetic, then return to their cages.
6. To monitor the levels of infection urethral swabs can be collected.
7. To collect swabs anesthetize as in **step 1**. Insert a Dacron swab approximately 2 cm into the urethra, rotate gently up to five times, remove and place in SPG for storage at −80 °C until needed [8, 15].

3.3 Mice: C. muridarum and C. pneumoniae Respiratory Infection (See Note 18)

1. Ear-tag mice and record weights prior to infection.
2. Anesthetize mice using 4% isoflurane until mice lose consciousness.
3. Pick up the mouse holding it so that the ventral side is facing upward and the head is at a downward angle.
4. Pipette 10 μL (5 μL per nare) of 1000 IFUs of *C. muridarum*, or 1–5 × 10^4 IFUs of *C. pneumoniae* onto the nostrils (*see* Fig. 3). Allow the animal to inhale the solution as they recover from anesthesia.
5. Return mice to the cage, weighing daily for weight loss (*see* **Note 19**).

3.4 Mice: C. muridarum and C. trachomatis Gastrointestinal Infections (See Note 20)

1. Anesthetize mouse through the delivery of 4% isoflurane until mouse loses consciousness.
2. Restrain the mouse by firmly holding the skin from the scruff of the neck down the back to immobilize the head.
3. Hold the mouse in a vertical position, so there is a straight line between the mouth and pharynx, into the esophagus.
4. It is recommended that the infectious inoculum is loaded into the syringe through the 20–22G Jorgensen feeding needle (these needles have a round end to avoid piercing the

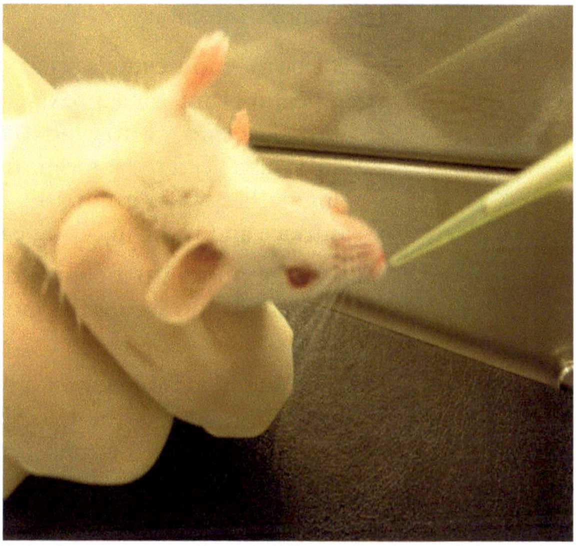

Fig. 3 Intranasal infection (or immunization) of mice

esophageal tissues). This will ensure that the needle is preloaded with the inoculum and the correct dosage is administered.

5. Pass the needle gently through the mouth and pharynx into the esophagus. When done correctly the needle should not need to be forced into position; it should slide in gently.
6. Inject the required dose of *Chlamydia* by applying gentle pressure to the syringe. The infectious dose for *C. muridarum* is 10^3–10^4 IFU, for *C. trachomatis* is 10^6–10^7 IFU, which is delivered in a volume of 30–50 μL [16–18].
7. Return the mouse to the cage for recovery.
8. To directly monitor the levels of infection in the gastrointestinal tissues animals need to be euthanized and tissues collected for sub-culturing or PCR (*see* **Note 4**). Another method to track the infection is to use luciferase-expressing *C. muridarum* [19] for infections. To detect the luciferase, 30 min prior to imaging inject 500 μL of D-luciferin (40 mg/mL) intraperitoneally. Mice can then be imaged using the Xenogen IVIS imaging system (PerkinElmer).

3.5 Ocular Infections

1. Anesthetize mice with ketamine (100 mg/kg) and xylazine (10 mg/kg) via the intraperitoneal route.
2. Using a micropipette and sterile tips place 5 μL of the infectious inoculum (5 × 10^4 IFU) onto the upper fornix of the eye and close the eye [20].
3. Return animals to their cages.

3.6 Systemic Infection (Mice)

Intravenous Infection

1. Place the mouse into a separate cage and warm the mouse using a heating lamp for approximately 5 min. Ensure that the lamp is not close enough to cage to either melt or cause ignition. The mice also need to be monitored for heat stress (panting, sweating, lethargy). Warming the mice will bring the tail vein to the surface and make injecting easier.
2. Place mouse into a commercial Tailvein Restrainer (Braintree Scientific) and secure the mouse with the tail exposed and free at the end.
3. The lateral tail veins should be used and are located on the left and right sides of the tail. Slightly turn the tail so one of the lateral veins is at the top.
4. Slightly bend the tail over the finger that is on the underside of the tail to create an approximate 45° angle with the tip of the tail pointing downward, while the base of the tail is close to vertical. Veins are easier to see in white mice and appear purple in color. In black/brown mice they are harder to see but are the darker lines down the tail.
5. Using a 29–30G needle and syringe loaded with 10^6 IFU of *Chlamydia*, insert the needle into the tail vein [21]. If the needle is in the vein, a small amount of blood will backflow into the syringe. Slowly inject the required volume. If there is resistance to the injection, the needle is not in the vein and needle insertion will need to be reattempted.
6. Slowly remove the needle and apply pressure to the puncture wound with sterile gauze or absorbent material until bleeding stops.
7. Return the mouse to the cage.

Retro-orbital Infection

1. Anesthetize mouse through delivery the of 4% isoflurane until mouse loses consciousness.
2. Remove the mouse from the isoflurane and place on the stomach. Gently retract the facial skin toward the body, causing the eye to protrude. Be sure not to apply too much pressure as this will restrict breathing.
3. Insert the needle with the bevel edge facing up, into the medial canthus of the eye at a 45° angle to the nose, into the vessels behind the eye. Be sure not to move the needle once inserted into the eye as this will rupture the vessels, causing bleeding and loss of the inoculum into the back of the eye.
4. Inject no more than a 100 μL volume containing the *Chlamydia* (10^6 IFU) [21].

5. Withdraw the needle and apply light pressure to the eye to control any bleeding. Once the animal has regained consciousness return to the cage.

4 Notes

1. Animal work requires ethical approval from the appropriate Institute/University and procedures should be performed in a BSL-2/PC2 approved facility. Infections of rodents should be performed in a facility isolated from noninfectious mice to prevent cross-contamination and infection of clean rodents with *Chlamydia* spp. (particularly *C. muridarum* which can spread to other mice as a respiratory infection). Detailed information regarding manipulation of animals etc. described in this protocol can be found here http://www.usp.br/bioterio/Artigos/Procedimentos%20experimentais/Routeadministration-4.pdf.
2. The use and purchase of anesthetics requires approval from national governing bodies and is pertinent to the country where the animal work is to be performed in.
3. Infection of female mice requires hormonal synchronization into diestrus which can be achieved via a single subcutaneous injection of 2.5 mg Depo Provera® diluted to 200 μL with PBS 1 week prior to challenge [6].
4. Table 1: PCR amplification of *C. trachomatis* serD, *C. pneumoniae* and *C. muridarum*, Primer sequences.

   Thermocycler Conditions:

   (PCR mastermix and Taq polymerase were from InVitrogen)
   95 °C-10 min; (95 °C-30 s/52 °C-30 s/74 °C-30 s) × 34; 74 °C-2 min; Hold at 4 °C.
5. In our hands, infection of male mice with *C. muridarum* does not resolve and viable *Chlamydiae* can be cultured from testicular homogenates up to 1-year postinfection [22]. It is possible to monitor chlamydial shedding in the urine by culture or PCR, but results are highly variable and inconsistent depending on the volume of voided urine, hydration of the animal, and inhibitors in the urine.
6. Respiratory infection of mice with human isolates of *C. pneumoniae* recovered from the brains of Alzheimer's patients also leads to brain infection in mice [23]. Infection resulted in the detection of Amyloid beta deposits in mouse brains, which increased in number, size, and density as the infection progressed.

Table 1
PCR amplification of *C. trachomatis* serD, *C. pneumoniae*, and *C. muridarum*, primer sequences

Gene	Dir	Sequence (5′–3′)	Anneal	Amplicon	$MgCl_2$ conc
Cpn *ompA*	F	cgctggcgtagcaacag	57	70	2.0–2.5 mM
	R	ggctcctacttgccattcatgataa	54		
Cpn 16S rRNA	F	gcaccttacctggacttgacatgt	56	82	1.5–2.5 mM
	R	ccatgcagcacctgtgtatctg	57		
Ctr *ompA*	F	tgccgctttgagttctgctt	56	75	2.0–2.5 mM
	R	gtcgatcataaggcttggttcag	55		
Ctr 16S rRNA	F	gcgaaggcgcttttctaatttat	55	76	1.5–2.5 mM
	R	ccagggtatctaatcctgtttgct	55		
Cmu *ompA*	F	gccgttttgggttctgctt	59	53	1.5–2.5 mM
	R	cgtcaatcataaggcttggttca	59		
Cmu 16S rRNA	F	tggttcagattgaacgctgg	59	121	1.5–2.5 mM
	R	ggccgtttccaaccgttatt	59		

7. Genital tract infection with *C. muridarum* results in a long-lasting (60 to >100 days) gastrointestinal infection [16, 24].
8. Guinea pigs can be infected with *C. caviae* for studies of conjunctivitis [25].
9. Genital infection of susceptible SKG mice, but not wild-type BALB/c mice, with *C. muridarum* results in arthritis, spondylitis, and psoriasiform dermatitis 5 weeks postinfection providing a model of reactive arthritis [26].
10. Intravenous infection of mice leads to a long-lasting infection only in the gastrointestinal tract [21].
11. *C. muridarum* is a natural pathogen of mice [5, 27] and readily replicates and disseminates in vivo. In female genital tract infections, vaginal inoculation of the lower reproductive tract is easily established and repeatable, and progresses to infect the uterine horns, oviducts (Fallopian tubes), and ovaries of the upper reproductive tract. Wild-type strains (Nigg and Weiss) will cause inflammation, scarring, occlusion, and swelling of the oviducts. Various doses of *C. muridarum* have been used with doses as low as 50 inclusion forming units (IFUs) routinely establishing an infection, with 500–5000 IFUs representing a natural male to female transmission dose [13]. Commonly, female mice are experimentally infected with 5×10^4 or up to 1×10^7 IFUs [6]. The peak of infection in females occurs within 6–9 days and resolves within 35 days. After 35 days, hydrosalpinx (oviduct scarring and blockage) is macroscopically visible and can be measured to determine the severity of occlusion. Hydrosalpinx does not resolve and mice remain infertile [28].

12. Vaginal infection of female mice with the human pathogen *C. trachomatis* requires higher doses of inoculum (10^7 IFUs), does not ascend from the vagina to the upper reproductive tract, causes minimal inflammation and self-resolves within a week. Postinfection inflammation only occurs following direct inoculation of *Chlamydia* into the uterine horns or ovarian bursa [29, 30].
13. As vaginal infection of mice does not ascend from the vagina to upper reproductive tract and cause infertility, direct inoculation of the uterus can be achieved by traversing the cervix [31].
14. Rats are not often used for genital chlamydial infections due to their very close similarities to the murine model. Much like mice, infection of female rats requires hormonal synchronization into diestrus, which can be achieved via two subcutaneous injections of 10 mg progesterone (Calbiochem) suspended in saline 1 week prior to challenge and the day of challenge [32, 33]. The intrauterine infection procedure requires recovery surgery; therefore, all materials and equipment must be sterile to prevent contamination of the surgical site and subsequent, unwanted infection of the rat. This procedure requires two persons to be present; one is primarily the "surgeon," while the other acts as the "non-sterile assistant," monitoring animal breathing and passing tools to the surgeon. Surgery should be performed with the rat on a heating pad to ensure body temperature does not drop during the procedure.
15. Guinea pigs have a 14–19-day estrus cycle that more closely mimics human females and thus are a more relevant model of human *C. trachomatis* infections [34, 35]. Unlike other rodents, guinea pigs do not require Depo Provera® pretreatment to achieve a chlamydial vaginal infection [36]. The stage of cycle at infection does, however affect the severity of upper reproductive tract inflammation [36]. Another advantage of the Guinea pig model is that females can be naturally infected by mating with infected males [8]. Guinea pigs are larger than mice and require two people to safely handle and experiment on.
16. Open the guinea pig cage and allow the animal to become alert to your presence as they are docile but easily frightened. Firmly, but gently grasp the animal around the shoulders with the left hand, placing the thumb on the back and index and middle fingers under the chest. Using the right hand, support the hind quarters and sit the animal vertically while crossing the front legs. Place the left thumb under the animal's jaw to prevent biting. Place the animal at a 45° angle allowing the second person to pull out the left hind leg and intraperitoneally inject anesthetic into the posterior quadrant of the abdomen.

17. A male model of urogenital chlamydial infection was established by Pal et al., and allows researchers to study immunopathogenicity and vaccine trials in mice [10]. Infection of the urethral glans leads to urethral, prostate, bladder, seminiferous vesicle, vas deferens, epididymal and testicular infection.
18. *Chlamydia muridarum* (formally *C. trachomatis* mouse *pneumonitis* (MoPn)) was originally isolated from mice suffering from respiratory infections [5, 27, 37]. *C. muridarum* infects the epithelia of the lung leading to fibrosis and collagen deposition, leading to cachexia and death if the animal is unable to recover from the infection. Respiratory infections of 10^3 IFUs are generally resolved within 10 days, with higher doses ($>10^6$) being fatal [38]. Infection of mice with the human pathogen *Chlamydia pneumoniae* can also induce immunopathology [39].
19. Depending on the decision of the Institute or University's animal ethics committee, mice should be euthanized if their weight drops >10% below the original preinfection weight.
20. Due to the acidic nature of the stomach, the pH of the delivery solution should not be alkaline, as it will not be well tolerated. When loading the feeding tube remember to take into account the volume that will remain in the needle after delivery.

Acknowledgments

KB is supported by NHMRC grants APP1083314 and APP1062198. KB and AJC are supported by NHMRC grant APP1145825. The authors thank Emily Bryan, Logan Trim, and David Van Der Heide for their assistance. CWA and AJC contributed equally.

References

1. WHO (2016) Sexually Transmitted Infections (STIs). Fact Sheets August 2016. Available from http://www.who.int/mediacentre/factsheets/fs110/en/
2. WHO (2018) Trachoma: Fact Sheet February 2018. Available from: http://www.who.int/news-room/fact-sheets/detail/trachoma
3. Zeidler H, Hudson AP (2016) Causality of Chlamydiae in arthritis and spondyloarthritis: a plea for increased translational research. Curr Rheumatol Rep 18(2):9
4. Choroszy-Krol I et al (2014) Infections caused by *Chlamydophila pneumoniae*. Adv Clin Exp Med 23(1):123–126
5. Nigg C (1942) An unidentified virus which produces pneumonia and systemic infection in mice. Science 95(2454):49–50
6. O'Meara CP, Andrew DW, Beagley KW (2014) The mouse model of Chlamydia genital tract infection: a review of infection, disease, immunity and vaccine development. Curr Mol Med 14(3):396–421
7. Murray ES (1964) Guinea pig inclusion conjunctivitis virus. I. Isolation and identification as a member of the psittacosis-lymphogranuloma-trachoma group. J Infect Dis 114:1–12
8. Rank RG et al (2003) Characterization of chlamydial genital infection resulting from sexual

transmission from male to female guinea pigs and determination of infectious dose. Infect Immun 71(11):6148–6154
9. Howard L, Orenstein NS, King NW (1974) Purification on renografin density gradients of *Chlamydia trachomatis* grown in the yolk sac of eggs. Appl Microbiol 27(1):102–106
10. Pal S, Peterson EM, de la Maza LM (2004) New murine model for the study of *Chlamydia trachomatis* genitourinary tract infections in males. Infect Immun 72(7):4210–4216
11. Mackern-Oberti JP et al (2011) Male rodent genital tract infection with Chlamydia muridarum: persistence in the prostate gland that triggers self-immune reactions in genetically susceptible hosts. J Urol 186(3):1100–1106
12. Motrich RD et al (2012) Male rat genital tract infection with *Chlamydia muridarum* has no significant consequence on male fertility. J Urol 187(5):1911–1917
13. O'Meara CP et al (2016) Induction of partial immunity in both males and females is sufficient to protect females against sexual transmission of Chlamydia. Mucosal Immunol 9 (4):1076–1088
14. Rank RG et al (1981) Cystitis associated with chlamydial infection of the genital tract in male guinea pigs. Sex Transm Dis 8(3):203–210
15. Wang Y et al (2010) Local host response to chlamydial urethral infection in male guinea pigs. Infect Immun 78(4):1670–1681
16. Igietseme JU, Portis JL, Perry LL (2001) Inflammation and clearance of *Chlamydia trachomatis* in enteric and nonenteric mucosae. Infect Immun 69(3):1832–1840
17. Erkkila L et al (2008) Intragastric primary infection sensitizes to lung reinfection in a *Chlamydia pneumoniae* mouse model. Vaccine 26(20):2503–2509
18. Wang L et al (2016) The *Chlamydia muridarum* organisms fail to auto-inoculate the mouse genital tract after colonization in the gastrointestinal tract for 70 days. PLoS One 11(5):e0155880
19. Campbell J et al (2014) Bioluminescence imaging of *Chlamydia muridarum* ascending infection in mice. PLoS One 9(7):e101634
20. Whittum-Hudson JA, O'Brien TP, Prendergast RA (1995) Murine model of ocular infection by a human biovar of *Chlamydia trachomatis*. Invest Ophthalmol Vis Sci 36 (10):1976–1987
21. Dai J et al (2016) Intravenous inoculation with *Chlamydia muridarum* leads to a long-lasting infection restricted to the gastrointestinal tract. Infect Immun 84(8):2382–2388
22. Kollipara A et al (2016) Chlamydia infection in males; an underappreciated problem, in 13th Congress of the International Society for Immunology of Reproduction (ISIR) and the European Society for Reproductive Immunology (ESRI). J Reprod Immunol 52
23. Little CS et al (2004) *Chlamydia pneumoniae* induces Alzheimer-like amyloid plaques in brains of BALB/c mice. Neurobiol Aging 25 (4):419–429
24. Zhang Q et al (2015) In vivo and ex vivo imaging reveals a long-lasting chlamydial infection in the mouse gastrointestinal tract following genital tract inoculation. Infect Immun 83 (9):3568–3577
25. Rank RG et al (2008) Chlamydiae and polymorphonuclear leukocytes: unlikely allies in the spread of chlamydial infection. FEMS Immunol Med Microbiol 54(1):104–113
26. Baillet AC et al (2015) High Chlamydia burden promotes tumor necrosis factor-dependent reactive arthritis in SKG mice. Arthritis Rheumatol 67(6):1535–1547
27. Nigg C, Eaton MD (1944) Isolation from normal mice of a pneumotropic virus which forms elementary bodies. J Exp Med 79(5):497–510
28. Ramsey KH, DeWolfe JL, Salyer RD (2000) Disease outcome subsequent to primary and secondary urogenital infection with murine or human biovars of *Chlamydia trachomatis*. Infect Immun 68(12):7186–7189
29. Tuffrey M et al (1986) Infertility in mice infected genitally with a human strain of *Chlamydia trachomatis*. J Reprod Fertil 78 (1):251–260
30. Tuffrey M et al (1986) Salpingitis in mice induced by human strains of *Chlamydia trachomatis*. Br J Exp Pathol 67(4):605–616
31. Gondek DC et al (2012) CD4+ T cells are necessary and sufficient to confer protection against *Chlamydia trachomatis* infection in the murine upper genital tract. J Immunol 189 (5):2441–2449
32. Kaushic C et al (1998) *Chlamydia trachomatis* infection in the female reproductive tract of the rat: influence of progesterone on infectivity and immune response. Infect Immun 66 (3):893–898
33. Kaushic C et al (2000) Effects of estradiol and progesterone on susceptibility and early immune responses to *Chlamydia trachomatis* infection in the female reproductive tract. Infect Immun 68(7):4207–4216
34. Barron AL et al (1979) Target tissues associated with genital infection of female guinea pigs by the chlamydial agent of guinea pig

inclusion conjunctivitis. J Infect Dis 139 (1):60–68

35. Rank RG, Sanders MM (1992) Pathogenesis of endometritis and salpingitis in a guinea pig model of chlamydial genital infection. Am J Pathol 140(4):927–936
36. Rank RG, Sanders MM, Kidd AT (1993) Influence of the estrous cycle on the development of upper genital tract pathology as a result of chlamydial infection in the guinea pig model of pelvic inflammatory disease. Am J Pathol 142(4):1291–1296
37. Gogolak FM (1953) Purification of murine pneumonitis virus from mouse lung. J Infect Dis 92(3):248–253
38. Gogolak FM (1953) A quantitative study of the infectivity of murine pneumonitis virus in mice infected in a cloud chamber of improved design. J Infect Dis 92(3):240–247
39. Kaukoranta-Tolvanen SS et al (1993) Experimental infection of *Chlamydia pneumoniae* in mice. Microb Pathog 15(4):293–302

Chapter 16

Enumeration of Viable Chlamydia from Infected Animals Using Immunofluorescent Microscopy

Steven Liang and James B. Mahony

Abstract

An appropriate means of quantitating infectious *Chlamydia* from infected animals is essential for the evaluation of vaccines. However, unlike methods involving culture, nonculture methods, including detection of antigen or DNA, are not able to differentiate between viable and nonviable organisms. As an obligate intracellular bacterium, *Chlamydia* replicates inside host cells by forming unique organelles called inclusions. Here, we describe the enumeration of viable *C. trachomatis* from infected mice by culturing vaginal swabs on McCoy cells and counting inclusions via immunofluorescent microscopy.

Key words Chlamydia, Immunofluorescent microscopy, Shedding, Vaccine efficacy, Viable *Chlamydia* enumeration

1 Introduction

An accurate, reproducible, and appropriate means of evaluating vaccine efficacy is crucial for the development of a human vaccine against *Chlamydia trachomatis*—the most common sexually transmitted bacterial infection in the world [1]. Primary outcome measures for assessing vaccine effectiveness in animal models include incidence and severity of infection, and associated pathology of the upper genital tract (UGT) [2]. Typically, in a vaccine study, a lower level of bacterial shedding in vaccinated animals, compared to unvaccinated control animals, is frequently used as the first screen of any potential vaccine candidate. Therefore, selecting a method of quantifying bacterial shedding appropriate for a given experimental design is critical for identifying promising vaccine candidates.

Bacterial load can be quantified using culture or non-culture methods. Non-culture quantification may involve antigen detection, nucleic acid detection, or cytologic examination [3]. With the exception of nucleic acid detection, these methods have lower sensitivity compared to culture methods, while having similar specificity [3, 4]. Nucleic acid amplification tests (NAAT) have high specificity

Amanda Claire Brown (ed.), *Chlamydia trachomatis: Methods and Protocols*, Methods in Molecular Biology, vol. 2042, https://doi.org/10.1007/978-1-4939-9694-0_16,

and can theoretically detect a single copy of chlamydial DNA, and can be modified to quantitate the number of genome copies in a sample using quantitative polymerase chain reaction (qPCR). However, methods involving only DNA detection, and other nonculture methods, are unable to differentiate between viable and nonviable organisms [3]. Since only live bacteria can cause incident infection, spread to the UGT and cause pathology [5], vaccines should aim to reduce the shedding of viable organisms, which can only be quantified by culture. As *Chlamydia* is an obligate intracellular bacterium that replicates within inclusion bodies in a host cell, culture methods involve infection of susceptible and permissive cells for 16–48 h and then enumerating inclusions, which are visualized by iodine staining, Giemsa staining, or immunofluorescent (IF) staining, with IF staining being the most specific method [6, 7].

Here, we describe the enumeration of infectious *C. trachomatis* from vaginal swabs using culture and IF staining, and offer suggestions to maximize efficiency and avoid pitfalls that can arise when performing this commonly used laboratory technique in *Chlamydia* vaccine research.

2 Materials

2.1 Processing of Swab Samples

1. Sucrose–phosphate–glutamate (SPG) Buffer (*see* **Note 1**): 250 mM sucrose, 10 mM sodium phosphate, 0.5 mM L-glutamic acid (pH 7.4 at 4 °C). Filter sterilize and store at 4 °C.
2. FLOQSwabs (Copan Diagnostics 516CS01.)
3. Glass Beads (1.0 mm) (BioSpec 11079110) (*see* **Note 2**).
4. Sterile Microcentrifuge Tubes with Screw Caps.
5. Vortex Mixer with Microtube Holder Attachment (Scientific Industries SI-0236 and SI-H524.)
6. Glass Pasteur Pipettes (VWR 14672-200.)
7. 10 mL Serological Pipettes (VWR 89130-898.)
8. Vacuum pump.

2.2 Preparation of McCoy Monolayers

1. McCoy B mouse fibroblasts: ATCC CRL-1696 (*see* **Note 3**). Store in liquid nitrogen (vapor phase).
2. T75 cell culture flasks.
3. McCoy Media: DMEM +10% fetal bovine serum (FBS). Store at 4 °C.
4. Trypsin.
5. Hemocytometer.
6. 96-Well Clear Flat-Bottom Cell Culture Plates with Lids (*see* **Note 4**)
7. Inverted microscope.

2.3 Infection of McCoy Monolayers

1. 96-well clear flat-bottom cell culture plates with lids containing attached 5×10^4 McCoy B Mouse Fibroblast cells per well.
2. *Chlamydia* Swab Sample (stored in SPG at −80 °C).
3. Infection Media: RPMI 1640 with L-glutamine, HEPES, 10% FBS, 50 μg/mL gentamicin, 1 μg/mL cycloheximide (*see* **Note 5**). Store at 4 °C.
4. Centrifuge with plate holder attachment.
5. 96-well U-bottom plates.
6. Inverted microscope.

2.4 Immuno-fluorescent Staining

1. Fixation Buffer (*see* **Note 6**) (BD Biosciences 554655.)
2. Staining Buffer (*see* **Note 7**) (BD Biosciences 554723.)
3. Washing Buffer: PBS with 0.1% Tween 20.
4. *Chlamydia*-Specific Primary Antibody (*see* **Note 8**) (Genetex GTX13943.)
5. Fluorophore-conjugated secondary antibody (*see* **Note 9**) (Jackson ImmunoResearch Laboratories 211-605-109.)

2.5 Immuno-fluorescent Microscopy

1. Fluorescent Microscope.

3 Methods

Certain *Chlamydia* species may cause human disease; apply standard of practice appropriate for the biosafety level of your pathogen.

3.1 Processing of Swab Samples

1. Use tweezers to take cotton filters out of two 10 mL serological pipettes and fit them snugly into the larger opening end of a glass Pasteur pipette and attach the other end into a rubber tube connected to a vacuum pump. Turn on vacuum pump to suction approximately 0.1 mL beads into the glass pipette, and then carefully detach the glass pipette from the rubber tube just until the vacuum is removed and beads can fall into a 2 mL sterile microcentrifuge tube.
2. On the day of sample collection, add 0.5 mL of SPG into each 2 mL sterile microcentrifuge tube containing 0.1 mL glass beads.
3. Gently insert FLOQSwabs into the vaginal vault until resistance is felt. Rotate the swab for 5–10 s. Withdraw the swab, unscrew your SPG microcentrifuge tube and place the cotton portion of the swab into the SPG. Use scissors to cut the shaft

of the swab just above the opening of the tube, taking care not to splash contents. Screw the cap tightly on the tube and keep the sample on ice until all samples are collected (*see* **Note 10**).

4. Insert the microcentrifuge tubes into a microtube holder attachment and vortex for 30 s on setting 10 to release bacteria from the swab (*see* **Note 11**).
5. Store samples at −80 °C.

3.2 Preparation of McCoy Monolayers

1. Maintain McCoy cells in McCoy Media at 37 °C/5% CO_2 in a T75 cell culture flask. Subculture as needed (*see* **Note 12**).
2. Remove McCoy Media and dissociate McCoy cell monolayer from cell culture flask by adding 3 mL of trypsin to the cells for 5–10 min at 37 °C (*see* **Note 13**).
3. Add 3 mL of McCoy Media to neutralize trypsin activity.
4. Count cells using a hemocytometer; seed 5×10^4 cells in at least 150 μL of McCoy Media per well into a 96-well flat bottom cell culture plate.
5. Allow cells to attach at 37 °C/5% CO_2 (*see* **Note 14**).

3.3 Infection of McCoy Monolayers

1. The next day check McCoy cells to ensure they are 90–100% confluent and there are no variations in confluency between wells.
2. Thaw *Chlamydia* samples in a 37 °C water bath, until they are almost completely thawed.
3. Make tenfold serial dilutions of *Chlamydia* samples (15 μL of sample into 135 μL of Infection Media) in a 96-well U-bottom plate (*see* **Note 15**).
4. Remove McCoy Media from the 96-well flat-bottom plate containing the monolayers and add 100 μL of your serially diluted samples (including an undiluted sample, if desired) into each well (*see* **Note 16**).
5. Centrifuge plate at 1500 × *g* for 1 h at 25 °C (*see* **Note 17**).
6. Remove inoculum and replace with at least 150 μL of Infection Media.
7. Incubate for 16–48 h at 37 °C/5% CO_2 (*see* **Note 18**).

3.4 Immunofluorescent Staining

1. Remove Infection Media and add at least 50 μL of Fixation Buffer to each well, to cover the monolayer, and leave on for 20 min.
2. Remove Fixation Buffer and wash each well twice with 100 μL of Washing Buffer.
3. Remove Washing Buffer and add at least 70 μL of Staining Buffer into each well.

4. Incubate for at least 1 h at room temperature or overnight at 4 °C.
5. Remove Staining Buffer, add 50 μL of *Chlamydia*-Specific Primary Antibody, and incubate for 30–60 min at 37 °C.
6. Remove primary antibodies and wash twice with 100 μL of Washing Buffer.
7. Perform all steps in the dark from this point forward (turn off fluorescent lights).
8. Remove Washing Buffer, add 50 μL of fluorophore-conjugated secondary antibody, and incubate for 30–60 min at 37 °C.
9. Remove secondary antibodies and wash three times with 100 μL of Washing Buffer.
10. Remove Washing Buffer and keep plate in the dark until ready to read under a fluorescent microscope (*see* **Note 19**).

3.5 Immunofluorescent Microscopy

1. For each sample, identify the well that has 10–100 inclusions (*see* **Note 20**).
2. Count the total number of inclusions in that well.
3. To calculate the number of infectious *Chlamydia* obtained from a swab, use the following formula:

   Inclusion forming unit (IFU)/Swab = (# inclusions per well) × (Dilution Factor) × (Sample Volume/Inoculum Volume)

   For example, if the swab is placed into 0.5 mL of SPG and 0.1 mL of a 10^{-4} dilution is used to infect a monolayer, yielding 20 inclusions, then according to the formula:

$$\begin{aligned} \text{IFU/Swab} &= 20 \times 10^4 \times 0.5/0.1 \\ &= 10^6 \text{ IFU/Swab} \end{aligned}$$

4 Notes

1. Avoid autoclaving solutions containing sucrose as high temperatures may affect its chemical structure. SPG is the most commonly used medium for long term storage of *Chlamydia*, but other formulations with similar effectiveness in preserving *Chlamydia* infectivity can also be used, such as those containing EMEM and FCS [8]. If desired, gentamicin (50 μg/mL) may be added to prevent bacterial contamination [9].
2. Although larger sized glass beads can be used, we have found that glass beads with a smaller diameter are more efficient at releasing bacteria from swabs.

3. McCoy cells are among the most commonly used cell lines for propagation of *Chlamydia*. Other cell types, including HeLa cells, can also be used with this protocol [10, 11].
4. To minimize the edge effect, where water loss from evaporation is more pronounced in wells near the edge of the plate, avoid using the outer wells for culture. Instead, add 100 μL of Infection Media to outer wells. Alternatively, specially designed 96-well plates with outer moats that can be filled with liquid are also available for purchase from many manufacturers.
5. Cycloheximide is an eukaryote protein synthesis inhibitor. Although not required, we recommend adding cycloheximide as it increases the size of inclusions, making them easier to count [12]. Gentamicin should always be added to prevent bacterial contamination.
6. Any formaldehyde based fixation buffer will work. Acetone is not recommended if using plastic plates.
7. If using an alternative staining buffer, make sure it contains a membrane permeabilizing reagent, such as saponin or Tween-20, as *Chlamydia* will be primarily intracellular.
8. The primary anti-*Chlamydia* antibody should not be made from the same species from which you obtained your samples. For example, if you obtained vaginal swabs from mice, we recommend using primary anti-*Chlamydia* antibodies made in any other species but mice. This is done to maximize signal-to-noise ratio, since if the primary anti-*Chlamydia* antibody was made in the same species from which you took your sample, then the secondary antibody can bind, in addition to the primary antibody, endogenous antibodies from the sample, thereby increasing background staining.
9. Ideally, to minimize the possibility of background staining, the secondary antibody should be made in the same species from which you obtained your samples, but can be made in other species.
10. *Chlamydia* loses its viability very quickly outside its host. In SPG at 4 °C, 95% of *C. pneumoniae* will survive after 4 h, while only 89% will survive at 22 °C [8].
11. Bacteria trapped in mucus on the swab may have trouble diffusing out into the solution, affecting the ability of *Chlamydia* to infect the McCoy monolayer and the accuracy of serial dilutions. This problem can be avoided by vortexing the swab sample with glass beads, thus releasing the bacteria into SPG.
12. McCoy cells from ATCC can be passaged for at least 43 times before *Chlamydia* growth is affected. Routinely check cells for

susceptibility to *Chlamydia* infection with a positive sample of known titer.

13. Cells could take longer to dissociate depending on the confluency of the monolayer and the specific dissociating reagent used. To help speed up this process, wash monolayer with sterile PBS to remove FBS before adding trypsin.
14. Cells will attach within a few hours and can be infected immediately after, if desired.
15. The number of serial dilutions needed depends on the number of viable *Chlamydia* that is shed during infection, which depends on the chlamydial species and strain used, the animal model, and the specific experimental design. Typically, during an intravaginal *C. trachomatis* infection in mice, no more than 10^7 viable bacteria will be shed at any given point. Therefore, dilutions beyond 10^{-6} should not be necessary. Use a U-bottom or V-bottom plate to optimize mixing and sample recovery.
16. Be careful not to aspirate McCoy cells—stop before the pipette reaches the bottom of the well, and pipette consistently in the same well position.
17. Centrifugation assisted infection significantly increases the sensitivity and accuracy of this method by enabling infection of host cells with all viable *Chlamydia* in the sample [13].
18. The chlamydial developmental cycle takes 24–72 h to complete depending on the chlamydial species [14, 15]. Make sure to

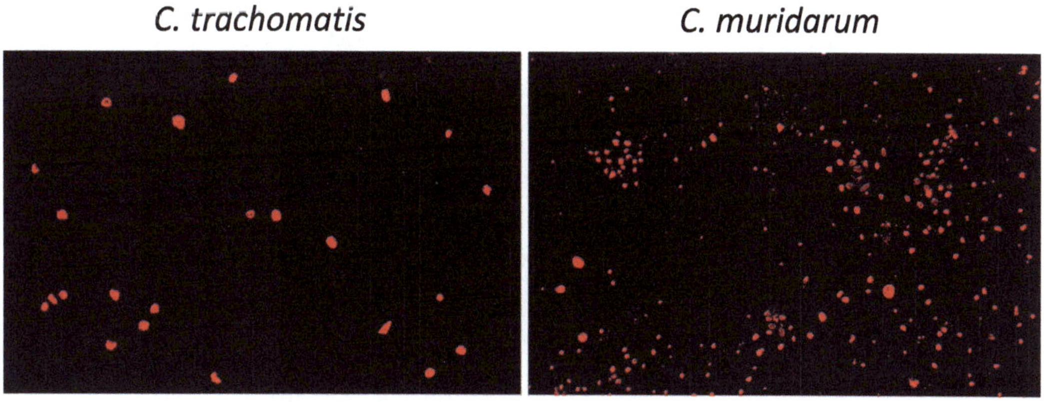

Fig. 1 *C. trachomatis* or *C. muridarum* (both at 5×10^3 IFU/mL) were used to infect confluent McCoy monolayers. At 42 h post infection, the monolayer was fixed, stained, and viewed under a fluorescent microscope. Although the two inoculums had the same number of infectious *Chlamydia*, a higher number of inclusions are seen in the *C. muridarum* infected monolayer compared to the *C. trachomatis* infected monolayer. This is explained by the fact that at 42 h post infection, *C. muridarum*, which has a 24-h developmental cycle, but not *C. trachomatis*, which has a 48-h developmental cycle, has completed its first developmental cycle and the newly formed elementary bodies have started to infect adjacent cells

stop the cycle before progeny bacteria are released from the infected monolayer. Newly formed elementary bodies can yield new inclusions, making it difficult to discriminate original inclusions from nascent inclusions (Fig. 1).

19. Depending on the fluorophore used, the fixation method, and the storage method, shelf life for stained plates may vary. Using materials and methods described in this protocol, plates can be stored for at least 4 weeks after staining.
20. Counting wells with too many or too few infected cells can yield inaccurate results.

References

1. World Health Organization (2015) Progress report of the implementation of the global strategy for prevention and control of sexually transmitted infections: 2006-2012
2. Zhong G, Brunham RC, de la Maza LM, Darville T, Deal C (2017) National Institute of Allergy and Infectious Diseases workshop report: "Chlamydia vaccines: the way forward". Vaccine
3. Black CM (1997) Current methods of laboratory diagnosis of *Chlamydia trachomatis* infections. Clin Microbiol Rev 10:160–184
4. Wu CH, Lee MF, Yin SC, Yang DM, Cheng SF (1992) Comparison of polymerase chain reaction, monoclonal antibody based enzyme immunoassay, and cell culture for detection of *Chlamydia trachomatis* in genital specimens. Sex Transm Dis 19:193–197
5. Brunham RC, Rey-Ladino J (2005) Immunology of Chlamydia infection: implications for a *Chlamydia trachomatis* vaccine. Nat Rev Immunol 5:149–161
6. Stamm WE, Tam M, Koester M, Cles L (1983) Detection of *Chlamydia trachomatis* inclusions in McCoy cell cultures with fluorescein-conjugated monoclonal antibodies. J Clin Microbiol 17:666–668
7. Stephens RS, Kuo CC, Tam MR (1982) Sensitivity of immunofluorescence with monoclonal antibodies for detection of *Chlamydia trachomatis* inclusions in cell culture. J Clin Microbiol 16:4–7
8. Maass M, Dalhoff K (1995) Transport and storage conditions for cultural recovery of *Chlamydia pneumoniae*. J Clin Microbiol 33:1793–1796
9. Wentworth BB (1973) Use of gentamicin in the isolation of subgroup A Chlamydia. Antimicrob Agents Chemother 3:698–702
10. Rota TR (1977) *Chlamydia trachomatis* in cell culture. II. Susceptibility of seven established mammalian cell types in vitro. Adaptation of trachoma organisms to McCoy and BHK-21 cells. In Vitro 13:280–292
11. Jenkin HM (1966) The continuous passage of agents of trachoma in cell culture. I. Characteristics of TW-3 and Bour strains of trachoma cultivated in serial passage in HeLa 229 cells. J Infect Dis 116:390–399
12. Ripa KT, Mardh PA (1977) Cultivation of *Chlamydia trachomatis* in cycloheximide-treated McCoy cells. J Clin Microbiol 6:328–331
13. Allan I, Pearce JH (1979) Modulation by centrifugation of cell susceptibility to chlamydial infection. J Gen Microbiol 111:87–92
14. Abdelrahman YM, Belland RJ (2005) The chlamydial developmental cycle. FEMS Microbiol Rev 29:949–959
15. Lyons JM, Ito JI Jr, Pena AS, Morre SA (2005) Differences in growth characteristics and elementary body associated cytotoxicity between *Chlamydia trachomatis* oculogenital serovars D and H and *Chlamydia muridarum*. J Clin Pathol 58:397–401

Chapter 17

Proximity Labeling of the *Chlamydia trachomatis* Inclusion Membrane

Macy G. Olson, Lisa M. Jorgenson, Ray E. Widner, and Elizabeth A. Rucks

Abstract

In the study of intracellular bacteria that reside within a membrane-bound vacuole, there are many questions related to how prokaryotic or eukaryotic transmembrane or membrane-associated proteins are organized and function within the membranes of these pathogen-containing vacuoles. Yet this host–pathogen interaction interface has proven difficult to experimentally resolve. For example, one method to begin to understand protein function is to determine the protein-binding partners; however, examining protein–protein interactions of hydrophobic transmembrane proteins is not widely successful using standard immunoprecipitation or coimmunoprecipitation techniques. In these scenarios, the lysis conditions that maintain protein–protein interactions are not compatible with solubilizing hydrophobic membrane proteins. In this chapter, we outline two proximity labeling systems to circumvent these issues to study (1) eukaryotic proteins that localize to the membrane-bound inclusion formed by *Chlamydia trachomatis* using BioID, and (2) chlamydial proteins that are inserted into the inclusion membrane using APEX2. BioID is a promiscuous biotin ligase to tag proximal proteins with biotin. APEX2 is an ascorbate peroxidase that creates biotin-phenoxyl radicals to label proximal proteins with biotin or 3,3′-diaminobenzidine intermediates for examination of APEX2 labeling of subcellular structures using transmission electron microscopy. We present how these methods were originally conceptualized and developed, so that the user can understand the strengths and limitations of each proximity labeling system. We discuss important considerations regarding experimental design, which include careful consideration of background conditions and statistical analysis of mass spectrometry results. When applied in the appropriate context with adequate controls, these methods can be powerful tools toward understanding membrane interfaces between intracellular pathogens and their hosts.

Key words *Chlamydia trachomatis*, Proximity labeling, Host–pathogen interactions, BioID, APEX2

1 Introduction

1.1 Overview

Members of *Chlamydiae* are obligate intracellular pathogens and the entirety of their characteristic and unique biphasic developmental cycle is completed within a vacuole, which is called the chlamydial inclusion. *Chlamydia* manipulates the host cell in order to create

---

Macy G. Olson and Lisa M. Jorgenson contributed equally to this work.

Amanda Claire Brown (ed.), *Chlamydia trachomatis: Methods and Protocols*, Methods in Molecular Biology, vol. 2042,
https://doi.org/10.1007/978-1-4939-9694-0_17,

and maintain the inclusion as a stable specialized niche to promote chlamydial growth and development. Although *Chlamydia* has undergone significant genomic reduction, they have maintained a significant portion of their coding capacity towards a type III secretion system and associated effector proteins [1]. The chlamydial type III secretion system is essential [2], which highlights the importance of this system in modulating chlamydial–host interactions. A significant portion of chlamydial type III effectors have large hydrophobic transmembrane domains and belong to the Inc family of proteins; these effectors reside within the inclusion membrane. Incs are characterized by having two or more hydrophobic transmembrane domains, with the amino-terminal and carboxy-terminal ends of the proteins exposed to the eukaryotic cytosol [3–5]; although, it is conceivable that the N- and C-termini of some Incs may be exposed to the lumen of the chlamydial inclusion. While these proteins are important to the intracellular development of *Chlamydia*, their high hydrophobicity makes them difficult proteins to purify and study. This chapter describes methods that allow for the study of protein–protein interactions of chlamydial membrane and host membrane proteins. Ostensibly, these methods will be applicable to any investigator trying to examine the molecular interactions of pathogen and host proteins, particularly within the context of membrane-associated interactions.

1.2 Introduction to *Chlamydia*

Chlamydia trachomatis is the chlamydial organism most commonly associated with human disease and one of the most common human bacterial pathogens. *C. trachomatis* serovars cause either blinding trachoma [6] or the most common bacterial sexually transmitted infection [7, 8]. Primary chlamydial infections in women are often innocuous, which increases the likelihood that they will go untreated and ascend from the cervix into the upper genital tract to cause pelvic inflammatory disease, ectopic pregnancy, and/or infertility. A hallmark of chlamydial infection is the ability of the pathogen to thrive within the host while limiting obvious signs of inflammation during an active infection [9]. Current prevention and treatment strategies have failed to reduce the incidence of new infections and subsequent sequelae.

Chlamydiae are obligate intracellular bacterial pathogens that, during a productive growth cycle, use a developmental cycle to alternate between infectious, metabolically quiescent elementary bodies (EBs) and non-infectious, metabolically active reticulate bodies (RBs) (*see* [10] for review). The two forms of *Chlamydia* reflect their distinct roles: the EB mediates attachment and internalization into a susceptible host cell and the RB grows and divides using a novel budding mechanism [11]. EB-differentiation into an RB, RB growth and division, and RB secondary differentiation into infectious EBs occurs within the chlamydial inclusion. The molecular events and signals required for differentiation between these morphologic forms are not understood and the secondary differentiation step is asynchronous, meaning that while some RBs undergo differentiation into EBs, other RBs continue to divide via budding. Genome-wide microarray and targeted transcriptional studies have helped define genes that may be important at critical stages such as the EB-to-RB (early) and RB-to-EB (late) transition [12–14].

Conceptually, host–pathogen interactions entail a pathogen directly engaging with host machinery or secretion of an effector to modulate host cell activity to help create a more hospitable environment for the pathogen. However, the inclusion membrane creates a barrier between *Chlamydia* and the host cell. Without direct access to the host cytosol, *Chlamydia* orchestrates host cell functions via the inclusion membrane. Both chlamydial protein synthesis and type III secretion are necessary for the early remodeling of the inclusion membrane [15, 16]. Early seminal studies attempting to identify components of the inclusion revealed the presence of chlamydial proteins, which were later defined as Incs, inserted into the inclusion membrane [17, 18]. Incs have one key identifying motif: a large hydrophobic region encoding at least two transmembrane domains. *Inc* genes encode a type III secretion signal that allows the bacterium to secrete these proteins for insertion into the inclusion membrane [15].

Type III secretion systems are virulence factors of Gram-negative pathogens [19] and, in *Chlamydia*, are linked to chlamydial survival and virulence [20]. There are type III effectors that are directly secreted into the host cell cytosol [21, 22]. Once Incs are type III secreted, they are inserted into the inclusion membrane, where they likely act as virulence proteins and contribute to chlamydial pathogenesis. Topological analyses indicate that both the N- and C-termini of Incs are exposed on the cytoplasmic side of the inclusion membrane [5], and a recent study using an anti-FLAG antibody showed that the C-terminus of IncD-FLAG is indeed exposed to the cytoplasm [3]. Given the similarity in structure, other Incs are thought to be positioned similarly in the chlamydial inclusion with their terminal ends being exposed to the cytosol. While Incs are obvious effectors to engage host cell proteins, only a handful have been shown to interact with a host cell component [23–25]. Despite the fundamental importance of the inclusion membrane to chlamydial fitness, specific molecular events, including Inc protein function, that are responsible for inclusion membrane composition and integrity are unknown.

The lack of mechanistic data concerning Incs and the inclusion membrane is due to two main reasons. First, the inclusion membrane is difficult to biochemically purify [26, 27]. A recent study reports the identification of a chlamydial inclusion interactome. This study relied on an extensive purification scheme that only captured 50% of the inclusions from 6×10^7 infected cells and failed to identify host–chlamydial or Inc–Inc interactions [26]. Labor-intensive schemes complicate attempts to characterize the inclusion membrane interactome over the course of the developmental cycle—key requirements in understanding the inclusion and its function. Second, Inc proteins are membrane proteins, which are notoriously difficult to study due to their hydrophobicity. Identification of interacting partners for Incs have typically relied on expressing the cytosolic domains of a given large Inc in yeast two-hybrid systems or purifying a recombinant version of this domain for pulldowns [23–25, 28–31]. A last and final consideration is that the addition of molecular tags may prevent Inc-fusion constructs from being properly type III secreted. To address two out of three of these obstacles, we have adapted several proximity labeling systems to understand protein–protein interactions at the chlamydial inclusion membrane. In this chapter, we will outline the origins of these systems, discuss how these methods have already been applied by the field to understand chlamydial–host interactions, and present our methods in adapting these systems to help understand chlamydial–host interactions.

1.3 Origins of BioID

The origins of the BioID proximity dependent labeling began with the examination of the impact of point mutations on the enzymatic activity of BirA, a bacterial biotin ligase. Biotinylation reactions by BirA occur in two steps: the first being the binding of BirA to biotinyl-5′-ATP [32], and the second is the recognition of the biotin acceptor tag, or BAT, where the biotin intermediate is transferred to a specific lysine [33]. Mutation of the arginine at position 118 to a glycine resulted in an alteration in BirA substrate specificity [34]. Further characterization of the mutant BirA, recognized that the BirA R118G (BirA∗) mutant maintained the ability to transfer biotin moieties to lysine residues of proteins in close proximity to the enzyme, but any specificity towards specific target proteins with known BATs was eliminated [35]. Most of these studies had been performed in *Escherichia coli*, but the fact that BirA∗ promiscuously biotinylated proteins in a proximity-dependent manner made it an attractive tool to adapt for cell biology studies. In the first cell biology-based study, Kyle Roux and colleagues created a molecular fusion between BirA∗ and a nuclear lamin protein [36]. Nuclear lamins are transmembrane proteins and important for the function of the nuclear envelope. Their hydrophobicity made them difficult proteins to study. Given that biotin has a strong affinity for streptavidin (K_D 10^{-13} to 10^{-15} M; [37]), they designed a BirA∗ construct to biotinylate potential lamin-interacting proteins, and then relied on the high affinity of biotin for streptavidin to purify these binding partners for identification by mass spectrometry [36].

Since its inception, the BioID system has been used in cell biology studies to identify regulators of centriole duplication and structure [38, 39] and to understand proteins involved in the insulin-like growth factor I receptor pathway [40]. In the field of cellular microbiology, the BioID system has been useful in determining protein targets of the *Legionella pneumophila* effector protein, PieE, [41], identifying components of the *Toxoplasma* inner membrane complex [42], components of the *Trypanosoma brucei* biolobe [43], and novel proteins on the *Plasmodium* parasitophorous vacuolar membrane [44]. Further, several laboratories, are adapting BioID in novel ways to advance the understanding of protein-protein complexes. These include the development of a "split BioID" system, that places the N-terminus or C-terminus of BirA* onto separate proteins. If the proteins interact, then the biotin ligase activity is restored, and biotinylation of surrounding proteins occurs [45]. A similar 2C-BioID system has been developed, which also recapitulates a two component or two-hybrid system within a native cellular environment [46]. In all cases, utilization of BirA*–protein fusion constructs have helped elucidate difficult to study protein–protein interaction networks.

1.4 Origins of APEX2

Another proximity-dependent biotinylation system to determine difficult to study protein–protein interaction networks is APEX2 [47]. APEX is a soybean ascorbate peroxidase, whose activity is similar to horseradish peroxidase (HRP). HRP has been useful in electron microscopy studies because, in the presence of H_2O_2, it can polymerize 3,3′-diaminobenzidine (DAB), which gives a localized contrast in activity when the samples are processed with osmium tetroxide (OsO_4) [48, 49]. Attempts to create fusion constructs with various mammalian proteins and HRP failed to create a system to tag neighboring protein networks. HRP has four structurally essential disulfide bonds and two Ca^{2+} binding sites for activity. The mammalian cytosol is a reducing environment that is also Ca^{2+} poor, which renders HRP inactive [50]. APX, the endogenous soybean ascorbate peroxidase and a class I cytosolic plant peroxidase, is 40% smaller in size than HRP, and lacks disulfide bonds and the necessity for Ca^{2+} cofactors for activity. The protein known as APEX, which stands for "enhanced APX," was molecularly engineered with point mutations within *APX* to increase enzymatic activity, and this original construct was used to label the mitochondria and endoplasmic reticulum with DAB to render greater structural resolution [51]. This same group substituted biotinyl-tyramide for DAB as a substrate and demonstrated that they could use APEX-mitochondrial protein constructs to selectively label, with biotin, other mitochondrial proteins within the inner and outer membranes of the mitochondrial matrix in a proximity dependent manner [52]. In these reactions, biotinyl-tyramide or, more commonly used, biotin-phenol is incubated with eukaryotic cells that are expressing APEX-fusion constructs. In the presence of H_2O_2, APEX reacts with the biotin-phenol, producing a biotin phenoxyl radical, that can then covalently attach (or react with) tyrosine, tryptophan, histidine, or cysteine amino acids within adjacent proteins [52]. As discussed below, these adjacent proteins include binding partners as well as proximal proteins.

The original APEX molecule was monomeric, and using directed evolution to select for improved activity, APEX2 was created. This new protein contained a point mutation at position 134 that resulted in a proline replacing an alanine, creating APEX2 (APEX, A134P) [47]. APEX2 is able to form dimeric complexes [47], which has been shown to increase the stability and activity of ascorbate peroxidases [53]. It also requires heme for activity [47], and heme is a molecule that is ubiquitous in eukaryotic cells, as it is required for oxidation–reduction reactions, amongst other metabolic processes [54].

Many APEX2 constructs have been generated and have contributed to studies that have created "organelle barcodes," which reveal the subcellular localization of difficult to study proteins [55], and spatiotemporally resolved protein interaction networks of G-protein coupled receptors [56]. The latter study also discussed the resolution of proteins that are directly involved in protein networks or those that are considered 'bystander' proteins, which are proteins that are in the area of the interactions but are not directly involved in their function or signaling networks. These studies have exploited the power and the limitations of the proximity-dependent labeling systems to further our understanding of protein–protein interaction dynamics and have increased our understanding of the spatial organization of many proteins within eukaryotic cells.

As it relates to understanding protein–protein interactions within bacteria or host–pathogen interactions, Eric Cascales's group used the APEX2 system to identify novel binding partners for a type VI secretion system protein (T6SS), TssA, which helped illuminate dynamics of T6SS apparatus assembly [57].

Relative to the field of *Chlamydia*, both the BioID and APEX2 systems have been employed. For example, the BioID system has been used to determine if specific eukaryotic proteins are recruited to the chlamydial inclusion. For example, to clarify if a eukaryotic protein, syntaxin 6, and a mutant of syntaxin 6 lacking a signal sequence trafficked to the chlamydial inclusion, a BirA∗-syntaxin 6 and its mutant was exogenously expressed in chlamydial infected HeLa cells [58]. Syntaxin 6 is a eukaryotic SNARE protein that localizes to the Golgi and traffics between the Golgi and the plasma membrane [59]. As the chlamydial inclusion develops within the proximity of the Golgi, localization of Golgi-resident or Golgi-associated proteins can be murky. Wild type BirA∗-syntaxin 6, but not the mutant, was recruited and, in the presence of exogenously added biotin, biotinylated the chlamydial inclusion membrane [58]. BioID has also been used to identify eukaryotic interacting proteins for a *Chlamydia psittaci* type III effector protein, SINC. In this study, BirA∗-SINC exogenously expressed in eukaryotic cells was determined to localize similarly as SINC that is expressed and secreted by *C. psittaci* and used to identify several candidates that might bind SINC during a chlamydial infection [60]. In another study, BioID was used to demonstrate that a chlamydial type III effector, TmeA, and a specific eukaryotic protein interacted within the host cytosol [61]. In this study, the eukaryotic protein, AHNAK, was fused to BirA∗ and exogenously expressed in HeLa cells. As a control, a eukaryotic protein, BirA∗--Perforin-2, which was known to interact with early inclusions [62], but not TmeA, was also expressed. Five hours postinfection, the chlamydial protein was analyzed for biotinylation, and was found to be biotinylated by BirA∗-AHNAK only, thereby confirming the proteins' interaction [61]. To date, BirA∗ has not been fused to a chlamydial protein and successfully expressed by *Chlamydia*. It is unknown if BirA∗ can be type III secreted by the chlamydial type III secretion apparatus. In contrast, it has been demonstrated that chlamydial Inc proteins with C-terminal APEX2 fusions are expressed and type III secreted by *C. trachomatis* [63, 64]. As this seminal study demonstrated, it is important to optimize Inc-APEX2 expression levels from *Chlamydia* to replicate, as close as possible, expression levels of endogenously expressed chlamydial proteins, and to limit possible negative effects of Inc overexpression on inclusion development [64]. Methods provided in this chapter outline our progress in these areas.

1.5 General Procedure

1.5.1 General Considerations

When embarking on one of these proximity labeling systems that use biotin to tag binding partners or adjacent proteins, there are several considerations that must be factored into experimental design. For each system, one must consider where the greatest source of background will come from, so it can be controlled for accordingly. Second, if expressing these constructs from *C. trachomatis*, one should consider how expression will be controlled (i.e., via an inducible promoter, endogenous promoter, from a chlamydial transformation plasmid, or by inserting the construct into the chlamydial chromosome). Further, one should consider whether addition of APEX2 or BirA∗ tags prevents normal localization (or secretion) of the protein, which will impact not only the proteins that are targeted by the fusion constructs but how the resulting data are interpreted.

Biotinylation is a natural posttranslational modification in eukaryotes and is also used in prokaryotes to modify protein function. As such, there are several cellular sources of natural background proteins that will be purified by any streptavidin pulldown of whole eukaryotic cell lysates. Some of these naturally biotinylated eukaryotic proteins include five mitochondrial and cytosolic carboxylases [65], mitochondrial proteins [66], mRNA processing proteins [67], and ribosomal proteins [68]. Also, given the high homology between conserved bacterial proteins and mitochondrial

Table 1
Comparison of BioID and APEX2 proximity-dependent labeling systems

	BioID (BirA∗)[a]	APEX2[b]
Enzyme	Promiscuous biotin ligase	Ascorbate peroxidase
Substrate	Biotin	Biotin-phenol
Intermediate	Biotinyl-5′-AMP	Biotin phenoxyl radical
Target amino acids	Lysine (Lys, K)	Tyrosine (Tyr, Y); Tryptophan (Trp, W); Histidine (His, H); Cysteine (Cys, C)
Half-life	Several minutes	~1 ms
Labeling time	5–24 h[c]	30 s to 1 min
Labeling radius	~40 nm	20 nm

[a][36]
[b][52]
[c][36, 61]

proteins, careful consideration of how to account for the presence of naturally biotinylated mitochondrial proteins so that false positive and false negative hits are not arbitrarily included or excluded, respectively.

Table 1 includes a direct comparison of the BioID and APEX2 proximity labeling systems. Part of the appeal of using BioID is that you can get a "roadmap" of where your protein has been over an extended period of time. If using a eukaryotic protein–BirA∗ fusion, one can also understand how chlamydial infection impacts protein interactions within a specific network/pathway. With the APEX2 system, you will receive a snapshot of protein interactions at specific time intervals. Specifically, after expression of a BirA∗--protein construct, one will add biotin to the tissue culture medium for 5–24 h before using the samples for downstream applications [36, 61]. In contrast, the labeling requirements of APEX2-protein constructs happen quickly. Biotin-phenol is added to the medium for 30 min, and the labeling reaction is catalyzed by the addition of H_2O_2 for 30 s to a minute [52, 56], followed by a quenching reaction that relies on a temperature change, peroxidase inhibitors, and a competitive substrate to reduce APEX2 activity [52].

1.6 Anticipated Results and Statistical Evaluation of Mass Spectrometry Results

In an APEX2 or BioID experiment, a single mass spectrometry data set can yield hundreds to thousands of protein identifications. These data sets are typically a list of peptides or proteins represented with mass-to-charge (m/z) ratios and their corresponding signal intensities, analyzed against a known and annotated protein database, like the *Homo sapiens* Swiss protein database. Very clearly, not all these proteins are true interacting partners, or even spatially related to the protein of interest. Conversely, the same results analyzed against a *Chlamydia trachomatis* Swiss protein database, will yield at maximum: 895 proteins; however, some of these protein identifications may be proteins that are highly homologous to mitochondrial proteins. Great care must be taken to ensure that proteins are not overrepresented or underrepresented in the final results. Statistical evaluation of these data is critical towards obtaining an unbiased and rigorously examined list of significant protein targets. Integral towards having a statistically significant dataset is having a minimum of five biological replicate experiments, which include all experimental and control conditions, and are processed identically for mass spectrometry.

Further, the type of statistical analysis used to properly evaluate these data is important. The current trend in the field is to use a statistical analysis program that uses Bayesian statistics [69, 70]. For our studies, we have used one such program, known as SAINT for "significance analysis of interactome" [69]. With this program, the spectral counts for a given peptide or protein are normalized to the length of the protein and the total number of spectra that were received in the experiment/purification. Further, the statistical output is used to estimate the Bayesian false discovery rate (BFDR), which gives the probability of whether a specific protein is likely to be a significant finding or part of background [69]. This method is an improvement upon standard *t*-tests of spectral counts between replicates and allows for the inclusion of proteins/peptides that appear weakly in controls (background) and strongly and definitively in the experimental conditions [70].

2 Materials

2.1 Construction of BioID Gene Fusions for Transfection and Expression in Eukaryotic Cells

1. Tabletop centrifuge with rotors or adaptors for 15 mL and 50 mL conical tubes.
2. Microcentrifuge to accommodate 1.5 mL microfuge tubes.
3. Standard cloning materials and vectors.
4. 37 °C and 42 °C water bath.
5. Plasmid Miniprep kit.
6. Plasmid Midiprep kit.
7. 30 °C or 37 °C incubator for agar plates.
8. 30 °C or 37 °C shaking incubator for liquid cultures.
9. Standard *E. coli* strain for cloning (10-β or DH5α).
10. PCR reagents, including a proofreading polymerase.
11. Ligation kit (e.g., NEBuilder HiFi Assembly Cloning Kit, NEB Cat# E5520S).
12. −80 °C freezer.
13. −20 °C freezer.
14. 4 °C refrigerator.

2.2 Construction of APEX2 Gene Fusions for Transformation and Expression from C. trachomatis

1. Tabletop centrifuge with rotors or adaptors for 15 mL and 50 mL conical tubes.
2. Microcentrifuge to accommodate 1.5 mL microfuge tubes.
3. Standard cloning materials and vectors.
4. 37 °C and 42 °C water bath.
5. Plasmid Miniprep kit.
6. Plasmid Midiprep kit.
7. 30 °C or 37 °C incubator for agar plates.
8. 30 °C or 37 °C shaking incubator for liquid cultures.
9. Standard *E. coli* strain for cloning (10-β or DH5 α).
10. dam^-/dcm^- *E. coli* for *Chlamydia* transformation.
11. −80 °C freezer.
12. −20 °C freezer.
13. 4 °C refrigerator.

2.3 Proximity Labeling Using BioID Constructs

1. HeLa 229 cells (ATCC, CCL-2.1) (*see* **Note 1**).
2. Mycoplasma test kit (e.g., "LookOut Mycoplasma Detection Kit," Sigma-Aldrich, MP0035) (*see* **Note 2**).
3. Standard tissue culture materials.
4. Biosafety cabinet.
5. 5% CO_2 37 °C incubator for tissue culture.
6. 37 °C water bath.
7. Refrigerated microcentrifuge (capable of 10,000 × *g*; 4 °C).
8. 6-well tissue culture plates.
9. 24-well tissue culture plates.
10. Round 12 mm glass coverslips.
11. 1× DMEM *without* biotin (Hyclone # SH30243.01 or Gibco DMEM #11-965-092).
12. Heat-inactivated fetal bovine serum.
13. BioID-fusion construct (*see* **Note 3**).
14. Biotin powder suitable for cell culture (stock 1 mM in serum free 1× DMEM, *see* **Note 4**).
15. 1× DMEM supplemented with 1% heat inactivated FBS.
16. Transfection reagent (*see* **Note 5**).
17. Antibody against Myc and appropriate secondary antibody.
18. Fluorescent streptavidin conjugate (e.g., Streptavidin–488 Jackson ImmunoResearch Laboratories Inc., 016-540-084).
19. Fixative (e.g., methanol or paraformaldehyde) and permeabilization reagent (e.g., 0.1% Triton X-100 or 0.5% saponin) (*see* **Note 6**).

2.4 Proximity Labeling Using APEX2 Constructs

1. HeLa 229 cells (ATCC, CCL-2.1) (*see* **Note 1**).
2. Mycoplasma test kit (e.g., "LookOut Mycoplasma Detection Kit," Sigma-Aldrich, MP0035) (*see* **Note 2**).
3. Standard tissue culture materials.
4. Biosafety cabinet.
5. 5% CO_2 37 °C incubator for tissue culture.
6. 37 °C water bath.
7. Refrigerated microcentrifuge (capable of 10,000 × *g*; 4 °C).
8. 6-well tissue culture plates.
9. 24-well tissue culture plates.
10. Round 12 mm glass coverslips.
11. 1× DMEM *without* biotin (Hyclone, SH30243.01 or Gibco DMEM, 11-965-092).
12. Heat inactivated fetal bovine serum.
13. 1× Phosphate buffered saline (PBS).
14. Anhydrous DMSO.
15. Antibiotics: cycloheximide, gentamicin, penicillin, or another selective antibiotic.
16. A clonal population of *C. trachomatis* transformants.
17. Fixative (e.g., methanol or paraformaldehyde) (*See* **Note 6**).
18. Permeabilization reagent (e.g., 0.1% Triton X-100 or 0.5% saponin).
19. Anhydrotetracycline.
20. Biotin phenol (biotinyl-tyramide) (Adipogen; stock 50 mM in anhydrous DMSO (*see* **Note 7**).
21. Hydrogen peroxide.
22. Quenching wash solution (10 mM sodium ascorbate, 10 mM sodium azide, 5 mM Trolox in dPBS) (*see* **Note 8**).
23. RIPA buffer A (50 mM Tris–HCl, pH 7.4; 150 mM NaCl, 0.1% SDS, 0.5% sodium deoxycholate) modified with the following:
    (a) 150 μM clastolactacystin β-lactone (*see* **Note 9**).
    (b) 10 mM sodium azide.
    (c) 10 mM sodium ascorbate.
    (d) 5 mM Trolox (*see* **Note 8**).
    (e) 5% Triton X-100.
    (f) 1% SDS.
    (g) 1× HALT +1× EDTA protease inhibitor (ThermoFisher, 87786).
    (h) 0.1 μL/mL Universal Nuclease (ThermoFisher, 88701).

24. Sonicator.
25. Cell scrapers.
26. 15 mL conical tubes (sterile).
27. 1.5 mL microcentrifuge tubes (sterile).
28. Tabletop centrifuge with rotors or adaptors for 15 mL and 50 mL conical tubes.

2.5 Affinity Purification of Biotinylated Proteins

1. EZQ Protein Quantification Kit (ThermoFisher, R33200).
2. RIPA buffer (50 mM Tris–HCl, pH 7.4; 150 mM NaCl, 0.1% SDS, 0.5% sodium deoxycholate).
3. Streptavidin magnetic beads (e.g., Pierce streptavidin magnetic beads, 88816) (*see* **Note 10**).
4. Wash buffer A: RIPA buffer (50 mM Tris–HCl, pH 7.4; 150 mM NaCl, 0.1% SDS, 0.5% sodium deoxycholate) modified with:
   (a) 500 mM NaCl.
   (b) 5% Triton X-100.
   (c) 1% SDS.
5. Wash buffer B: 2 M urea in 10 mM Tris–HCl (pH 7.4).
6. 2× Sample buffer (both laboratory prepared and mass spectrometry grade) (*see* **Note 11**).
7. β-mercaptoethanol.
8. Biotin (3mM).
9. 2× Laemmli sample buffer (Use premade/high grade Laemmli sample buffer (e.g., 2× Sample Buffer, Bio-Rad, 1610737) if intended for mass spectrometry.) (*see* **Note 11**).
10. 1.5 mL microcentrifuge tubes (sterile).
11. Magnetic rack and individual magnet (*see* **Note 12**).
12. Tube rotator (1.5 mL tube compatible).

2.6 Confirmation of Protein Biotinylation and Mass Spectrometry Identification

1. Gel electrophoresis apparatus.
2. Standard denaturing gel (e.g., Bio-Rad 4–20% gradient Criterion gels, 5671093.
3. Western blot transfer apparatus.
4. Western blotting reagents: Polyvinylidene membrane (PVDF), blocking reagents (e.g., bovine serum albumin or milk), antibodies (e.g., streptavidin conjugate, anti-FLAG antibody), staining trays, wash solution (e.g., PBS + 0.1% Tween 20).
5. Streptavidin–680 in 5% bovine serum albumin in PBS + 0.1% Tween 20 (PBST) (*see* **Note 13**).
6. Gel imaging system.

7. Coomassie stain (Coomassie G-250, 10% methanol, 5% acetic acid in tissue culture grade water) (*see* **Note 14**).
8. Destain solution (10% methanol, 5% acetic acid in tissue culture grade water).
9. 30% methanol (*see* **Note 15**).
10. Nonautoclaved tubes.
11. Scalpel blades or razor for cutting SDS-PAGE protein bands.
12. Mass spectrometry core facility (*see* **Note 16**).
13. Significance Analysis of Interactome (SAINT) or similar appropriate statistical analysis program to accurately assess background biotinylated proteins from the proximal proteins identified and prevent bias (*see* **Note 17**).

2.7 DAB Staining for Electron Microscopy

1. Plastic coverslips suitable for electron microscopy (e.g., Thermo Scientific™ Nunc™ Thermanox™ coverslips, cat. no. 174985).
2. 6-well cell tissue culture plate.
3. HeLa 229 cells (*see* **Note 1**).
4. Mycoplasma test kit (e.g., "LookOut Mycoplasma Detection Kit," Sigma-Aldrich, MP0035) (*see* **Note 2**).
5. DMEM +10% FBS.
6. Cycloheximide (*see* **Note 18**).
7. Penicillin G (*see* **Note 18**).
8. Enumerated *Chlamydia* strains.
9. Anhydrotetracycline.
10. Dulbecco's phosphate-buffered saline.
11. Fixing solution (2% paraformaldehyde, 2% glutaraldehyde, 0.1 M sodium cacodylate fixative): Combine 8 mL of 25% glutaraldehyde, 12 mL of 16% paraformaldehyde, 50 mL of 0.2 M sodium cacodylate, and 30 mL of distilled H_2O.
12. Wash buffer (0.1 M sodium cacodylate): Combine 100 mL of 0.2 M sodium cacodylate, pH 7.4 with 100 mL of distilled H_2O.
13. Quenching/blocking buffer (20 mM glycine in 0.1 M sodium cacodylate, 2 mM $CaCl_2$ blocking buffer). As described in [71], dissolve 75.1 mg glycine in 50 mL of 0.1 M sodium cacodylate, 2 mM $CaCl_2$. To make 0.1 M sodium cacodylate, 2 mM $CaCl_2$, dissolve 32.103 g of sodium cacodylate trihydrate in 450 mL Milli-Q™ H_2O. Decrease the pH to 7.4 by adding HCl drop by drop. Dissolve 0.441 g of calcium chloride, dihydrate. Bring the solution up to 500 mL with Milli-Q™ H_2O. This is 0.3 M sodium cacodylate, 6 mM $CaCl_2$.

Then, combine 167 mL of this solution and 333 mL of Milli-Q™ H_2O to give 0.1 M sodium cacodylate, 2 mM $CaCl_2$.

14. 3,3′-Diaminobenzidine (DAB) stock solution (5 mg/mL). As described in [51, 71], dissolve 50 mg of DAB in 10 mL of 0.1 M HCl. Make 1 mL aliquots, freeze these by placing in dry ice, and store at −80 ° C.
15. Working DAB solution (0.5 mg/mL DAB, 0.1 M sodium cacodylate, 2 mM $CaCl_2$ solution): As described in Martell et al. 2017 [71], combine 1 mL of 5 mg/mL DAB stock solution, 3.33 mL of 0.3 M sodium cacodylate, 6 mM $CaCl_2$ (see above), and 5.67 mL of Milli-Q™ H_2O.
16. 0.5 mg/mL DAB, 3 mM H_2O_2, 0.1 M sodium cacodylate, 2 mM $CaCl_2$ solution: As described in Martell et al. 2017 [71], mix solutions together as for the previous solution. Add 3 μL 30% (weight/weight) H_2O_2. The concentration of H_2O_2 must be determined empirically. The DAB solutions with and without H_2O_2 should be made fresh just before adding them to the samples. Add H_2O_2 last to avoid beginning the polymerization reaction before the solution has been applied to the samples.
17. Electron Microscopy Facility.

3 Methods

Biotinylation is a natural post-translational modification in eukaryotes. As such, there are several cellular sources of natural background proteins that will be purified by any streptavidin pulldown of whole eukaryotic cell lysates. Some of these naturally biotinylated eukaryotic proteins include five mitochondrial and cytosolic carboxylases [65], mitochondrial proteins [66], mRNA processing proteins [67], and ribosomal proteins [68]. Therefore, during the experimental design steps, several negative controls for background biotinylated proteins should be considered. Table 2 outlines the controls that we use in our experiments.

The methodology for implementing a proximity labeling system in *Chlamydia* can be broken down into three major steps. First, clone the gene of interest in frame with either BioID or APEX2 into the desired vector (e.g., mammalian or *C. trachomatis* vector). Second, transfect or transform the vector encoding the BioID or APEX2 fusions (e.g., gene X-BioID, gene Y-APEX2) into the eukaryotic cells or *Chlamydia*, respectively and then optimize the expression of the fusion protein in your system. Third, affinity-purify the biotinylated proteins from the solubilized lysate and confirm biotinylation by western blotting.

Table 2
Negative controls for BioID, APEX2, and DAB labeling

BioID (BirA∗) controls	Biotin	BirA∗	Reason
Uninfected cells	+	+	For comparison to infected cells to determine if infection with *Ct* causes any changes in protein–protein interactions
WT L2	−	+	Negative control to determine endogenous biotinylation during transfection
WT L2	+	−	Negative control to determine endogenous biotinylation during infection with WT *Ct*
Uninfected cells	+	−	Negative control to determine endogenous biotinylation
APEX2 controls	**aTc[a]**	**BP[b]**	**Reason**
APEX2 fusion control 1	+	−	Negative control for background biotinylation
WT L2	+	+	Negative control for background biotinylation
Uninfected cells	+	+	Negative control for background biotinylation
DAB controls	**DAB[c]**	**H_2O_2**	**Reason**
WT L2	+	+	Negative control to check for background DAB labeling
APEX2 transformant	−	−	Negative control to make sure there is no DAB labeling in the absence of DAB and H_2O_2

[a]Anhydrotetracycline (aTc)
[b]Biotin-phenol (BP)
[c]3,3′-Diaminobenzidine (DAB)

3.1 Construction of BioID Gene Fusions for Transfection and Expression in Eukaryotic Cells

This section will briefly describe the steps involved in constructing BioID gene fusions and assumes that the scientist is proficient in standard cloning techniques.

1. Design primers to amplify your gene of interest, and perform PCR using standard methodologies and a proofreading polymerase.
2. Purify PCR products using a commercial PCR purification or gel extraction kit.
3. Clone in frame into the BioID plasmid, pcDNA3.1 mycBioID (Addgene plasmid #35700) [36] using standard cloning methods (*see* **Note 3**).
4. Transform into a standard *E. coli* cloning strain, isolate the plasmid, and verify the DNA sequence. The isolated plasmid will be used to transfect HeLa cells to allow for biotin labeling of proteins proximal to your protein of interest with the addition of exogenous biotin.

3.2 Construction of APEX2 Gene Fusions for Transformation and Expression from C. trachomatis

This section will briefly describe the steps involved in constructing APEX2 gene fusions and assumes that the scientist is proficient in standard cloning techniques.

1. Amplify the gene of interest from *Chlamydia* genomic DNA (or other template as needed), and *APEX2* from pcDNA3 APEX2-NES (Addgene cat#49386) [47].
2. Purify PCR products using a commercial PCR purification or gel extraction kit.
3. Clone in frame into the desired chlamydial expression vector (e.g., pASK-mKate [72], or pBomb4 [3]) followed by *APEX2* (*see* **Note 19**).
4. Transform into standard *E. coli* cloning strains.
5. Isolate plasmids and confirm by DNA sequencing analysis.
6. Transform sequence verified clones into *dam*$^-$/*dcm*$^-$ *E. coli* [73, 74] and purify plasmids.
7. Transform demethylated plasmids into *C. trachomatis* following the chlamydial transformation protocol as previously published [73, 74]. In brief: Approximately 24 h prior to transformation, seed McCoy cells in a 6-well tissue culture plate. Plate enough wells for one well per sample intended for transformation. The next day resuspend purified *C. trachomatis* elementary bodies that lack the endogenous plasmid (-pL2) in calcium chloride ($CaCl_2$) and add 2 μg of demethylated plasmid DNA. Incubate the mixture at room temperature for 30 min. Rinse the 6-well plate of McCoy cells with Hanks Buffered Salt Solution (HBSS) and replace with 2 mL of HBSS. After incubation, add the EB–pDNA mixture dropwise to a single well of a 6-well tissue culture plate per condition. Centrifuge the plates at 400 × *g* for 15 min at room temperature then incubate at 37 °C + 5% CO_2 for 15 min. Replace the HBSS with DMEM +10% FBS and place the 6-well plate back at 37 °C + 5% CO_2 incubator. At 8 hpi, aspirate the media and replace with DMEM +10% FBS containing the appropriate antibiotic and cycloheximide. Passage samples every 48 h by scraping and infecting onto a fresh monolayer of McCoy cells until mature inclusions are observed.
8. Finally, test the *C. trachomatis* transformants for the appropriate localization of the constructs containing APEX2 by indirect immunofluorescence microscopy, and optimize the expression of each construct. For example, infect HeLa cell monolayers with *C. trachomatis* L2 transformants and induce with variable amounts of anhydrotetracycline. Fix and stain for indirect immunofluorescence to determine induction conditions that yield localization that most closely matches endogenous protein localization according to the literature. In addition, it is

best to perform subsequent biotinylation experiments with a clonal population of *C. trachomatis* transformants. This can be done by plaque purification (*see* Chapter 12, Subheading 3.4) or by limiting dilution.

3.3 Proximity Labeling Using BioID Constructs (See Note 5)

1. Seed HeLa 229 cells (*see* **Notes 1** and **2**) in a 6-well plate at the appropriate cell density determined from troubleshooting in 3 mL per well of 1× DMEM supplemented with 10% heat inactivated FBS.
   (a) Place one round 12 mm glass coverslip in 2 separate wells of the 6-well plate to monitor transfection and biotin-labeling efficiency via immunofluorescence.
2. Approximately 24 h after seeding cells and before transfecting, wash the cells twice with 1 mL per well of prewarmed 1× DMEM supplemented with 1% FBS (*we use 1% FBS to limit exogenous biotin contamination from the FBS*) and replace media with 2 mL of 1× DMEM containing 1% heat inactivated FBS.
3. Transfect with the BioID-fusion construct using preferred transfection method in 500 μL and allow to incubate at 37 °C and 5% CO_2 for 4 h.
4. Aspirate transfection medium and replace with 2 mL of DMEM +1% FBS + 50 μM Biotin (*see* **Note 4**) per well (or not as a control) to begin the labeling. *See* Table 2 for the appropriate negative controls.
   (a) Make sure to use separate pipet tips and Pasteur pipettes between samples to prevent cross-contamination.
5. Allow the cells to recover for 2 h.
6. Infect the cells with wild-type *Chlamydia* 6 h posttransfection (4-h transfection + 2-h recovery) by centrifugation at 400 × *g* for 15 min at room temperature, then incubate 37 °C 5% CO_2.
7. At the timepoint post infection you are interested in, remove glass coverslips from the two wells and place them into a new 24-well plate containing 500 μL of appropriate fixative and permeabilize with appropriate permeabilization reagent (e.g., 0.1% Triton X-100 or 0.5% saponin), *see* **Note 6**. Process the coverslips by indirect immunofluorescence assay [75] to monitor transfection efficiency (use an antibody against Myc) and biotin labeling (use a fluorescent streptavidin conjugate (e.g., Streptavidin–488 Jackson ImmunoResearch Laboratories Inc., 016-540-084)).
   (a) If the transfection efficiency is too low or biotinylation is not sufficient, discard the samples and repeat.
   (b) It is also important to monitor localization of the BirA*-protein fusion in uninfected cells to ensure the

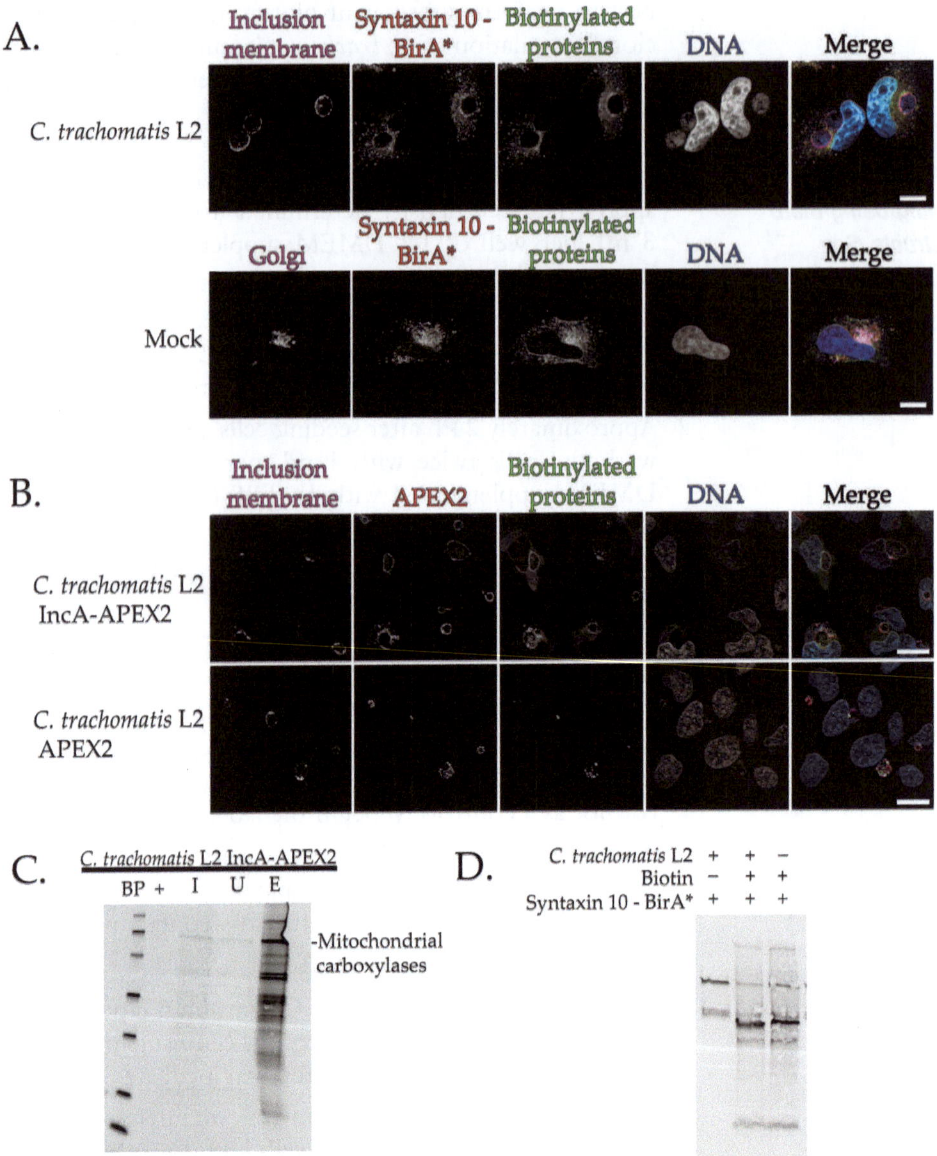

Fig. 1 Confirmation of BioID and APEX2 specific biotinylation using indirect immunofluorescence assays and western blotting. (**a**) HeLa cells transfected with Syntaxin10-BirA∗ were mock-infected or infected with *C. trachomatis* L2 in media supplemented with 50 μM biotin to catalyze the labeling of proximal proteins with biotin. Cells were fixed at 23 h postinfection and processed for immunofluorescence to visualize the inclusion membrane in infected cells (anti-IncA, pink) or the Golgi in uninfected cells (anti-Giantin, pink), expression of Syntaxin10-BirA∗ (anti-Myc, red), biotinylated proteins (streptavidin–488 conjugate, green) and DNA (DAPI, blue). Images were taken at 100× magnification using a Zeiss Apotome 2.1. Scale bar = 10 μm. (**b**) HeLa cells infected with *C. trachomatis* L2 transformed with IncA-APEX2 or APEX2 only and induced with anhydrotetracycline (aTc) at 7 hpi. Biotin-phenol was added 30 min prior to the biotin labeling step at 24 hpi. Biotinylation was catalyzed by the addition of 3 mM H_2O_2 for 1 min and washed with a quenching solution. Coverslips were fixed and processed for immunofluorescence to visualize the inclusion membrane (anti-CT223, pink), expression of the construct (anti-Flag, red), biotinylated proteins (streptavidin–488

addition of BirA∗ is not disrupting the protein's normal localization.

(c) *See* Fig. 1a for an example of an indirect immunofluorescence assay of a BioID labeling experiment using a host protein normally found within the Golgi, syntaxin 10, which has been previously shown to localize to the chlamydial inclusion [76] with a C-terminal BirA∗ fusion. In infected cells, the exogenously expressed syntaxin 10–BirA∗ can be seen around the inclusion, while in mock infected cells the syntaxin 10-BirA∗ is found in the Golgi. Biotinylated proteins are colocalized with syntaxin 10-BirA∗ in both infected and uninfected cells.

8. Continue with sample collection and affinity purification (*see* Subheading 3.4, **step 8**). (The processing for BioID and APEX2 cell lysate converges at this step.)

3.4 Proximity Labeling Using APEX2 Constructs

1. Seed HeLa 229 cells (*see* **Notes 1** and **2**) in a 6-well plate (1×10^6 cells/well) in biotin-free media (DMEM) supplemented with 1% heat inactivated FBS. *We decreased the FBS to deplete biotin levels in HeLa cells.* Seed one 6-well plate per condition/control sample. Place one glass coverslip in two separate wells of the 6-well plate to determine the expression level of the APEX2 construct and biotin labeling efficiency by immunofluorescence. It is important to include the appropriate negative controls for endogenous biotinylated proteins in both the host cell and *Chlamydia* (*see* Table 2).
2. After the biotin limiting step (~24 h), aspirate the media and infect the HeLa cells with the appropriate *C. trachomatis* L2 APEX2 transformant or negative controls in DMEM +10% FBS (1 μg/mL cycloheximide, 10 μg/mL gentamicin, and the appropriate selective antibiotic). *We have found more robust biotinylation using 10% FBS than with 1% FBS with limited additional background biotinylation.*

Fig. 1 (continued) conjugate, green), and DNA (DAPI, blue). Coverslips were imaged using a Zeiss with Apotome 2.1 63× Scale bar = 20 μm. (**c**) Lysates were collected from *C. trachomatis* L2 IncA-APEX2 infected HeLa cells treated with biotin phenol and the biotinylation reaction was catalyzed to label proximal proteins. The total solubilized lysate (input; I), unbound fraction (U), and eluate fraction (E) from the streptavidin affinity purification were separated by SDS-PAGE and transferred to a PVDF membrane. To determine if biotinylated proteins were affinity-purified, membranes were blotted with a streptavidin–680 conjugate. Images were taken using an Azure c600. (**d**) Lysates were collected from HeLa cells transfected with Syntaxin10–BirA∗ and infected with wild-type *C. trachomatis* L2 (or not), in the presence (or not) of 50 μM Biotin to induce biotin-labeling of proteins proximal to syntaxin 10 with BirA∗ (as shown in part A). Lysates were affinity-purified using streptavidin magnetic beads and eluates were resolved by SDS-PAGE, transferred to a PVDF membrane, and were blotted for biotinylated proteins using a streptavidin–680 conjugate. Images were taking using an Azure c600

3. Infect the HeLa cells by rocking, or by centrifugation at 400 × *g* for 15 min at room temperature, then transfer the plates to an incubator at 37 °C + 5% CO_2.
4. Induce expression of constructs at a predetermined time post-infection (e.g., 7–10 hpi), by adding an optimized concentration of anhydrotetracycline (aTc) directly to the tissue culture media. *Ensure that induction conditions allow for the appropriate localization of the APEX2 fusion construct as overexpression can cause mislocalization.*
5. Add biotin-phenol (1.5 mM final concentration) to each well 30 min before the desired reaction time (e.g., 23.5 hpi for 24 hpi biotinylation time) and incubate for 30 min at 37 °C + 5% CO_2. Keep the biotin-phenol stock solution ≤50 mM in anhydrous DMSO for optimal solubility (*see* **Note 7**).
6. To catalyze the biotin labeling reaction, aspirate the media and add 2 mL/well of 3 mM H_2O_2 in PBS. Incubate the samples at room temperature for 1 min, with rocking. *We have found that labeling two to four plates at a time is manageable. If there are greater than five 6-well plates in the experiment, stagger the infection times.*
7. After the labeling step, aspirate the H_2O_2 solution. Quench the labeling reaction with 3 × 1 mL/well washes using the quenching wash solution (10 mM sodium ascorbate, 10 mM sodium azide, 5 mM Trolox in dPBS (*see* **Note 8**)). Use gentle rocking during washes (approximately 15–30 s). After the washes, add 2 mL/well of PBS and then remove the coverslips for fixation to confirm construct expression via indirect immunofluorescence microscopy. This can be done using a fluorescent streptavidin conjugate, anti-FLAG antibody (FLAG is in N-terminus of APEX2), and other appropriate cell markers using the methods described above (*see* Subheading 3.3, **step** 7) (*see* **Note 6**). An example is provided in Fig. 1b of biotinylation at the inclusion membrane using an Inc fused to APEX2 compared to APEX2 only which lacks the type III secretion signal and thus biotinylation appears to localize with the individual *Chlamydiae*.
8. Immediately after the coverslips are removed, scrape the remaining cells and pipet into a 15 mL conical tube on ice (12 mL total volume per 6-well plate). Pellet the cells by centrifugation at 900 × *g* for 10 min at 4 °C. *Keep these steps at 4 °C and move quickly to limit protease activity.*
9. Aspirate the supernatant and resuspend the pellet in 1 mL of modified RIPA buffer A per sample (The modified RIPA contains a protease inhibitor cocktail and clastolactacystin β-lactone, which specifically inhibits the chlamydial protease-

like activity factor (CPAF) [77] (*see* **Note 9**). Transfer the lysate to a 1.5 mL tube for sonication. *These tubes work well for the sonication step to prevent excessive foaming of the sample.*

10. To solubilize the lysate, sonicate samples 3 times for 20 s at 20% amplitude. *Keep the microtip near the bottom of the 1.5 mL tube to prevent foaming.*
11. After sonication, incubate the samples on ice for 90 min. Vortex the samples every 30 min.
12. Clarify the lysate by centrifuging at 14,000 × *g* for 10 min at 4 °C. Transfer the supernatant (soluble protein fraction) into a fresh tube and resuspend the pellet (insoluble protein fraction) in 400 μL modified RIPA. *If a large pellet remains, sonicate the samples (see* **step 10**) *again. Using the above described conditions, the pellet is almost nonexistent.* Save each fraction and make a small aliquot (e.g., 50 μL). Store samples at −80 °C until protein quantification, western blot confirmation of protein biotinylation, and affinity purification steps are performed.

3.5 Affinity Purification of Biotinylated Proteins

1. Using the small aliquot, quantify the protein from each sample condition collected. *We use the EZQ Protein Quantification Kit (ThermoFisher) as it is compatible with high detergent levels in the modified RIPA lysis buffer and only requires a few microliters of sample (<5 μL).*
2. Normalize the solubilized lysate prior to SDS-PAGE and western blot analysis (e.g., 1 mg/mL).
3. Analyze the solubilized lysate by standard denaturing gel and western blot techniques, *see* Chapter 5 Western Blotting: An Introduction [78]; for gel electrophoresis and western blotting conditions, in brief: Load the normalized lysate mixed with sample buffer into a 4–20% gradient polyacrylamide gel and electrophorese at 200 V for 30 min, or as desired. To transfer for western blotting, soak the polyvinylidene (PVDF) membrane in methanol, then place in transfer buffer. Assemble the gel sandwich by placing one blotting pad at the base of the transfer cassette. Then place the polyacrylamide gel on the blotting pad, followed by the membrane. Place a second blotting pad on top of the membrane and close the cassette and load it into the wet transfer system, with the membrane closest to the positive electrode and transfer as desired (e.g., 100 V for 48 min with cold pack).
   (a) Confirm that biotinylated proteins are present in the soluble fraction and not in the insoluble fraction by blotting the membrane with a streptavidin conjugate (e.g., streptavidin–680 conjugate; Li-Cor Biosciences) in blocking buffer (e.g., 5% BSA in PBST (PBS + 0.1% Tween 20)) (*see* **Note 13**). After incubating with the streptavidin

conjugate, wash the membrane with PBST three times for 5 min at room temperature with rocking. Then perform a final wash step in PBS only (5 min) and image using the preferred imaging system.

4. Affinity-purify the biotinylated proteins from solubilized lysate using streptavidin beads (*see* **Note 10**). Vortex the beads to resuspend them, then aliquot 100 μL of the streptavidin magnetic beads (10 mg/mL stock) per sample condition into a 1.5 mL tube for affinity purification.
5. To equilibrate the streptavidin magnetic beads, add 1 mL of RIPA (modified with 5% Triton X-100 and 1% SDS) to each tube.
6. Rotate the beads for 5 min at room temperature using a tube rotator.
7. Pellet the beads using a magnetic rack or handheld magnet and remove the supernatant.
8. Repeat once more for a total of two washes. Remove the liquid *immediately* before adding soluble lysate to prevent the beads from drying out (*see* **Note 12**).
9. After the beads are equilibrated, add the solubilized lysate (1 mg/mL; 1 mL total volume/tube) and rotate the samples for 1.5–2 h at room temperature or overnight at 4 °C.
10. After the affinity-purification incubation step, pellet the beads using the magnetic rack and *SAVE the unbound fraction*. By western blot, confirm that the biotinylated proteins were affinity-purified from the solubilized lysate and do not remain in the unbound fraction (Fig. 1c).
11. Perform the following washes to remove nonspecific proteins from the streptavidin beads. Incubate in the indicated wash buffer for 5 min at room temperature on a rotator. Pellet the beads between each wash using the magnetic rack (*see* **Note 12**):
    (a) 2 washes × wash buffer A (1 mL RIPA buffer (plus 1% SDS, 5% Triton X-100, 500 mM NaCl))
    (b) 1 wash × wash buffer B (1 mL 2 M urea in 10 mM Tris–HCl (pH 7.4))
    (c) 2 washes × 1 mL RIPA buffer (unmodified).
12. After the final wash, remove all residual liquid and elute the biotinylated proteins directly in 2× Laemmli sample buffer (containing 5% β-mercaptoethanol, 3 mM biotin) (*see* **Note 11**). Elute the biotinylated proteins from the streptavidin magnetic beads with 40 μL of sample buffer per tube.
13. Rotate the beads for 5 min with the elution buffer, then boil at 95 °C for 4 min (*see* **Note 20**).

14. Pellet the beads using the magnet and transfer the eluate to a new tube. Save the leftover beads in case they are needed to test/troubleshoot elution conditions (*see* **Note 10**).

3.6 Confirmation of Protein Biotinylation and Mass Spectrometry Identification (See Note 21)

1. Electrophorese until the desired separation is reached (e.g., 200 V ~50 min for a midi gel).
2. Transfer to membrane for blotting following typical transfer methods, in brief: Soak the PVDF membrane in methanol, then place in transfer buffer. Assemble the gel sandwich by placing one blotting pad at the base of the transfer cassette. Then place the polyacrylamide gel on the blotting pad, followed by the membrane. Place a blotting pad on top of the membrane and close the cassette and load it into the wet transfer system. Place the side with the membrane closest to the positive electrode and transfer as desired (e.g., 100 V for 48 min with cold pack).
3. Blot for biotinylated proteins using a streptavidin conjugate (Fig. 1c, d) (e.g., streptavidin 680 in 5% BSA in PBST for 2 h at room temperature or overnight) (*see* **Note 13**).

Gel Electrophoresis for Mass Spectrometry Identification of Biotinylated Proteins

4. Use precast gels to reduce keratin contamination (e.g., 4–20% gradient midi Criterion gels, Bio-Rad, 5671093) for samples intended for mass spectrometry analysis.
5. Electrophorese the samples for 10–15 min, or about ~5.0 cm into the gel. *This reduces the area of gel sections which helps the efficiency of the digestion in downstream processing steps for mass spectrometry.*
    (a) While the gel is running, use 30% methanol (*see* **Note 15**) to remove keratin and other contaminants to clean the Coomassie gel staining container.
6. Remove gel from the cassette and add the Coomassie stain solution; incubate for 1 h at room temperature (*see* **Note 14**).
7. Destain the Coomassie gel overnight at 4 °C with gentle rocking using the destain solution. Perform additional wash steps as needed to remove excess Coomassie, leaving as little stain remaining as possible.
8. Cut the gel pieces in a biosafety cabinet (or area with dead space, away from vents which are a major source for keratin contamination) and place in tubes that have not been autoclaved (*see* **Note 15**).
9. Submit the gel samples to the mass spectrometry core facility for processing and mass spectrometry identification of affinity-purified biotinylated proteins (*see* **Note 16**).
10. Use the appropriate statistical analysis program to accurately assess background biotinylated proteins from the proximal

proteins identified and prevent bias. For example, Significance Analysis of Interactome (SAINT; [69]) uses Bayesian statistics to calculate the probability that a protein is real in test samples compared to designated control samples (*see* **Note 17**).

3.7 DAB Staining for Electron Microscopy

Day 1: Seed Cells

1. Place 25 mm plastic coverslips in the wells of a 6-well tissue culture plate.
2. Seed 1.0×10^6 HeLa 229 cells per well in DMEM +10% FBS (*see* **Notes 1** and **2**).
3. Incubate in a CO_2 incubator set at 37 °C and 5% CO_2 for approximately 1 day.

Day 2: Infect and Induce

4. The next day, verify that the number of cells doubled using a light microscope.
5. Aspirate the old cell culture growth media from the wells.
6. Immediately replace the old media with fresh DMEM +10% FBS supplemented with 1 μg/mL cycloheximide and 1–2 U/mL penicillin G for wells to be infected with transformed *Chlamydia* strains and 1 μg/mL cycloheximide only for wells to be infected with wild type *Chlamydia* (*see* **Note 18**). Replacing the media immediately after aspiration is important to ensure cells do not dry out and become stressed.
7. Thaw *Chlamydia* strains to be used for infection on ice (*see* **Note 22**) using controls indicated in Table 2.
8. Infect with a multiplicity of infection of 0.75 *Chlamydia* cells per host cell for transformed strains and 0.4 for the wild type strain. The transformed *Chlamydia* strains are used at a higher multiplicity of infection than the wild type strain because they may not infect as efficiently.
9. Tilt the plate back and forth and side to side to distribute the *Chlamydia* organisms throughout the cell layer. Centrifuge the plate at $400 \times g$ for 15 min. The plate is centrifuged to aid attachment to and infection of the host cells by the *Chlamydia* cells. Incubate the plate at 37 ° C and 5% CO_2 for 7 h.
10. At 7 h postinfection (hpi), pipette anhydrotetracycline into the wells to induce expression of APEX2 constructs. The optimal concentration must be determined empirically. In one example, APEX2 expression is induced at 7 hpi to maximize construct expression. The final concentration of anhydrotetracycline we used was 5.0 nM for some constructs and lower concentrations for one of the constructs [64].
11. Tilt the plate back and forth and side to side to distribute the anhydrotetracycline throughout the wells.
12. Incubate the plate at 37 °C and 5% CO_2 until 24 hpi.

Day 3: Process for TEM (*See* **Notes 23** and **24**)

13. At 24 hpi, aspirate the media from the wells.
14. Wash the wells once with DPBS that has been prewarmed in a 37 ° C water bath.
15. Aspirate the DPBS.
16. Fix the samples by applying a solution of room temperature 2% paraformaldehyde, 2% glutaraldehyde, 0.1 M sodium cacodylate buffer to the wells.
17. Place the plate on ice for 1 h to fix the cells.
18. Aspirate the fixative from the wells.
19. Wash the wells five times for 2 min each time with cold 0.1 M sodium cacodylate. Keep the plate on ice during the washes.
20. Block the samples by applying a solution of cold 20 mM glycine in 0.1 M sodium cacodylate, 2 mM $CaCl_2$ for 5 min. Keep the plate on ice.
21. Wash the wells five times for 2 min each time with cold 0.1 M sodium cacodylate. Keep the plate on ice during the washes.
22. Pretreat the samples with a solution of DAB that does not contain H_2O_2 by applying a solution of cold 0.5 mg/mL DAB, 0.1 M sodium cacodylate, 2 mM $CaCl_2$ for 30 min (*see* **Note 25**). Keep the plate on ice. As a negative control, do not apply this solution to one of the samples infected with a *Chlamydia* strain transformed with an APEX2 construct. Rather, apply a solution of cold 0.1 M sodium cacodylate instead.
23. Aspirate the solutions from the wells.
24. Treat the samples with a solution containing both DAB and H_2O_2 by applying a solution of cold 0.5 mg/mL DAB, 3 mM H_2O_2, 0.1 M sodium cacodylate, 2 mM $CaCl_2$ for 30 min. Keep the plate on ice. For the negative control that is not supposed to receive DAB or H_2O_2, apply a solution of cold 0.1 M sodium cacodylate instead.
25. Aspirate the solutions from the wells. Wash the wells five times for 2 min each time with cold 0.1 M sodium cacodylate. Keep the plate on ice.
26. The samples are then processed by an EM Facility. Briefly, the cells are postfixed with osmium tetroxide, stained with Toluidine Blue (to see the cells during processing), dehydrated with a series of increasing ethanol concentrations, embedded, sectioned, stained with uranyl acetate, stained with Reynold's lead citrate, and visualized via TEM using an FEI Tecnai G2 Spirit transmission electron microscope. *See* Fig. 2 for TEM images of DAB staining in HeLa cells infected with APEX2 fusion constructs.

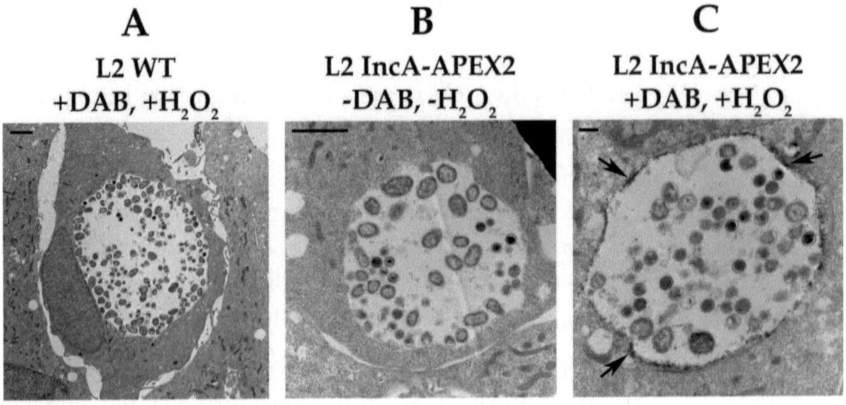

Fig. 2 DAB staining of the inclusion membrane by *C. trachomatis* serovar L2 *IncA-APEX2*. Wild type *C. trachomatis* serovar L2 (L2 WT)-infected HeLa 229 cells treated with DAB and H_2O_2 (**a**) and *C. trachomatis* serovar L2 *IncA-APEX2* (L2 IncA-APEX2) -infected HeLa 229 cells not treated with DAB or H_2O_2 (**b**) do not demonstrate DAB staining of the inclusion membrane. L2 IncA-APEX2-infected HeLa 229 cells treated with DAB and H_2O_2 (**c**) demonstrate DAB staining at the inclusion membrane as evidenced by the presence of the black polymer. Arrows point to examples of the DAB polymer. Scale bars (upper left corner) represent 2 μm in (**a**) and (**b**) and 0.5 μm in (**c**)

4 Notes

1. HeLa cells at higher passage numbers tend to demonstrate decreased biotinylation compared to lower passage number HeLa cells. For best results, use HeLa cells between passage 3 and passage 16 after cells are cultured from frozen stocks. In addition, when transfecting HeLa cells with exogenous DNA (i.e., BioID constructs), all media should be antibiotic free as addition of antibiotics to tissue culture medium can negatively affect the transfection efficiency. If HeLa cells are routinely passaged in media containing antibiotics, passage the cells in medium without antibiotics at least twice before starting the BioID labeling experiments.
2. Routinely test your cells for *Mycoplasma* contamination using a *Mycoplasma* test kit. We test our cells with each fresh vial and/or once a month.
3. When creating BioID (BirA∗) fusions, consideration must be made when choosing to position BirA∗ on either the N or C termini of your protein of interest. Thus, it is important to know the domains of your protein of interest and understand that the addition of BirA∗ could disrupt its function/localization. For example, we have created a construct containing *BirA∗* fused to the N-terminus of *syntaxin 6* [58], a host *trans*-Golgi network (TGN) protein, since it contains a C-terminal transmembrane domain. This positioning would

help ensure the addition of BirA* would not disrupt its normal transmembrane localization within TGN-derived membranes.

4. The stock of 1 mM biotin in serum free 1× DMEM for BioID labeling experiments is stable for 8 weeks stored at 2–8 °C. Dilutions to 50 μM in 1× DMEM + 1% FBS should be made the day of the biotin-labeling experiment for best results.
5. Prior to affinity purification experiments using BioID constructs, it is critical to troubleshoot transfection conditions using your preferred transfection reagent (amount of DNA, amount of transfection reagent, and cell density) in a 6-well tissue culture plate following the recommendations from the manufacturer. Different constructs will have different parameters for transfection, so it is important to troubleshoot each construct individually to maximize transfection efficiency. Each experimental condition uses a full 6-well plate to ensure proteins are at a high enough concentration for affinity purification and identification via mass spectrometry.
6. Use consideration when determining what type of fixation solution and permeabilization reagent for immunofluorescence assays to use that will best capture the proteins in which you are interested. We use a 4% paraformaldehyde solution for 15 min at room temperature followed by permeabilization with 0.1% Triton X-100 in 1× PBS for 5 min at room temperature since the proteins we are interested in are normally localized within the Golgi and we want to preserve organelle/Golgi structure. Other techniques using methanol as a fixative can disrupt the structure of lipid membranes/organelles. Furthermore, cell components have been shown to be inside the inclusion after fixation when they are not present during live-cell imaging [79].
7. Both the solubilization reagent used and the stock concentration of biotin-phenol are important to obtain efficient biotinylation during the labeling reaction step. We have observed the most efficient biotinylation using anhydrous DMSO (AnDMSO) to solubilize biotin-phenol immediately prior to the experiment (or ≤3 months at −20 °C). We also observe the best results when our stock biotin-phenol kept at 50 mM in AnDMSO rather than higher concentrations, which may affect the solubility of biotin-phenol.
8. Trolox can be difficult to make as it requires several steps and has limited solubility (max solubility 100 mM). To make 100 mM Trolox: In a 15 mL conical tube mix 0.1 g Trolox with 0.43 mL of methanol and vortex vigorously until Trolox is in solution. After Trolox is in solution, add 3.2 mL of water and the sample will change color accompanied by an expansion in volume (it will have a cement like consistency). To get the

Trolox back into solution, add 0.36 mL of 1 M sodium hydroxide and mix well by vortexing, adding more in small amounts as necessary. When the solution reaches pH 9.0–9.5, the Trolox will go into solution and should be slightly yellow. Store at −20 °C until use. Scale up as desired for larger volumes of 100 mM Trolox. We routinely make 45 mL of 100 mM Trolox at a time.

9. Clastolactacystin β-lactone is soluble in methyl acetate and anhydrous DMSO (AnDMSO). Methyl acetate can be harmful to the cells. If purchased in methyl acetate, evaporate the methyl acetate with nitrogen gas and resuspend Clastolactacystin β-lactone in AnDMSO. Be aware that high DMSO concentrations can affect eukaryotic cell viability. To reduce this possibility, we resuspend Clastolactacystin β-lactone at 23 mM in AnDMSO. In an aqueous solution, Clastolactacystin β-lactone is only stable for 1 day, so dilute in RIPA buffer the day of sample collection, but it is stable in AnDMSO at −20 °C.
10. We prefer to use Pierce streptavidin magnetic beads (cat# 88816) as they have a high binding capacity. Other brands will likely work, but you will have to optimize the input concentration and elution conditions. In most cases boiled beads cannot be reused for affinity purification experiments.
11. Use premade/high grade Laemmli sample buffer (e.g., 2× Sample Buffer, Bio-Rad, 1610737) if intended for mass spectrometry.
12. To minimize the loss of beads during the pipetting steps, hold a strong magnet directly next to the tube while removing supernatant (e.g., washes/unbound fraction).
13. We have found that 5% milk in PBST greatly reduces background for visualization of most antibodies; however, streptavidin conjugates work best using 5% BSA in PBST.
14. Do not exceed 1 h for the Coomassie staining. Overstaining the gel can interfere with the downstream processing steps prior to the mass spectrometry analysis.
15. Autoclaving is significant source for keratin contamination. Before adding the gel slices, rinse the empty tubes with 30% methanol. This is as an extra wash step to remove contaminants.
16. Work with your preferred core facility to design a digestion and mass spectrometry protocol. We have found a double digestion with Trypsin and AspN proteases improves the detection of membrane proteins. We also use a more sensitive mass spectrometer (Orbitrap Fusion™ Lumos™; Thermo Scientific) to increase the resolution of peptides detected.

17. SAINT is freely available and uses output files from Scaffold (Scaffold, http://www.proteomesoftware.com/products/free-viewer) to construct the SAINT files for statistical analysis.
18. Cycloheximide inhibits host cell protein synthesis; the media is supplemented with it to aid the growth of transformed *Chlamydia* strains. Penicillin G is added to select for the transformants as the plasmid used for transformation contains the gene for beta-lactamase. Consequently, untransformed *Chlamydia* organisms will present as large, aberrant cells in the presence of penicillin G while transformed organisms will show normal morphology. For this reason, penicillin G is not added to the media of the wild type-infected sample. The concentration of penicillin G must be determined empirically.
19. *APEX2* can either be positioned at the N-terminus or the C-terminus of the gene of interest based on functions of the expressed proteins. For example, inclusion membrane proteins (Incs) have an N-terminal type III secretion signal [15], so *APEX2* was placed at the C-terminus to avoid interfering with secretion.
20. Perform a quick centrifugation step before loading the eluate sample into a polyacrylamide gel.
21. This section describes gel electrophoresis and western blotting methods to confirm sample biotinylation and preparation of samples for mass spectrometry analysis. After confirming biotinylation by western blot, to identify the biotinylated proteins, we recommend separating the sample briefly by electrophoresis and cutting the Coomassie-stained gel into sections. This is preferred over digesting the biotinylated proteins directly off the beads or sending the entire eluate for mass spectrometry analysis; both can cause significant streptavidin background. Sectioning the gel enhances the resolution of peptide identification during mass spectrometry analysis. This is especially important when looking for chlamydial proteins in complex eukaryotic cell background.
22. Be sure to have one cell sample infected with wild type, untransformed *Chlamydia* as a negative control, one cell sample infected with a strain of *Chlamydia* transformed with an APEX2 construct that will not receive DAB or H_2O_2 treatment later as another negative control, and one or more strains of *Chlamydia* transformed with APEX2 constructs as the experimental samples.
23. The TEM protocol is adapted from Martell et al. 2017 [71].
24. Every time a solution is placed in the wells, lift the coverslips slightly using a sharp object, such as a needle, to allow that solution to flow underneath the coverslips.

25. Pretreatment with a solution containing DAB but not H_2O_2 may not be necessary depending on the situation. In our experience, we have seen better DAB labeling when samples are pretreated in comparison to samples that are not. Pretreatment allows the DAB molecules to diffuse into the cells [71].

Acknowledgments

We would like to thank Scot Ouellette for critical reading of the manuscript. We would also like to thank Tom Bargar and Nicholas Conoan of the Electron Microscopy Core Facility (EMCF) at the University of Nebraska Medical Center for technical assistance. The EMCF is supported by state funds from the Nebraska Research Initiative (NRI) and the University of Nebraska Foundation, and institutionally by the Office of the Vice Chancellor for Research. This work is supported by NIH/NIAID R01 AI114670 (E. Rucks) and UNMC Department of Pathology and Microbiology start-up funds (E. Rucks).

References

1. Stephens RS, Kalman S, Lammel C, Fan J, Marathe R, Aravind L, Mitchell W, Olinger L, Tatusov RL, Zhao Q, Koonin EV, Davis RW (1998) Genome sequence of an obligate intracellular pathogen of humans: *Chlamydia trachomatis*. Science (New York, NY) 282 (5389):754–759
2. Muschiol S, Bailey L, Gylfe A, Sundin C, Hultenby K, Bergstrom S, Elofsson M, Wolf-Watz H, Normark S, Henriques-Normark B (2006) A small-molecule inhibitor of type III secretion inhibits different stages of the infectious cycle of *Chlamydia trachomatis*. Proc Natl Acad Sci U S A 103(39):14566–14571. https://doi.org/10.1073/pnas.0606412103
3. Bauler LD, Hackstadt T (2014) Expression and targeting of secreted proteins from *Chlamydia trachomatis*. J Bacteriol 196 (7):1325–1334. https://doi.org/10.1128/jb.01290-13
4. Hackstadt T, Scidmore-Carlson MA, Shaw EI, Fischer ER (1999) The *Chlamydia trachomatis* IncA protein is required for homotypic vesicle fusion. Cell Microbiol 1:119–130
5. Rockey DD, Grosenbach D, Hruby DE, Peacock MG, Heinzen RA, Hackstadt T (1997) *Chlamydia psittaci* IncA is phosphorylated by the host cell and is exposed on the cytoplasmic face of the developing inclusion. Mol Microbiol 24(1):217–228
6. Schachter J (1999) Infection and disease epidemiology. In: Stephens RS (ed) Chlamydia: intracellular biology, pathogenesis, and immunity. American Society for Microbiology, Washington, DC, pp 139–169
7. cdc.gov (2005). http://www.cdc.gov/ncidod/DBMD/diseaseinfo/chlamydiapneumonia_t.htm
8. Datta SD, Sternberg M, Johnson RE, Berman S, Papp JR, McQuillan G, Weinstock H (2007) Gonorrhea and chlamydia in the United States among persons 14 to 39 years of age, 1999 to 2002. Ann Intern Med 147:89–97
9. Darville T, Hiltke TJ (2010) Pathogenesis of genital tract disease due to *Chlamydia trachomatis*. J Infect Dis 201:S114–S125
10. Abdelrahman YM, Belland RJ (2005) The chlamydial developmental cycle. FEMS Microbiol Rev 29(5):949–959. https://doi.org/10.1016/j.femsre.2005.03.002
11. AbdelRahman Y, Ouellette SP, Belland RJ, Cox JV (2016) Polarized cell division of *Chlamydia trachomatis*. PLoS Pathog 12(8):e1005822. https://doi.org/10.1371/journal.ppat.1005822
12. Belland RJ, Zhong G, Crane DD, Hogan D, Sturdevant D, Sharma J, Beatty WL, Caldwell HD (2003) Gemomic transcriptional profiling of the developmental cycle of *Chlamydia*

trachomatis. Proc Natl Acad Sci U S A 100:8478–8483

13. Nicholson TL, Olinger L, Chong K, Schoolnik G, Stephens RS (2003) Global stage-specific gene regulation during the developmental cycle of *Chlamydia trachomatis*. J Bacteriol 185:3179–3189
14. Shaw EI, Dooley CA, Fischer ER, Scidmore MA, Fields KA, Hackstadt T (2000) Three temporal classes of gene expression during the *Chlamydia trachomatis* developmental cycle. Mol Microbiol 37:913–925
15. Fields KA, Mead DJ, Dooley CA, Hackstadt T (2003) *Chlamydia trachomatis* type III secretion: evidence for a functional apparatus during early-cycle development. Mol Microbiol 48:671–683
16. Scidmore MA, Fischer ER, Hackstadt T (2003) Restricted fusion of *Chlamydia trachomatis* vesicles with endocytic compartments during the initial stages of infection. Infect Immun 71:973–984
17. Bannantine JP, Stamm WE, Suchland RJ, Rockey DD (1998) *Chlamydia trachomatis* IncA is localized to the inclusion membrane and is recognized by antisera from infected humans and primates. Infect Immun 66 (12):6017–6021
18. Scidmore-Carlson MA, Shaw EI, Dooley CA, Fischer ER, Hackstadt T (1999) Identification and characterization of *Chlamydia trachomatis* early operon encoding four novel inclusion membrane proteins. Mol Microbiol 33:753–765
19. Hueck CJ (1998) Type III protein secretion systems in bacterial pathogens of animals and plants. Microbiol Mol Biol Rev 62(2):379–433
20. Betts-Hampikian HJ, Fields KA (2010) The chlamydial type III secretion mechanism: revealing cracks in a tough nut. Front Microbiol 1:114. https://doi.org/10.3389/fmicb.2010.00114
21. Clifton DR, Fields KA, Grieshaber SS, Dooley CA, Fischer ER, Mead DJ, Carabeo RA, Hackstadt T (2004) A chlamydial type III translocated protein is tyrosine-phosphorylated at the site of entry and associated with recruitment of actin. Proc Natl Acad Sci U S A 101 (27):10166–10171. https://doi.org/10.1073/pnas.0402829101
22. Hower S, Wolf K, Fields KA (2009) Evidence that CT694 is a novel *Chlamydia trachomatis* T3S substrate capable of functioning during invasion or early cycle development. Mol Microbiol 72(6):1423–1437. https://doi.org/10.1111/j.1365-2958.2009.06732.x
23. Derre I, Swiss R, Agaisse H (2011) The lipid transfer protein CERT interacts with the *Chlamydia* inclusion protein IncD and participates to ER-*Chlamydia* inclusion membrane contact sites. PLoS Pathog 7:e1002092
24. Lutter EI, Barger AC, Nair V, Hackstadt T (2013) *Chlamydia trachomatis* inclusion membrane protein CT228 recruits elements of the myosin phosphatase pathway to regulate release mechanisms. Cell Rep 3 (6):1921–1931. https://doi.org/10.1016/j.celrep.2013.04.027
25. Scidmore MA, Hackstadt T (2001) Mammalian 14-3-3beta associates with the *Chlamydia trachomatis* inclusion membrane via its interaction with IncG. Mol Microbiol 39:1638–1650
26. Aeberhard L, Banhart S, Fischer M, Jehmlich N, Rose L, Koch S, Laue M, Renard BY, Schmidt F, Heuer D (2015) The proteome of the isolated *Chlamydia trachomatis* containing vacuole reveals a complex trafficking platform enriched for retromer components. PLoS Pathog 11(6):e1004883. https://doi.org/10.1371/journal.ppat.1004883
27. Saka HA, Thompson JW, Chen Y-S, Kumar Y, Dubois LG, Moseley MA, Valdivia RH (2011) Quantitative proteomics reveals metabolic and pathogenic properties of *Chlamydia trachomatis* developmental forms. Mol Microbiol 82:1185–1203
28. Delevoye C, Nilges M, Dehoux P, Paumet F, Perrinet S, Dautry-Varsat A, Subtil A (2008) SNARE protein mimicry by an intracellular bacterium. PLoS Pathog 4:e100022
29. Elwell CA, Jiang S, Kim JH, Lee A, Wittmann T, Hanada K, Melancon P, Engel JN (2011) *Chlamydia trachomatis* co-opts GBF-1 and CERT to acquire host sphingomyelin for distinct roles during intracellular development. PLoS Pathog 7:e1002198
30. Mital J, Lutter EI, Barger AC, Dooley CA, Hackstadt T (2015) *Chlamydia trachomatis* inclusion membrane protein CT850 interacts with the dynein light chain DYNLT1 (Tctex1). Biochem Biophys Res Commun 462 (2):165–170. https://doi.org/10.1016/j.bbrc.2015.04.116
31. Weber MM, Faris R (2018) Subversion of the endocytic and secretory pathways by bacterial effector proteins. Front Cell Dev Biol 6:1. https://doi.org/10.3389/fcell.2018.00001
32. Lane MD, Rominger KL, Young DL, Lynen F (1964) The enzymatic synthesis of holotranscarboxylase from apotranscarboxylase and (+)-biotin. II Investigation of the reaction mechanism. J Biol Chem 239:2865–2871

33. Beckett D, Kovaleva E, Schatz PJ (1999) A minimal peptide substrate in biotin holoenzyme synthetase-catalyzed biotinylation. Protein Sci 8(4):921–929. https://doi.org/10.1110/ps.8.4.921
34. Kwon K, Beckett D (2000) Function of a conserved sequence motif in biotin holoenzyme synthetases. Protein Sci 9(8):1530–1539. https://doi.org/10.1110/ps.9.8.1530
35. Choi-Rhee E, Schulman H, Cronan JE (2004) Promiscuous protein biotinylation by *Escherichia coli* biotin protein ligase. Protein Sci 13 (11):3043–3050. https://doi.org/10.1110/ps.04911804
36. Roux KJ, Kim DI, Raida M, Burke B (2012) A promiscuous biotin ligase fusion protein identifies proximal and interacting proteins in mammalian cells. J Cell Biol 196(6):801–810. https://doi.org/10.1083/jcb.201112098
37. Green NM (1963) Avidin. 1. the use of (14-C) biotin for kinetic studies and for assay. Biochem J 89:585–591
38. Firat-Karalar EN, Rauniyar N, Yates JR 3rd, Stearns T (2014) Proximity interactions among centrosome components identify regulators of centriole duplication. Curr Biol 24 (6):664–670. https://doi.org/10.1016/j.cub.2014.01.067
39. Firat-Karalar EN, Stearns T (2015) Probing mammalian centrosome structure using BioID proximity-dependent biotinylation. Methods Cell Biol 129:153–170. https://doi.org/10.1016/bs.mcb.2015.03.016
40. Bareja A, Hodgkinson CP, Soderblom E, Waitt G, Dzau VJ (2018) The proximity-labeling technique BioID identifies sorting nexin 6 as a member of the insulin-like growth factor 1 (IGF1)-IGF1 receptor pathway. J Biol Chem 293(17):6449–6459. https://doi.org/10.1074/jbc.RA118.002406
41. Mousnier A, Schroeder GN, Stoneham CA, So EC, Garnett JA, Yu L, Matthews SJ, Choudhary JS, Hartland EL, Frankel G (2014) A new method to determine in vivo interactomes reveals binding of the *Legionella pneumophila* effector PieE to multiple rab GTPases. MBio 5 (4). https://doi.org/10.1128/mBio.01148-14
42. Chen AL, Kim EW, Toh JY, Vashisht AA, Rashoff AQ, Van C, Huang AS, Moon AS, Bell HN, Bentolila LA, Wohlschlegel JA, Bradley PJ (2015) Novel components of the Toxoplasma inner membrane complex revealed by BioID. MBio 6(1):e02357-14. https://doi.org/10.1128/mBio.02357-14
43. Morriswood B, Havlicek K, Demmel L, Yavuz S, Sealey-Cardona M, Vidilaseris K, Anrather D, Kostan J, Djinovic-Carugo K, Roux KJ, Warren G (2013) Novel bilobe components in *Trypanosoma brucei* identified using proximity-dependent biotinylation. Eukaryot Cell 12(2):356–367. https://doi.org/10.1128/ec.00326-12
44. Schnider CB, Bausch-Fluck D, Bruhlmann F, Heussler VT, Burda PC (2018) BioID reveals novel proteins of the Plasmodium parasitophorous vacuole membrane. mSphere 3(1). https://doi.org/10.1128/mSphere.00522-17
45. Schopp IM, Bethune J (2018) Split-BioID—Proteomic analysis of context-specific protein complexes in their native cellular environment. J Vis Exp (134). https://doi.org/10.3791/57479
46. Chojnowski A, Sobota RM, Ong PF, Xie W, Wong X, Dreesen O, Burke B, Stewart CL (2018) 2C-BioID: an advanced two component BioID system for precision mapping of protein interactomes. iScience 10:40–52. https://doi.org/10.1016/j.isci.2018.11.023
47. Lam SS, Martell JD, Kamer KJ, Deerinck TJ, Ellisman MH, Mootha VK, Ting AY (2015) Directed evolution of APEX2 for electron microscopy and proximity labeling. Nat Methods 12(1):51–54. https://doi.org/10.1038/nmeth.3179
48. Connolly CN, Futter CE, Gibson A, Hopkins CR, Cutler DF (1994) Transport into and out of the Golgi complex studied by transfecting cells with cDNAs encoding horseradish peroxidase. J Cell Biol 127(3):641–652
49. Li J, Wang Y, Chiu SL, Cline HT (2010) Membrane targeted horseradish peroxidase as a marker for correlative fluorescence and electron microscopy studies. Front Neural Circuits 4:6. https://doi.org/10.3389/neuro.04.006.2010
50. Hopkins C, Gibson A, Stinchcombe J, Futter C (2000) Chimeric molecules employing horseradish peroxidase as reporter enzyme for protein localization in the electron microscope. Methods Enzymol 327:35–45
51. Martell JD, Deerinck TJ, Sancak Y, Poulos TL, Mootha VK, Sosinsky GE, Ellisman MH, Ting AY (2012) Engineered ascorbate peroxidase as a genetically encoded reporter for electron microscopy. Nat Biotechnol 30 (11):1143–1148. https://doi.org/10.1038/nbt.2375
52. Rhee HW, Zou P, Udeshi ND, Martell JD, Mootha VK, Carr SA, Ting AY (2013) Proteomic mapping of mitochondria in living cells via spatially restricted enzymatic tagging. Science (New York, NY) 339(6125):1328–1331. https://doi.org/10.1126/science.1230593

53. Mandelman D, Schwarz FP, Li H, Poulos TL (1998) The role of quaternary interactions on the stability and activity of ascorbate peroxidase. Protein Sci 7(10):2089–2098. https://doi.org/10.1002/pro.5560071005
54. Khan AA, Quigley JG (2011) Control of intracellular heme levels: heme transporters and heme oxygenases. Biochim Biophys Acta 1813(5):668–682. https://doi.org/10.1016/j.bbamcr.2011.01.008
55. Lee S-Y, Kang M-G, Park J-S, Lee G, Ting Alice Y, Rhee H-W (2016) APEX fingerprinting reveals the subcellular localization of proteins of interest. Cell Rep 15(8):1837–1847. https://doi.org/10.1016/j.celrep.2016.04.064
56. Lobingier BT, Huttenhain R, Eichel K, Miller KB, Ting AY, von Zastrow M, Krogan NJ (2017) An approach to spatiotemporally resolve protein interaction networks in living cells. Cell 169(2):350–360.e12. https://doi.org/10.1016/j.cell.2017.03.022
57. Santin YG, Doan T, Lebrun R, Espinosa L, Journet L, Cascales E (2018) In vivo TssA proximity labelling during type VI secretion biogenesis reveals TagA as a protein that stops and holds the sheath. Nat Microbiol 3(11):1304–1313. https://doi.org/10.1038/s41564-018-0234-3
58. Kabeiseman EJ, Cichos KH, Moore ER (2014) The eukaryotic signal sequence, YGRL, targets the chlamydial inclusion. Front Cell Infect Microbiol 4:129. https://doi.org/10.3389/fcimb.2014.00129
59. Watson RT, Pessin JE (2000) Functional cooperation of two independent targeting domains in syntaxin 6 is required for its efficient localization in the *trans*-golgi network of 3T3L1 adipocytes. J Biol Chem 275:1261–1268
60. Mojica SA, Hovis KM, Frieman MB, Tran B, Hsia RC, Ravel J, Jenkins-Houk C, Wilson KL, Bavoil PM (2015) SINC, a type III secreted protein of *Chlamydia psittaci*, targets the inner nuclear membrane of infected cells and uninfected neighbors. Mol Biol Cell 26(10):1918–1934. https://doi.org/10.1091/mbc.E14-11-1530
61. McKuen MJ, Mueller KE, Bae YS, Fields KA (2017) Fluorescence-reported allelic exchange mutagenesis reveals a role for *Chlamydia trachomatis* TmeA in invasion that is independent of host AHNAK. Infect Immun 85(12). https://doi.org/10.1128/iai.00640-17
62. Fields KA, McCormack R, de Armas LR, Podack ER (2013) Perforin-2 restricts growth of *Chlamydia trachomatis* in macrophages. Infect Immun 81(8):3045–3054. https://doi.org/10.1128/iai.00497-13
63. Moore ER, Ouellette SP (2014) Reconceptualizing the chlamydial inclusion as a pathogen-specified parasitic organelle: an expanded role for Inc proteins. Front Cell Infect Microbiol 4:157. https://doi.org/10.3389/fcimb.2014.00157
64. Rucks EA, Olson MG, Jorgenson LM, Srinivasan RR, Ouellette SP (2017) Development of a proximity labeling system to map the *Chlamydia trachomatis* inclusion membrane. Front Cell Infect Microbiol 7:40. https://doi.org/10.3389/fcimb.2017.00040
65. Papageorgiou DN, Demmers J, Strouboulis J (2013) NP-40 reduces contamination by endogenous biotinylated carboxylases during purification of biotin tagged nuclear proteins. Protein Expr Purif 89(1):80–83. https://doi.org/10.1016/j.pep.2013.02.015
66. Hollinshead M, Sanderson J, Vaux DJ (1997) Anti-biotin antibodies offer superior organelle-specific labeling of mitochondria over avidin or streptavidin. J Histochem Cytochem 45(8):1053–1057. https://doi.org/10.1177/002215549704500803
67. Rodriguez P, Braun H, Kolodziej KE, de Boer E, Campbell J, Bonte E, Grosveld F, Philipsen S, Strouboulis J (2006) Isolation of transcription factor complexes by in vivo biotinylation tagging and direct binding to streptavidin beads. In: Bina M (ed) Gene mapping, discovery, and expression: methods and protocols. Humana Press, Totowa, NJ, pp 305–323. https://doi.org/10.1385/1-59745-097-9:305
68. de Boer E, Rodriguez P, Bonte E, Krijgsveld J, Katsantoni E, Heck A, Grosveld F, Strouboulis J (2003) Efficient biotinylation and single-step purification of tagged transcription factors in mammalian cells and transgenic mice. Proc Natl Acad Sci U S A 100(13):7480–7485. https://doi.org/10.1073/pnas.1332608100
69. Choi H, Larsen B, Lin ZY, Breitkreutz A, Mellacheruvu D, Fermin D, Qin ZS, Tyers M, Gingras AC, Nesvizhskii AI (2011) SAINT: probabilistic scoring of affinity purification-mass spectrometry data. Nat Methods 8(1):70–73. https://doi.org/10.1038/nmeth.1541
70. Webb-Robertson BJ, Matzke MM, Datta S, Payne SH, Kang J, Bramer LM, Nicora CD, Shukla AK, Metz TO, Rodland KD, Smith RD, Tardiff MF, McDermott JE, Pounds JG, Waters KM (2014) Bayesian proteoform modeling improves protein quantification of global proteomic measurements. Mol Cell Proteomics 13(12):3639–3646. https://doi.org/10.1074/mcp.M113.030932

71. Martell JD, Deerinck TJ, Lam SS, Ellisman MH, Ting AY (2017) Electron microscopy using the genetically encoded APEX2 tag in cultured mammalian cells. Nat Protoc 12 (9):1792–1816. https://doi.org/10.1038/nprot.2017.065
72. Wickstrum J, Sammons LR, Restivo KN, Hefty PS (2013) Conditional gene expression in using the tet system. PLoS One 8(10): e76743. https://doi.org/10.1371/journal.pone.0076743
73. Wang Y, Kahane S, Cutcliffe LT, Skilton RJ, Lambden PR, Clarke IN (2011) Development of a transformation system for *Chlamydia trachomatis*: restoration of glycogen biosynthesis by acquisition of a plasmid shuttle vector. PLoS Pathog 7(9):e1002258. https://doi.org/10.1371/journal.ppat.1002258
74. Mueller KE, Fields KA (2015) Application of beta-lactamase reporter fusions as an indicator of effector protein secretion during infections with the obligate intracellular pathogen *Chlamydia trachomatis*. PLoS One 10(8): e0135295. https://doi.org/10.1371/journal.pone.0135295
75. Willingham MC (1999) Fluorescence labeling of surface antigens of attached or suspended tissue-culture cells. In: Javois LC (ed) Immunocytochemical methods and protocols. Humana Press, Totowa, NJ, pp 113–119. https://doi.org/10.1385/1-59259-213-9:113
76. Lucas AL, Ouellette SP, Kabeiseman EJ, Cichos KH, Rucks EA (2015) The trans-Golgi SNARE syntaxin 10 is required for optimal development of *Chlamydia trachomatis*. Front Cell Infect Microbiol 5:68. https://doi.org/10.3389/fcimb.2015.00068
77. Chen AL, Johnson KA, Lee JK, Sutterlin C, Tan M (2012) CPAF: a chlamydial protease in search of an authentic substrate. PLoS Pathog 8(8):e1002842. https://doi.org/10.1371/journal.ppat.1002842
78. Kurien BT, Hal Scofield R (2015) Western blotting: methods and protocols, Methods in molecular biology, vol 1312. Humana Press, New York. https://doi.org/10.1007/978-1-4939-2694-7
79. Kokes M, Valdivia RH (2015) Differential translocation of host cellular materials into the *Chlamydia trachomatis* inclusion lumen during chemical fixation. PLoS One 10(10): e0139153. https://doi.org/10.1371/journal.pone.0139153

Chapter 18

High-Throughput Screening for Novel Inhibitors of Intracellular Pathogens, Including *Chlamydia trachomatis*

Amanda Claire Brown and Nicole L. Kushner

Abstract

High-throughput drug screening (HTS) is a powerful tool that can be used rapidly to identify new potential bacterial inhibitors and/or compounds which enhance host cell control of pathogens, which can then go on to be developed as novel therapeutics. Typically screening is commonly done in artificial culture medium; however, obligate intracellular pathogens, such as *Chlamydia trachomatis*, cannot be tested this way. Intracellular screening methods allow for such pathogens to undergo HTS, while still giving reliable and consistent data. Plus, as well as identifying new potential bacterial inhibitors, it is also able to detect compounds which enhance host cell control of pathogens, to allow for host-directed therapies to be developed.

Key words High throughput screen (HTS), *Chlamydia trachomatis*, Intracellular screen, Fluorescence, Intracellular drug screen

1 Introduction

C. trachomatis is the most common bacterial agent in sexually transmitted infections (STI), globally accounting for more than 100 million infections per year [1, 2]. *C. trachomatis* causes pelvic inflammatory disease in women, which has severe long-term consequences including ectopic pregnancy, infertility and chronic pelvic pain; in addition serovar A–C causes the blinding disease, trachoma, which affects millions of people worldwide [3, 4]. The loss of productivity and morbidity resulting from *C. trachomatis* infection has been recognized as placing a heavy economic burden on society [5, 6]. Drug resistance is a real concern amongst all infectious disease, and Chlamydia is no exception, with multiple drug resistant strains being responsible for treatment failure [7]. Therefore the need to discover and develop new and effective treatments is very timely. However due to its obligate intracellular lifestyle, typical high throughput drug screens (HTS), where

Amanda Claire Brown (ed.), *Chlamydia trachomatis: Methods and Protocols*, Methods in Molecular Biology, vol. 2042, https://doi.org/10.1007/978-1-4939-9694-0_18, © Springer Science+Business Media, LLC, part of Springer Nature 2019

compound libraries are tested for efficacy against bacteria grown in synthetic media, are not applicable.

The method presented in this chapter describes an intracellular HTS protocol, which we have routinely used to screen over 1.5 million compounds against *Mycobacterium tuberculosis (Mtb)*, and which is suitable for screening for new drugs with anti-chlamydial activity.

2 Materials

In addition to the materials listed below standard lab equipment, including multichannel pipettes, CO_2 incubators, centrifuges etc. are required. We give the details for 384-well format, although it can be scaled to others, as required. To maintain consistency we recommend the use of automated liquid handling devices, such as the Multidrop™ Combi Reagent Dispenser and PerkinElmer Janus®, especially in the case of the 384-well format, and higher. Sterile, nuclease-free, aerosol barrier pipette tips should be used for all liquid manipulations, and standard molecular microbiology protocols, such as wearing of gloves/PPE, minimizing aerosol production, taking care to not cross contaminate samples, etc., should be followed.

Please note that certain *Chlamydia* species may cause human disease; apply local standards of practice appropriate for the biosafety level of the pathogen you are working with.

2.1 Preparation of Chlamydia

1. Suitable eukaryotic cells (e.g., McCoy B Mouse Fibroblasts: ATCC CRL-1696) (*see* **Note 1**). Store in liquid nitrogen (vapor phase).
2. *Chlamydia* strain of interest, carrying a fluorescent reporter (*see* **Note 2**).
3. Cryovials.
4. Sucrose–phosphate–glutamate (SPG) Buffer: 250 mM sucrose, 10 mM sodium phosphate, and 0.5 mM L-glutamic acid (pH 7.4 at 4 °C). Filter sterilize and store at 4 °C.

2.2 Preparation of Host Cells

1. Suitable eukaryotic cells (e.g., McCoy B Mouse Fibroblasts: ATCC CRL-1696) (*see* **Note 1**). Store in liquid nitrogen (vapor phase).
2. T75 Cell Culture Flasks.
3. Suitable eukaryotic cell media (e.g., McCoy Media: DMEM with 10% fetal bovine serum (FBS)) (*see* **Note 1**). Store at 4 °C.
4. Sterile PBS.
5. Trypsin, if required.

6. Hemocytometer.
7. Inverted microscope.
8. 384-well black plates, with clear flat bottoms and lids (e.g., Corning 3712 (VWR)) (*see* **Note 3**).
9. Automated cell dispenser, such as the Multidrop™ Combi Reagent Dispenser 5840300 (ThermoFisher) (*see* **Note 4**).

2.3 Infection and Drug Addition

1. Quantified *Chlamydia* stock (*see* **Notes 5** and **6**).
2. Suitable eukaryotic cell media (e.g., McCoy Media: DMEM with 10% fetal bovine serum (FBS)) (*see* **Note 1**). Store at 4 °C.
3. Seeded eukaryotic cell 384-well "assay plates" from 2.2 (*see* **Note 7**).
4. Automated liquid handling device (e.g., PerkinElmer Janus® or similar) (*see* **Note 4**).
5. Compound library (*see* **Note 8**).
6. Axygen sterile 384 deep well, polypropylene plates (P-384-120-SQ-C-S, VWR).
7. Sterile DMEM, without added FBS.
8. Sterile 30 μL 384 filter tips.

2.4 Visualization and Identification of "Hits"

1. ATc (Sigma. Cat no. 37919), if required; dissolved in 100% ethanol, store at −20 °C. Before addition to plates dilute in sterile PBS.
2. Plate reader, such as a PerkinElmer Envision.

3 Methods

3.1 Preparation of Chlamydia

1. Propagate *Chlamydia* strain of interest, carrying a fluorescent reporter, in chosen cell line (*see* **Notes 1** and **2**).
2. Isolate *Chlamydia* by scraping cells and resuspending in sterile microfuge tubes containing 0.5 mL SPG buffer and ~10 1 mM sterile glass beads.
3. Vortex tubes for 2 min.
4. Briefly pulse spin in a microfuge to settle aerosols.
5. Remove supernatant and pool.
6. Mix the pooled supernatants and aliquot into cryovials.
7. Store at −80 °C.
8. Thaw 1 vial to quantify *Chlamydia* before HTS (*see* **Note 5**).

3.2 Preparation of Host Cells

1. Grow chosen cell line to 70–90% competency (*see* **Note 1**).
2. Harvest cells, as per cell line chosen requirements (*see* **Note 1**).

3. Count cells using hemocytometer.
4. Dilute cells to 1,000,000/mL (*see* **Note 6**) using cell culture media.
5. Seed 30 μL cell suspension in 384-well black, clear flat bottomed cell culture plates, to give 30,000 cells/well; this is now the "assay plate"—prepare two assay plates for each drug plate (*see* **Notes 4** and 7).

3.3 Infection and Drug Addition

1. Thaw pre-quantified *Chlamydia* stocks (*see* **Note 5**).
2. Dilute *Chlamydia* in cell culture media to give chosen MOI, for example—if chosen MOI is 1, and host cell number per well is 30,000, you will need ~30,000 *Chlamydia* cells in 10 μL (= 3,000,000/mL) as your infection stock (*see* **Note 6**).
3. Add 10 μL *Chlamydia* infection stock to each assay plate well, using PerkinElmer Janus®, or equivalent (*see* **Note 4**).
4. Centrifuge assay plates at 500 × *g* for 30 min, and incubate at 37 °C with 5% CO_2 for 1 h, or until bacteria have entered host cells.
5. Prepare drug plates by adding 0.5 μL compound (dissolved in 100% DMSO) at 1 mM to each well (*see* **Notes 4** and **8–10**).
6. Dilute drugs with 24.5 μL DMEM (no added FBS), to give 25 μL at 30 μM drug/well (*see* **Notes 4** and **8**), each drug plate will serve two assay plates (duplicate plates).
7. Transfer 10 μL/well drug to the equivalent well in the assay plate, using PerkinElmer Janus®, or equivalent (*see* **Notes 4** and **11**).
8. Incubate at 37 °C with 5% CO_2.

3.4 Visualization and Identification of "Hits"

1. If using a tet-inducible strain add 10 μL ATc (*see* **Note 12**) at the most suitable point of the incubation—typically after the test compounds have had sufficient time to act on the bacteria—this should be defined empirically by the user (*see* **Note 11**), and then read after 24, 48 and 72 h, using a plate reader, such as the PerkinElmer Envision (Fig. 1a).
2. If using a constitutive expressing strain the assay plates can be read daily, using a plate reader, such as the PerkinElmer Envision, and returned to the incubator until the reads become saturated (Fig. 1a).
3. We use variance across the plate to identify hits, *z*-scores are calculated as $z = x - m/s$ where m = median of sample plate, and s = sample standard deviation; a *z*-score of −2 or lower is considered to indicate a "hit" (Fig. 1b, c).

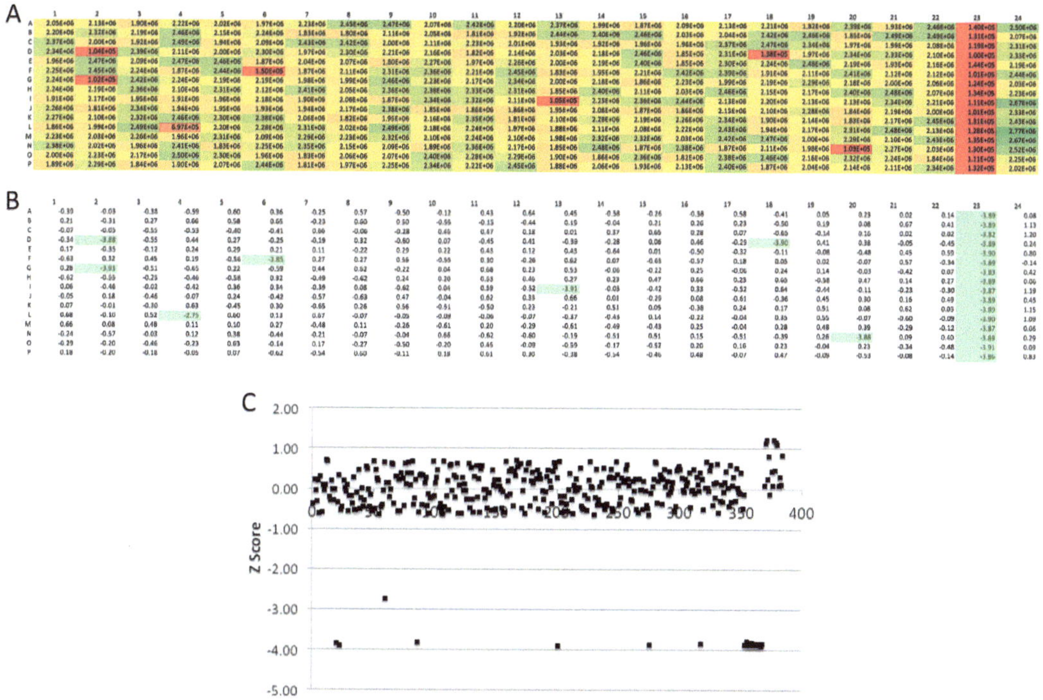

Fig. 1 Analysis of fluorescent readout data from a representative 384-well plate. (**a**) Raw fluorescent data as obtained from plate read. (**b**) Data from A transformed into *z*-scores, calculated as $z = x - m/s$ where $m =$ median of sample plate, and $s =$ sample standard deviation; a *z*-score of -2 or lower was considered to indicate a "hit." (**c**) Scatter plot of *z* scores, *x*-axis is compound number (1-384), data points at and below -2 are considered as hits

4 Notes

1. *Chlamydia* can be grown in a variety of different host cells, including McCoy B Mouse Fibroblasts and HeLa cells, and the most appropriate line for screening should be determined by the user. The growth media used should be the one most appropriate for the cell line chosen. Typically when screening for inhibitors against Mtb we do not include any routine antibiotics, such as pen-strep, in the growth media, as we do not want to include the action of these in our screening, therefore good aseptic technique is essential.
2. The fluorescent reporter chosen needs to be user defined. For *Mtb* screening we have successfully used both constitutive- and inducible- strains expressing fluorescent proteins, and both have their advantages and disadvantages. We favor mKOrange as a reporter, over mCherry, as, in our hands, with *Mtb*, it has greater sensitivity and a lower background, allowing us to perform day zero reads for our constitutive reporter strain.

We use a highly expressed promoter in both incidences, with the inducible strain being held under the control of the tet-on system (expression only occurs after the addition of tetracycline/ATc to the test wells). Constitutively expressing fluorescent strains allow for regular monitoring, however fluorescent proteins are extremely stable and are not easily degraded, so a reduction in signal can be difficult to identify. For constitutively expressing strains we typically include known positive and negative controls on each plate, and perform day zero (background) plate reads. Inducible strains, such as those under the control the tet-on systems, have an increased sensitivity, as there is no residual fluorescent protein present from dead bacteria, however the concentration of ATc to use, and the time point at which to add it, needs to be defined by titration by the user prior to starting a screen.

3. Here we describe the procedure for screening in a 384-well format, however this can be translated to 96- or 1536-well, as desired. Typically in scaling to a 96-well format we multiply the volumes given for 384-well by three (not four), as we have found this gives the most suitable ratio of liquid to well volume—however, this is just our preference. For 96-well plates we find that Corning 3603 work very well.
4. For plate manipulations we use automated liquid handling devices, namely a Multidrop™ Combi Reagent Dispenser for seeding cells and drug dilutions, and a PerkinElmer Janus for bacterial- and drug- additions, however these can also be carried out using a multichannel pipette, although there is an increased lack of consistency and the potential for user based error.
5. After isolation *Chlamydia* can be quantified by a number of different methods, including qPCR, however, although convenient PCR does not provide any information regarding the number of viable bacteria. For a more accurate quantification host cells should be infected with a dilution series created from one vial of isolated bacteria. There are numerous methods to visualize and quantify the infection, including fluorescent labeled antibodies; however, as the *Chlamydia* strain in this instance is carrying a fluorescent protein marker, this can be used to identify and enumerate the bacteria via confocal microscopy, following the addition of ATc, as required based on the reporter type chosen.
6. Identifying the MOI most suitable for the screening needs to be empirically defined, based on the specific strains and cell lines used.

7. It should be defined by the user how long the host cells should be incubated after seeding, prior to infection; typically overnight seems to work well.
8. We typically screen initially at a single concentration (5 μM final concentration), in duplicate plates; biological replicates are the most useful. Any hits are then rescreened in triplicate and in eight-point dose response.
9. Controls—it can be helpful to include both drug-positive and drug-negative controls. The drug-positive control we typically use is 10 μM rifampicin, which allows us to identify the margin signals obtained from "dead" bacteria vs the drug-negative controls. Rifampicin has a slight auto-fluorescence, which is why we use this concentration and no higher; the sensitivity to rifampicin should be empirically determined for any particular *Chlamydia* strain and host cell combination. Drug-negative controls typically use sterile PBS in the place of the drug, to keep volumes consistant. Controls also allow for a QC step between runs. Controls can be included in with drug plates (e.g., in columns 23 and 24) or prepared as a dedicated control drug plate which is included with each run, and processed in the same way as the other drug plates.
10. DMSO is toxic to cells at high concentrations (>1%), the final DMSO concentration for the dilutions described here is 0.16%. The sensitivity to DMSO should be empirically determined for any particular *Chlamydia* strain and host cell combination, if DMSO toxicity is a particular concern DMSO-positive and negative controls should be included.
11. We typically use the same set of tips for 10 plates, washing them with 70% ethanol and rinsing with sterile PBS, and then blotting on sterile 3 M blotting paper, before the next drug addition.
12. For *Mtb* screens we routinely use 200 ng/mL ATc final concentration, this amount has been previously observed to have no detrimental effect on *C. trachomatis* [8], but should be tested empirically in the form of a titration series with drug-positive and drug-negative wells, over a time course to establish the best incubation period for the assay as well as the optimal ATc concentration.

References

1. WHO. Global incidence and prevalence of selected curable sexually transmitted infections—2008. https://www.who.int/reproductivehealth/publications/rtis/stisestimates/en/
2. WHO. Prevalence and incidence of selected sexually transmitted infections. https://www.who.int/reproductivehealth/publications/rtis/9789241502450/en/
3. Mylonas I (2012) Female genital *Chlamydia trachomatis* infection: where are we heading? Arch Gynecol Obstet 285:1271–1285
4. Mariotti SP, Pascolini D, Rose-Nussbaumer J (2009) Trachoma: global magnitude of a

preventable cause of blindness. Br J Ophthalmol 93:563–568

5. Blandford JM, Gift TL (2006) Productivity losses attributable to untreated chlamydial infection and associated pelvic inflammatory disease in reproductive-aged women. Sex Transm Dis 33:S117–S121
6. Burton MJ, Mabey DCW (2009) The global burden of trachoma: a review. PLoS Negl Trop Dis 3:e460
7. Somani J, Bhullar VB, Workowski KA, Farshy CE, Black CM (2000) Multiple drug-resistant *Chlamydia trachomatis* associated with clinical treatment failure. J Infect Dis 181 (4):1421–1427. https://doi.org/10.1086/315372
8. Wickstrum J, Sammons LR, Restivo KN, Hefty PS (2013) Conditional gene expression in *Chlamydia trachomatis* using the tet system. PLoS One 8(10):e76743. https://doi.org/10.1371/journal.pone.0076743

Index

Amanda Claire Brown (ed.), *Chlamydia trachomatis: Methods and Protocols*, Methods in Molecular Biology, vol. 2042, https://doi.org/10.1007/978-1-4939-9694-0, © Springer Science+Business Media, LLC, part of Springer Nature 2019

P

R

S

T

U

V

W

MIX
Papier aus verantwortungsvollen Quellen
Paper from responsible sources
FSC® C105338

If you have any concerns about our products,
you can contact us on
ProductSafety@springernature.com

In case Publisher is established outside the EU,
the EU authorized representative is:
Springer Nature Customer Service Center GmbH
Europaplatz 3, 69115 Heidelberg, Germany

Printed by Libri Plureos GmbH
in Hamburg, Germany